Day and Thomson's Series.

HIGHER ARITHMETIC;

OR THE

SCIENCE AND APPLICATION OF NUMBERS

COMBINING THE

ANALYTIC AND SYNTHETIC MODES OF INSTRUCTION.

DESIGNED FOR

ADVANCED CLASSES IN SCHOOLS AND ACADEMIES.

BY JAMES B. THOMSON, A.M.,

AUTHOR OF MENTAL ARITHMETIC; EXERCISES IN ARITHMETICAL ANALYSIS; PRACTICAL ARITHMETIC; EDITOR OF DAY'S SCHOOL ALGEBRA; LEGENDRE'S GEOMETRY, ETC.

TWENTY FIFTH EDITION.

NEW-YORK:
PUBLISHED BY NEWMAN AND IVISON,
178 FULTON STREET,

CINCINNATI: MOORE & ANDERSON. AUBURN: J. C. IVISON & CO
CHICAGO: S. C. GRIGGS & CO DETROIT: A. McFARREN.

1853.

DAY AND THOMSON'S MATHEMATICAL SERIES.

FOR SCHOOLS AND ACADEMIES.

I. MENTAL ARITHMETIC; or, *First Lessons in Numbers;*—for Beginners. This work commences with the *simplest* combinations of numbers, and *gradually* advances to more difficult combinations, as the mind of the learner expands and is prepared to comprehend them.

II. PRACTICAL ARITHMETIC;—Uniting the *Inductive* with the *Synthetic* mode of Instruction; also illustrating the principles of CANCELATION. The design of this work is to make the pupil thoroughly acquainted with the *reason* of every operation which he is required to perform. It abounds in *examples,* and is *eminently practical.*

III. KEY TO PRACTICAL ARITHMETIC;—Containing the answers with numerous suggestions, &c.

IV. HIGHER ARITHMETIC; *or, the Science and Application of Numbers;*—For advanced Classes. This work is complete in itself, commencing with the fundamental rules, and extending to the highest department of the science.

V. KEY TO HIGHER ARITHMETIC;—Containing all the answers, with many suggestions, and the solution of the more difficult questions.

VI. THOMSON'S DAY'S ALGEBRA;—This work is designed to be a *lucid* and *easy* transition from the study of Arithmetic to the higher branches of Mathematics. The number of examples is much increased; and the work is every way adapted to the improved methods of instruction in Schools and Academies.

VII. KEY TO THOMSON'S DAY'S ALGEBRA;—Containing the answers, the solution of the more difficult problems, &c.

VIII. THOMSON'S LEGENDRE'S GEOMETRY;—With practical notes and illustrations. This work has received the approbation of many of the most eminent Teachers and Practical Educators.

IX. PLANE TRIGONOMETRY, AND THE MENSURATION OF HEIGHTS AND DISTANCES; *with a summary view of the Nature and Use of Logarithms;*—Adapted to the method of instruction in Schools and Academies.

X. ELEMENTS OF SURVEYING;—Adapted both to the wants of the learner and the practical Surveyor. (Published soon.)

Entered according to Act of Congress, in the year 1847,

BY JAMES B. THOMSON,

In the Clerk's Office of the Northern District of New York.

Stereotyped by Thomas B. Smith, 216 William St. N. Y.

PREFACE.

THE *Higher Arithmetic* which is now presented to the public, is the third and last of a *series of Arithmetics* adapted to the wants of different classes of pupils in Schools and Academies. The *title* of each explains the *character* of the work. The series is constructed upon the principle, that "there is a place for everything, and everything should be in its proper place." Each work forms an *entire treatise* in itself; the examples in each are all *different* from those in the others, so that pupils who study the series, will not be obliged to purchase the same matter *twice*, nor to solve the *same problems* over again.

The *Mental Arithmetic*, is designed for children from six to eight years of age. It is divided into progressive lessons of convenient length, beginning with the *simplest combinations* of numbers, and advancing by *gradual* steps, to more difficult operations, as the mind of the learner expands and is prepared to *comprehend* them.

The *Practical Arithmetic* embraces all the subjects requisite for a *thorough business education*. The principles and rules are carefully *analyzed* and *demonstrated;* the examples for practice are *numerous*, and the observations and notes contain much information pertaining to *business matters*, not found in other works of the kind. This is the FIRST SCHOOL BOOK in which the *Standard Units* of Weights and Measures adopted by the Government in 1834, were published.

The *Higher Arithmetic* is designed to give a full development of the *philosophy* of Arithmetic, and its *various applications* to commercial purposes. Its plan is the following:

1. The work is *complete* in itself. It commences with notation, and illustrating the different *properties* of numbers, the principles of *Cancelation*, and various other methods of *contraction*, extends to the higher operations in mercantile affairs, and the more abstruse departments of the science.

2. Great pains have been taken to render the *definitions* and *rules clear, concise, exact, comprehensive.*

3. It has been a cardinal point never to *anticipate* a principle; and never to *use* one principle in the explanation of another, until it has itself been *explained* or *demonstrated.*

4. Nothing is taken for granted which requires *proof.* Every principle therefore has been *reinvestigated*, and *carefully analyzed.*

5. The principles are arranged *consecutively*, and the *dependence* of each on those that precede it, is pointed out by references. Treated in this manner, the science of Arithmetic presents a series of principles and propositions alike *harmonious* and *logical;* and the study of it cannot fail to exert the happiest influence in *developing* and *strengthening* the reasoning powers of the learner.

6. The rules are *demonstrated* with care, and the reasons of every operation *fully illustrated.*

7. The examples are *copious* and *diversified;* calling every principle into exercise, and making its application thoroughly understood.

8. In the arrangement of subjects, the *natural order* of the science has been carefully followed. Common Fractions have therefore been placed immediately after Division, for *two reasons. First,* they *arise* from division, and a connexion so *intimate* should *not be severed* without cause. *Second,* in Reduction and the Compound Rules, it is often necessary to multiply and divide by fractions, to add and subtract them, also to carry for them, unless perchance the examples are constructed for the *occasion,* and with *special reference* to *avoiding* these difficulties.

For the same reason Federal Money, which is based upon the decimal notation, is placed after Decimal Fractions; Interest, Commission, &c., after Percentage. To require a pupil to *understand a rule* before he is acquainted with the *principles* upon which it is based, is compelling him to raise a *superstructure,* before he is permitted to lay a *foundation.*

9. In preparing the *Tables* of *Weights* and *Measures,* no effort has been spared to ascertain those in *present use* in our country; and rejecting such as are *obsolete,* we have introduced the *Standard Units* adopted by the Government, together with the methods of determining and applying those standards.

10. Great labor has also been expended in preparing full and accurate Tables of Foreign Weights and Measures, and Moneys of Account, and in comparing them with those of the United States.

Such is a brief outline of the present work. In a word, it is designed to be an *auxiliary* to the teacher, a *lucid* and *comprehensive text-book* for the pupil, and an *acceptable acquisition* to the counting-room. It contains many illustrations and principles not found in other works before the public, and much is believed to be gained in the method of *reasoning* and *analysis.* No labor has been spared to render it worthy of the marked favor with which the former productions of the author have been received.

J. B. THOMSON.

New York, August, 1847.

SUGGESTIONS

ON THE

MODE OF TEACHING ARITHMETIC.

I. Qualifications.—The chief qualifications requisite in teaching Arithmetic, as well as other branches, are the following:

1. A *thorough knowledge* of the subject.
2. A *love* for the employment.
3. An *aptitude* to teach. These are *indispensable to success.*

II. Classification.—*Arithmetic*, like reading, grammar, &c., should be taught in *classes.*

1. This method saves much time, and thus enables the teacher to devote more attention to *oral illustrations.*
2. The action of mind upon mind, is a *powerful stimulant* to exertion, and cannot fail to create a *zest* for the study.
3. The mode of analyzing and reasoning of one scholar, will often *suggest new ideas* to others in the class.
4. In the classification, those should be put together who possess as nearly equal capacities and attainments as possible. If any of the class learn quicker than others, they should be allowed to take up an extra study, or be furnished with additional examples to solve, so that the whole class may advance together.
5. The number in a class, if practicable, should not be less than six, nor over twelve or fifteen. If the number is less, the recitation is apt to be deficient in animation; if greater, the turn to recite does not come round sufficiently often to keep up the interest.

III. Apparatus.—The *Black-board* and *Numerical Frame* are as indispensable to the teacher, as tables and cutlery are to the house-keeper. Not a recitation passes without use for the black-board. If a principle is to be demonstrated or an operation explained, it should be done upon the *black-board*, so that all may see and understand it at once.

To illustrate the increase of numbers, the process of adding, subtracting, multiplying, dividing, &c., to young scholars, the *Numerical Frame* furnishes one of the most simple and convenient methods ever invented.

Every one who ciphers will of course have a *slate.* Indeed, it is desirable that every scholar in school, even to the very youngest, should be furnished with a slate, so that when their lessons are learned each one may busy himself in writing and drawing various familiar objects. *Idleness* in school is the parent of *mischief*, and *employment* is the best antidote against *disobedience.*

Geometrical diagrams and *solids* are also highly useful in illustrating many points in arithmetic, and no school should be without them.

IV. Recitations.—The *first* object in a recitation, is to secure the *attention* of the class. This is done chiefly by throwing *life* and *variety* into the exercise. Children *loathe* dullness, while animation and variety are *their delight.*

2. Every example should be *analyzed;* the "why and the wherefore" of every step in the solution should be required, till each member of the class becomes perfectly familiar with the process of reasoning and analysis.

3. To ascertain whether each pupil has the right answer, it is an excellent method to name a question, then call upon some one to give the answer, and before deciding whether it is right or wrong, ask how many in the class agree with it. The answer they give by raising their hand, will show at once how many are right. The explanation of the process may now be made.

V. OBJECTS OF THE STUDY.—When properly studied, two important ends are attained. 1st. *Discipline* of mind, and the *development* of the reasoning powers. 2d. *Facility* and *accuracy* in the application of numbers to business calculations.

VI. THOROUGHNESS.—The motto of every teacher should be *thoroughness.* Without it, the great ends of the study of Arithmetic are *defeated.*

1. In securing this object, much advantage is derived from *frequent reviews.*

2. Every operation should be *proved.* The intellectual discipline and habits of accuracy thus secured, will richly reward the student for his time and toil.

3. Not a recitation should pass without *practical exercises* upon the blackboard or slates, besides the lesson assigned.

4. After the class have solved the examples under a rule, each one should be required to give an *accurate account* of its principles with the *reason* for each step, either in his own language or that of the author.

5. *Mental Exercises* in arithmetic are *exceedingly useful* in making ready and accurate arithmeticians; hence, the practice of connecting *mental* with *written exercises*, throughout the whole course, is strongly recommended.

VII. SELF-RELIANCE.—The *habit* of *self-reliance* in study, is confessedly *invaluable.* Its power is proverbial; I had almost said, *omnipotent.* "Where there is a *will*, there is a *way.*"

1. To acquire this habit, the pupil, like a child learning to walk, must be taught to *depend upon himself.* Hence,

2. When assistance is required, it should be given *indirectly;* not by taking the slate and solving the example for him, but by explaining the *meaning* of the question, or illustrating the *principle* on which the operation depends, by supposing a more familiar case. Thus the pupil will be able to solve the question himself, and his eye will sparkle with the consciousness of victory.

3. The pupil should be encouraged to study out different solutions, and to adopt the most *concise* and *elegant.*

4. Finally, he should learn to perform examples *independent* of the answer. Without this attainment the pupil receives but little or no *discipline* from the study, and acquires no *confidence* in his own *abilities.* What though he comes to the recitation with an occasional wrong answer; it were better to solve one question *understandingly* and *alone*, than to copy a *score* of answers from the book. What would the study of mental arithmetic be worth, if the pupil had the answers before him? What is a young man good for in the *counting-room*, who cannot perform arithmetical operations without looking to the *answer?* Every one pronounces him *unfit* to be trusted with *business calculations.*

CONTENTS.

	PAGE.
INTRODUCTION,	13
SECTION I.	
NOTATION,	20
Numeration,	24
Formation of different systems of Notation,	29
SECTION II.	
GENERAL RULE for Addition,	33
Different methods of proving Addition,	34
Counting-room Exercises,	37
Adding numbers expressed by the Roman Notation,	40
SECTION III.	
GENERAL RULE for Subtraction,	44
Different methods of proving Subtraction,	44
Subtracting numbers expressed by the Roman Notation,	47
SECTION IV.	
GENERAL RULE for Multiplication,	53
Different methods of proving Multiplication,	54
Contractions in Multiplication, six methods,	56
SECTION V.	
GENERAL RULE for Division,	71
Different methods of Proving Division,	72
Contractions in Division, nine methods,	74
General principles in Division,	81
CANCELATION,	83
Applications of the fundamental rules,	85
SECTION VI.	
PROPERTIES OF NUMBERS,	89
Method of finding the prime numbers in any series,	93
Table of prime numbers from 1 to 3413.	94
Numbers changed from the decimal to other scales of Notation,	95
Analysis of Composite Numbers,	97
Greatest Common Divisor, two methods,	100
Least Common Multiple, three methods,	102
SECTION VII.	
FRACTIONS,	107
General principles pertaining to Fractions,	109
Reduction of Fractions,	111
Addition of Fractions,	119
Subtraction of Fractions,	122
Multiplication of Fractions, common method,	124
Multiplication of Fractions by Cancelation,	130
Contractions in Multiplication of Fractions, four methods,	131—132

1*

PAGE
Division of Fractions, common method, - - - - - 133
Division of Fractions by Cancelation, - - - - - 136
Contractions in Division of Fractions, three methods, - - 138
Complex Fractions reduced to Simple ones, - - - 139

SECTION VIII.

COMPOUND NUMBERS, - - - - - - - - 144
Tables of Compound Numbers, - - - - - - 144
The Standard for Gold and Silver Coin of the United States, - 145
The Standard UNIT of Weight of the United States, - - 147
Origin of Weights and Measures, - - - - - - 147
The Avoirdupois Pound of the United States and Great Britain, - 148
The Standard UNIT of Length of the United States, - - - 150
" " " of New York and Great Britain, - 151
The Standard UNIT of Liquid Measure of the United States, - - 154
" " of Dry Measure of the United States, - 155
Method of determining whether a given year is Leap Year, - - 157
French Money, Weights and Measures, - - - - 161
Foreign Weights and Measures, compared with United States, - 163
GENERAL RULE for Reduction, - - - - - - 166
Applications of Reduction, - - - - - - - 169
Cubic Measure reduced to Dry, or Liquid Measure, &c., - - 172
Longitude reduced to Time, &c., - - - - - - 175
Compound Numbers reduced to Fractions, - - - 176
Fractional Compound Numbers reduced to whole numbers, - - 179
Compound Addition, - - - - - - - 181
Compound Subtraction, - - - - - - - 183
Compound Multiplication, - - - - - - - 186
Compound Division, - - - - - - - - 189

SECTION IX.

DECIMAL FRACTIONS, their origin, &c., - - - - - 191
Method of reading Decimals, - - - - - - - 193
Addition of Decimals, - - - - - - - 195
Subtraction of Decimals, - - - - - - - 197
Multiplication of Decimals, - - - - - - 199
Contractions in Multiplication of Decimals, two cases, - - 201
Division of Decimals, - - - - - - - 205
Contractions in Division of Decimals, - - - - 208
Decimals reduced to Common Fractions, - - - - 210
Common Fractions reduced to Decimals, - - - - 211
Compound Numbers reduced to Decimals, - - - 215

SECTION X.

ADDITION of Circulating Decimals, - - - - - 222
Subtraction of Circulating Decimals, - - - - - - 223
Multiplication of Circulating Decimals, - - - - - 224
Division of Circulating Decimals, - - - - - 225

PAGE

SECTION XI.

Federal Money, - - - - - - - - 226
Addition of Federal Money, - - - - - - - 229
Subtraction of Federal Money, - - - - - - 231
Multiplication of Federal Money, - - - - - - 232
Division of Federal Money, - - - - - - 234
Counting-room Exercises, Contractions, &c., - - - - 237
To find the cost of articles bought and sold by the 100 or 1000, - 238
Bills, Accounts, &c., - - - - - - - - 239

SECTION XII.

Percentage, Percentage Table, &c., - - - - - 241
Applications of Percentage, - - - - - - - 244
Commission, Brokerage, and Stocks, - - - - - 244
Commission deducted in advance, and the balance invested, - - 247
Interest, - - - - - - - - - - 249
General method for computing Interest, - - - - - 252
Second method, multiplying the Prin. by the Int. of $1 for the time, 255
To compute Interest by half the number of months, - - - 257
To compute Interest by the number of days, - - - - 258
Applications of Interest, remarks on Promissory Notes, &c., - - 258
Forms of Negotiable Notes, &c., - - - - - - 260
Partial Payments; United States Rule, - - - - - 261
Connecticut Rule, - - - - - - - - 263
Vermont Rule, - - - - - - - - - 264
Problems in Interest, - - - - - - - 265
Compound Interest, - - - - - - - - 270
Discount, - - - - - - - - - - 272
Bank Discount, - - - - - - - - 274
To find what sum must be discounted to produce a given amount, 277
Insurance, four Cases, - - - - - - - - 278
The sum to be insured to recover the premium and the property, - 282
Life Insurance, - - - - - - - - - 282
Profit and Loss, four Cases, - - - - - - - 283
Duties, Specific and Advalorem, - - - - - - 289
Assessment of Taxes, - - - - - - - - 292
To find what sum must be assessed to raise a given net amount, - 295
Formation of Tax Bills, - - - - - - - 295
Rate Bills for Schools, - - - - - - - 297

SECTION XIII.

Analysis, - - - - - - - - - - 298
Analytic solutions of questions in Simple Proportion, - - 299
" " " Barter, - - - - - 301
" " " Partnership, - - - - 302
" " " General Average, - - - 303
" " " Alligation, - - - 304

PAGE
Analytic solutions of questions in Compound Proportion, 307
" " " Position, 308
" " " Practice, 309

SECTION XIV.

Ratio, general principles pertaining to it, 313
Simple Proportion, 321
Simple Proportion and its Proof by Cancelation, 325
Compound Proportion, 328
Compound Proportion and its Proof by Cancelation, 330
Conjoined Proportion, 332

SECTION XV.

Duodecimals, its principles, &c., 334
Multiplication of Duodecimals, 334

SECTION XVI.

Equation of Payments, 338
Partnership, 340
General Average, 343
Exchange of Currencies, 345
Foreign Coins and Moneys of Account, 348
Exchange, Form of Bills of Exchange, &c., 351
Arbitration of Exchange, 355
Alligation, Medial, and Alternate, 356

SECTION XVII.

Involution, and Evolution, 360
Properties of Squares and Cubes, 365
Extraction of the Square Root, 367
Demonstration of the Square Root, 368
Applications of the Square Root, 370
Extraction of the Cube Root, Horner's Method, 374
Demonstration of the Cube Root, 376
Extraction of Higher Roots, 378

SECTION XVIII.

Arithmetical Progression, 381
Geometrical Progression, 384
Annuities, 386
Permutations and Combinations, 388

SECTION XIX.

Application of Arithmetic to Geometry, 389
Mensuration of Surfaces and Solids, 389
Measurement of timber, 392
Gauging of Casks, 393
Tonnage of Vessels, 393
Mechanical Powers, 394
Miscellaneous Examples, 395

INTRODUCTION.

ART. **1.** Anything which can be *multiplied, divided,* or ***measured,*** is called QUANTITY. Thus, lines, weight, time, number, &c., **are** quantities.

OBS. 1. A *line* is a quantity, because it can be measured in feet and inches; *weight* can be measured in pounds and ounces; *time*, in hours and minutes numbers can be multiplied, divided, &c.

2. Color, and the operations of the mind, as love, hatred, desire, choice, &c., cannot be multiplied, divided, or measured, and therefore cannot properly be called quantities.

2. MATHEMATICS is the *science of Quantity.*

3. The fundamental branches of Mathematics are, ***Arithmetic,*** *Algebra,* and *Geometry.*

4. *Arithmetic* is the science of *Numbers.*

5. *Algebra* is a general method of solving problems, and of investigating the relations of quantities by means of *letters* and *signs.*

OBS. *Fluxions*, or the *Differential* and *Integral Calculus*, may be considered as belonging to the higher branches of Algebra.

6. *Geometry* is that branch of Mathematics which treats **of** *Magnitude.*

7. The term *magnitude* signifies that which is ***extended,*** **or** which has one or more of the three dimensions, ***length, breadth,*** **and** *thickness.* Thus, lines, surfaces, and solids are *magnitudes.*

QUEST.—1. What is Quantity? Give some examples of quantity. *Obs.* Why is a line a quantity? Weight? Time? Numbers? Are color and the operations of the mind quantities? Why not? 2. What is Mathematics? 3. What are the fundamental branches of mathematics? 4. What is Arithmetic? 5. Algebra? 6. Geometry? 7. What is meant by magnitude?

OBS. 1. A *line* is a magnitude, because it can be extended in length; a *surface*, because it has length and breadth; a *solid*, because it has length, breadth, and thickness.

2. *Motion*, though a quantity, is not, strictly speaking, a magnitude; for it has neither length, breadth, nor thickness.

3. The term *magnitude* is sometimes, though inaccurately, used as synonymous with *quantity*.

8. *Trigonometry* and *Conic Sections* are branches of Mathematics, in which the principles of Geometry are applied to *triangles*, and the *sections* of a *cone*.

9. Mathematics are either *pure* or *mixed*.

In *pure* mathematics, quantities are considered, independently of any substances actually existing.

In *mixed* mathematics, the relations of quantities are investigated in connection with some of the properties of matter, or with reference to the common transactions of business. Thus, in Surveying, mathematical principles are applied to the measuring of land; in Optics, to the properties of light; and in Astronomy, to the heavenly bodies.

OBS. The science of *pure mathematics* has long been distinguished for the clearness and distinctness of its principles, and the irresistible conviction which they carry to the mind of every one who is once made acquainted with them. This is to be ascribed partly to the nature of the subjects, and partly to the *exactness* of the *definitions*, the *axioms*, and the *demonstrations*.

10. A *definition* is an explanation of what is meant by a *word*, or *phrase*.

OBS. It is essential to a *complete* definition, that it *perfectly distinguishes* the thing defined, from everything else.

11. A *proposition* is something proposed to be *proved*, or required to be *done*, and is either a *Theorem*, or a *Problem*.

12. A *theorem* is something to be *proved*.

13. A *problem* is something to be *done*, as a question to be solved.

QUEST.—*Obs.* Why is a line a magnitude? A surface? A solid? Is motion a magnitude? Why not? 9. Of how many kinds are mathematics? In pure mathematics how are quantities considered? How in mixed mathematics? *Obs.* For what is the science of pure mathematics distinguished? 10. What is a definition? *Obs.* What is essential to a complete definition? 11. What is a proposition? 12. A theorem? 13. A problem?

OBS. 1. In the statement of every proposition, whether theorem or problem, certain things must be given, or assumed to be true. These things are called the *data* of the proposition.

2. The operation by which the answer of a problem is found, is called a *solution*.

3. When the given problem is so easy, as to be obvious to every one without explanation, it is called a *postulate*.

14. One proposition is *contrary*, or contradictory to another, when what is *affirmed* in the one, is *denied* in the other.

OBS. A proposition and its contrary, can never *both* be true. It cannot be true, that two given lines are equal, and that they are *not* equal, at the same time.

15. One proposition is the *converse* of another, when the order is inverted; so that, what is *given* or *supposed* in the first, becomes the *conclusion* in the last; and what is *given* in the last, is the *conclusion*, in the first. Thus, it can be proved, first, that if the *sides* of a triangle are equal, the *angles* are equal; and secondly, that if the *angles* are equal, the *sides are* equal. Here, in the first proposition, the equality of the *sides is given*, and the equality of the *angles inferred;* in the second, the equality of the *angles* is given, and the equality of the *sides* inferred.

OBS. In many instances, a proposition and its converse are both true, as in the preceding example. But this is not always the case. A circle is a figure bounded by a curve; but a figure bounded by a curve is not necessarily a circle.

16. The *process of reasoning* by which a proposition is shown to be true, is called a *demonstration*.

OBS. A demonstration is either *direct* or *indirect*.

A *direct* demonstration commences with certain principles or data which are admitted, or have been proved to be true; and from these, a series of other truths are deduced, each depending on the preceding, till we arrive at the truth which was required to be established.

An *indirect* demonstration is the mode of establishing the truth of a proposition by proving that the supposition of its *contrary*, involves an absurdity.

QUEST.—*Obs.* What is meant by the data of a proposition? By the solution of a problem? What is a postulate? 14. When is one proposition contrary to another? *Obs.* Can a proposition and its contrary both be true? 15. When is one proposition the converse of another? *Obs.* Can a proposition and its converse both be true? 16. What is a demonstration? *Obs.* Of how many kinds are demonstrations? What is a direct demonstration? An indirect demonstration?

This is commonly called *reductio ad absurdum*. The former is the more common method of conducting a demonstrative argument, and is the most satisfactory to the mind.

17. A *Lemma* is a subsidiary truth or proposition, demonstrated for the purpose of using it in the demonstration of a theorem, or the solution of a problem.

18. A *Corollary* is an inference or principle deduced from a preceding proposition.

19. A *Scholium* is a remark made upon a preceding proposition, pointing out its *connection, use, restriction,* or *extension.*

20. An *Hypothesis* is a *supposition,* made either in the statement of a proposition, or in the course of a demonstration.

AXIOMS.

21. An *Axiom* is a self-evident proposition; that is, a proposition whose *truth* is so *evident* at sight, that no process of reasoning can make it plainer. The following axioms are among the most common:

1. Quantities which are equal to the *same* quantity, are equal to each other.
2. If the same or equal quantities are *added* to equals, the *sums* will be equal.
3. If the same or equal quantities are *subtracted* from equals, the *remainders* will be equal.
4. If the same or equal quantities are *added to unequals,* the *sums* will be unequal.
5. If the same or equal quantities are *subtracted* from *unequals,* the *remainders* will be unequal.
6. If equal quantities are *multiplied* by the same or equal quantities, the *products* will be equal.
7. If equal quantities are *divided* by the same or equal quantities, the *quotients* will be equal.
8. If the same quantity is both *added* to and *subtracted* from another, the *value* of the latter will not be altered.

QUEST.—17. What is a lemma? 18. What is a corollary? 19. What is a scholium? 20. What is an hypothesis? 21. What is an axiom? Name some of the most common axioms.

9. If a quantity is both *multiplied* and *divided* by the same or ar equal quantity, its *value* will not be altered.

10. The *whole* of a quantity is *greater* than a *part.*

11. The *whole* of a quantity is equal to the sum of *all its parts.*

SIGNS.

22 *Addition* is represented by the sign (+), which is called *plus.* It consists of two lines, one horizontal, the other perpendicular, forming a cross, and shows that the numbers between which it is placed, are to be added together. Thus, the expression 6+8, signifies that 6 is to be added to 8. It is read, "6 plus 8," or "6 added to 8."

OBS.—The term *plus* is a Latin word, originally signifying "more," hence "added to."

23. *Subtraction* is represented by a short horizontal line (—), which is called *minus.* When placed between two numbers, it shows that the number after it is to be subtracted from the one before it. Thus, the expression 9—4, signifies that 4 is to be subtracted from 9; and is read, "9 minus 4," or "9 less 4."

OBS.—The term *minus* is a Latin word, signifying *less.*

24. *Multiplication* is usually denoted by two oblique lines crossing each other (×), called the *sign of multiplication.* It shows that the numbers between which it is placed, are to be multiplied together. Thus, the expression (9×6), signifies that 9 and 6 are to be multiplied together, and is read, "9 multiplied by 6," or simply, "9 into 6." Sometimes multiplication is denoted by a *point* (.) placed between the two numbers or quantities. Thus, 9.6 denotes the same as 9×6.

OBS. It is better to denote the multiplication of figures by a *cross* than by a *point;* for the latter is liable to be confounded with the *decimal* point.

24. *a.* When two or more numbers are to be subjected to the same operation, they must be connected by a line (——) placed

QUEST.—22. What is the sign of addition called? Of what does it consist? What does it show? *Obs.* What is the meaning of the term plus? 23. How is subtraction represented? What is the sign of subtraction called? What does it show? *Obs.* What does the term minus signify? 24. How is multiplication usually denoted? What does the sign of multiplication show? In what other way is multiplication sometimes denoted?

over them, called a *vinculum*, or by a parenthesis ( ). Thus the expression (12+3)×2, shows that the sum of 12 and 3, is to be multiplied by 2, and is equal to 30. But 12+3×2, signifies that 3 only is to be multiplied by 2, and that the product is to be added to 12, which will make 18.

25. *Division* is expressed in two ways:

First, by a horizontal line between two dots (÷), called the *sign of division*, which shows that the number before it, is to be divided by the number *after* it. Thus, the expression 24÷6, signifies that 24 is to be divided by 6.

Second, division is often expressed by placing the divisor *under* the dividend, in the form of a fraction. Thus, the expression $\frac{35}{7}$, shows that 35 is to be divided by 7, and is equivalent to 35÷7.

26. The *equality* between two numbers or quantities, is represented by two parallel lines (=), called *the sign of equality*. Thus, the expression 5+3=8, denotes that 5 added to 3 are equal to 8. It is read, "5 plus 3 equal 8," or "the sum of 5 plus 3 is equal to 8." So 7+5=16—4=12.

QUEST.—24. *a.* When two or more numbers are to be subjected to the same operation what must be done? 25. In how many ways is division expressed? What is the first? What does this sign show? What is the second? 26. How is the equality between two numbers or quantities represented?

ARITHMETIC.

SECTION I.

NOTATION AND NUMERATION.

Art. **27.** Any single thing, as a peach, a rose, a book, is called a *unit*, or *one*; if another single thing is put with it, the collection is called *two*; if another still, it is called *three*; if another, *four*; if another, *five*, &c.

The terms, *one, two, three*, &c., by which we express *how many single things* or *units* are under consideration, are the *names of numbers*. Hence,

28. Number signifies a *unit*, or a *collection of units*.

Obs. 1. Numbers are divided into two classes, *abstract* and *concrete*.

When they are applied to particular objects, as *two* pears, *five* pounds, *ten* dollars, &c., they are called concrete numbers.

When they do *not* refer to any particular object, as when we say *four* and *five* are *nine*, they are called abstract numbers.

2. *Whole* numbers are often called *integers*.

3. Numbers have various properties and relations, and are applied to various computations in the practical concerns of life. These properties and applications are formed into a system, called *Arithmetic*.

29. Arithmetic *is the science of numbers.*

Obs. 1. The term *Arithmetic* is derived from the Greek word *arithmētikē*, which signifies the *art* of *reckoning* by numbers.

2. The aid of Arithmetic is required to make and apply calculations not only in *business transactions*, but in almost every *department of mathematics.*

Quest.—27. What is a single thing called? If another is put with it, what is the collection called? If another, what? What are the terms one, two, three, &c.? 28. What does number signify? *Obs.* Into how many classes are numbers divided? When are they called concrete? When abstract? To what are numbers applied? 29. What is Arithmetic? *Obs.* In what is the aid of arithmetic required?

Numbers are expressed by *words*, by *letters*, and by *figures*.

NOTATION.

30. *The art of expressing numbers by letters or figures, is called* NOTATION. There are two methods of notation in use, the *Roman* and the *Arabic*.

31. The Roman method employs seven capital letters, viz: I, V, X, L, C, D, M. When standing alone, the letter I, denotes *one;* V, *five;* X, *ten;* L, *fifty;* C, *one hundred;* D, *five hundred;* M, *one thousand.* To express the intervening numbers from one to a thousand, or any number larger than a thousand, we resort to repetitions and various combinations of these letters. The method of doing this will be easily learned from the following

TABLE.

I	denotes	one.	XXX	denote	thirty.
II	"	two.	XL	"	forty.
III	"	three.	L	"	fifty.
IV	"	four.	LX	"	sixty.
V	"	five.	LXX	"	seventy.
VI	"	six.	LXXX	"	eighty.
VII	"	seven.	XC	"	ninety.
VIII	"	eight.	C	"	one hundred.
IX	"	nine.	CI	"	one hundred and one.
X	"	ten.	CX	"	one hundred and ten.
XI	"	eleven.	CC	"	two hundred.
XII	"	twelve.	CCC	"	three hundred
XIII	"	thirteen.	CCCC	"	four hundred.
XIV	"	fourteen.	D	"	five hundred.
XV	"	fifteen.	DC	"	six hundred.
XVI	"	sixteen.	DCC	"	seven hundred.
XVII	"	seventeen.	DCCC	"	eight hundred.
XVIII	"	eighteen.	DCCCC	"	nine hundred.
XIX	"	nineteen.	M	"	one thousand.
XX	"	twenty.	MM	"	two thousand.
XXI	"	twenty-one.	MDCCCLV,		one thousand eight hundred and fifty-five.
XXII	"	twenty-two, &c.			

QUEST.—29. How are numbers expressed? 30. What is notation? How many methods are there in use? 31. What is employed by the Roman method?

OBS. 1. This method of expressing numbers was invented by the Romans, and is therefore called the *Roman Notation*. It is now seldom used, except to denote chapters, sections, and other divisions of books and discourses.

2. The letters C and M, are the initials of the Latin words *centum*, and *mille*, the former of which signifies a *hundred*, and the latter a *thousand*; for this reason it is supposed they were adopted to represent these numbers.

31. *a.* It will be perceived from the Table above, that every time a letter is repeated, its *value* is repeated. Thus I, standing alone, denotes *one*; II, *two ones*, or *two*, &c. So X denotes *ten*; XX, *twenty*, &c.

When a letter of a *less* value is placed *before* a letter of a *greater* value, the less *takes away* its own value from the greater; but when placed *after*, it *adds* its own value to the greater.

32. A line or bar (—) placed over a letter, increases its value a *thousand times*. Thus, V denotes five, $\overline{\text{V}}$ denotes five thousand; X, ten; $\overline{\text{X}}$, ten thousand, &c.

OBS. 1. In the early periods of this notation, four was written IIII, instead of IV; nine was written VIIII, instead of IX; forty was written XXXX, instead of XL, &c.

The former method is more convenient in performing arithmetical operations in addition and subtraction; while the latter is shorter and better adapted to ordinary purposes.

2. A *thousand* was originally written CIↃ, which, in later times, was changed into M; *five hundred* was written IↃ instead of D. Annexing Ↄ to IↃ increased its value ten times. Thus, IↃↃ denoted *five thousand*; IↃↃↃ, *fifty thousand*, &c.

3. Prefixing C and annexing Ↄ to the expression CIↃ, makes its value *ten* times greater: thus, CCIↃↃ denotes *ten thousand*; CCCIↃↃↃ, a *hundred thousand*. According to Pliny, the Romans carried this mode of notation no further. When they had occasion to express a larger number, they did it by repetition. Thus, CCCIↃↃↃ, CCCIↃↃↃ, expressed *two hundred thousand*, &c.

33. The common method of expressing numbers is by the *Arabic Notation*. The Arabic method employs the following *ten characters* or *figures*, viz:

1	2	3	4	5	6	7	8	9	0
one,	two,	three,	four,	five,	six,	seven,	eight,	nine,	zero.

QUEST.—*Obs.* Why is this method called Roman? 31. *a.* What is the effect of repeating a letter? If a letter of less value is placed before another of greater value, what is the effect? If placed after, what? 32. When a line or bar is placed over a letter, how does it affect its value? 33. What is the common way of expressing numbers? How many characters does this method employ?

The first nine are called *significant* figures, because each one always has a value, or denotes some number. They are also called *digits*, from the Latin word *digitus*, which signifies a finger.

The last one is called a *cipher*, or *naught*, because when standing *alone* it has *no value*, or signifies *nothing*.

OBS. 1. It must not be inferred, however, that the cipher is *useless*; for when placed on the right of any of the significant figures, it increases their value. It may therefore be regarded as an *auxiliary* digit, whose office, it will be seen hereafter, is as important as that of any other figure in the system.

2. Formerly all the Arabic characters were indiscriminately called *ciphers*; hence the process of calculating by them was called *ciphering*; on the same principle that calculating by *figures* is called *figuring*.

34. It will be seen that *nine* is the *greatest* number that can be expressed by *any single* figure in the Arabic system of Notation.

All numbers *larger* than nine are expressed by combining together two or more of these ten figures, and assigning different values to them, according as they occupy different places. For example, ten is expressed by combining the 1 and 0, thus 10; eleven by two 1s, thus 11; twelve by 1 and 2, thus 12; twenty, thus 20; thirty, thus 30; &c. A hundred is expressed by combining the 1 and two 0s, thus 100; two hundred, thus 200; a thousand by combining the 1 and three 0s, thus 1000; two thousand, thus 2000; ten thousand, thus 10,000; a hundred thousand, thus 100,000; a million, thus 1,000,000; ten millions, thus 10,000,000; &c. Hence,

35. The digits 1, 2, 3, &c., standing alone, or in the right hand place, respectively denote *units* or *ones*, and are called *units* of the *first order*.

When they stand in the *second* place, they express *tens*, or *ten ones*; that is, their value is *ten times* as much as when standing

---

QUEST.—What are the first nine called? Why? What else are they called? What is the last one called? Why? *Obs.* Is the cipher useless? What may it be regarded? What is the origin of the term ciphering? 34. What is the greatest number that can be expressed by one figure? How are larger numbers expressed? 35. What do the digits, 1, 2, 3, &c., denote, when standing alone, or in the right hand place? What are they then called? What do they denote when standing in the second place?

in the first or right hand place, and they are called *units* of the *second order.*

When occupying the *third* place, they express *hundreds;* that is, their value is *ten times* as much as when standing in the second place, and they are called *units* of the *third order.*

When occupying the *fourth* place, they express *thousands;* that is, their value is *ten times* as much as when standing in the third place, and they are called *units* of the *fourth order,* &c. Thus, it will be seen that,

Ten units make *one ten, ten tens* make *one hundred,* and *ten hundreds* make *one thousand;* that is, *ten* in an *inferior* order are equal to *one* in the next *superior* order. Hence, universally,

36. *Numbers increase from right to left in a tenfold ratio; consequently each removal of a figure one place towards the left, increases its value ten times.*

Note.—1. The number which forms the *basis,* or which expresses the *ratio* of increase in a system of Notation, is called the RADIX of that system. Thus, the radix of the Arabic notation is *ten.*

2. The reason that numbers increase from *right* to *left,* instead of left to right, is probably owing to the ancient practice of writing from the right hand to the left.

37. The different values which the same figures have, are called *simple* and *local* values.

The *simple* value of a figure is the value which it expresses when it stands alone, or in the right hand place. Hence the simple value of a figure is the number which its name denotes.

The *local* value of a figure is the increased value which it expresses by having other figures placed on its right. Hence the local value of a figure depends on its locality, or the place which

---

QUEST.—What is their value then? What are they called? What is a figure called when it occupies the third place? What is its value then? What is it called when in the fourth place? What is its value? How many units are required to make one ten? How many tens make a hundred? How many hundreds make a thousand? How many of an inferior order are required to make one of the next superior order? 36. What is the general law by which numbers increase? What effect has it upon the value of a figure to remove it one place towards the left? *Note.* What is the number called which forms the basis or the ratio of increase in a system of notation? What is the radix of the Arabic notation? Why do numbers increase from right to left? 37. What are the different values of the same figure called? What is the simple value of a figure? What the local?

it occupies in relation to other numbers with which it is connected. (Art. 35.)

OBS. 1. This system of notation is called *Arabic*, because it is supposed to have been invented by the Arabs.

2. It is also called the *decimal system*, because numbers increase in a tenfold ratio. The term *decimal* is derived from the Latin word *decem*, which signifies ten.

3. The early history of the Arabic notation is veiled in obscurity. It is the opinion of some whose judgment is entitled to respect, that it was invented by the philosophers of India. It was introduced into Europe from Arabia about the eighth century, and about the eleventh century it came into general use, both in England and on the continent. The application of the term *digit* to the significant figures, affords strong presumptive evidence that the system had its origin in the ancient mode of counting and reckoning by means of the *fingers;* and that the idea of employing *ten* characters, instead of *twelve* or *any other* number, was suggested by the number of fingers and thumbs on both hands. (Art. 33.)

NUMERATION.

38. *The art of reading numbers when expressed by figures, is called* NUMERATION.

The pupil will easily learn to read the largest numbers from the following scheme, called the

NUMERATION TABLE.

Tredecillions.	Duodecillions.	Undecillions.	Decillions.	Nonillions.	Octillions.	Septillions.	Sextillions.	Quintillions.	Quadrillions.	Trillions.	Billions.	Millions.	Thousands.	Units.
685,	876,	389,	764,	391,	827,	218,	649,	853,	123,	234,	579,	793,	465,	623.
XV.	XIV.	XIII.	XII.	XI.	X.	IX.	VIII.	VII.	VI.	V.	IV.	III.	II.	I.

39. The different orders of numbers are divided into *periods* of three figures each, *beginning* at the *right hand.* The *first* period, which is occupied by units, tens, hundreds, is called *units'*

QUEST.—Upon what does the local value of a figure depend? *Obs.* Why is this system of notation called Arabic? What else is it sometimes called? Why? What do you say of its early history? When was it introduced into Europe? What is the probable origin of the system? Why were ten characters, rather than any other number, adopted? 38. What is Numeration? 39. How are the orders of numbers divided? What is the first period called? By what is it occupied?

period; the *second* is occupied by thousands, tens of thousands, hundreds of thousands, and is called *thousands'* period; the *third* is occupied by millions, tens of millions, hundreds of millions, and is called *millions'* period; the *fourth* is occupied by billions, tens of billions, hundreds of billions, and is called *billions'* period; and so on, the orders of each successive period being *units, tens,* and *hundreds.*

The figures in the table are read thus: 685 tredecillions, 876 duodecillions, 389 undecillions, 764 decillions, 391 nonillions, 827 octillions, 218 septillions, 649 sextillions, 853 quintillions, 123 quadrillions, 234 trillions, 579 billions, 793 millions, 465 thousand, 6 hundred and twenty-three.

Note.—1. The terms *thirteen, fourteen, fifteen,* &c., are obviously derived from three and ten, four and ten, five and ten, which by contraction become thirteen, fourteen, fifteen, and are therefore significant of the numbers which they denote. The terms *eleven* and *twelve,* are generally regarded as primitive words; at all events, there is no perceptible analogy between them and the numbers which they represent. Had the terms *oneteen* and *twoteen* been adopted in their stead, the names would then have been significant of the numbers one and ten, two and ten; and their etymology would have been similar to that of the succeeding terms.

The terms *twenty, thirty, forty,* &c., were formed from two tens, three tens, four tens, which were contracted into twenty, thirty, forty, &c.

The terms *twenty-one, twenty-two, twenty-three,* &c., are compounded of twenty and one, twenty and two, &c. All the other numbers as far as ninety-nine, are formed in a similar manner.

2. The terms *hundred, thousand* and *million* are primitive words, and bear no analogy to the numbers which they denote. The numbers between a hundred and a thousand are expressed by a repetition of the numbers below a hundred. Thus we say one hundred and one, one hundred and two, one hundred and three, &c.

3. The terms *billion, trillion, quadrillion,* &c., are formed from million and the Latin numerals *bis, tres, quatuor,* &c. Thus, prefixing *bis* to *million,* by a slight contraction for the sake of euphony, it becomes *billion;* prefixing *tres* to *million,* it is easily contracted into trillion, &c. The Latin word *bis* signifies two; *tres,* three; *quatuor,* four; *quinque,* five; *sex,* six; *septem,* seven; *octo,* eight; *novem,* nine; *decem,* ten; *undecim,* eleven; *duodecim,* twelve; *tredecim,* thirteen.

QUEST.—What is the second period called? By what occupied? What is the third called? By what occupied? What is the fourth called? By what occupied? What is the fifth called? By what occupied? Repeat the Numeration Table, beginning at the right hand.

Higher periods than those in the Table, may be easily formed by following the above analogy.

4. The foregoing law, which assigns superior values to these ten characters, according to the order or place which they occupy and the use of so many derivative and compound words in forming the names of numbers, saves an inconceivable amount of time and labor in learning Notation and Numeration, as well as in their application.

40. To read numbers which are expressed by figures.

Point them off into periods of three figures each; then, beginning at the left hand, read the figures of each period in the same manner as those of the right hand period are read, and at the end of each period, pronounce its name.

OBS. 1. The learner must be careful, in *pointing off* figures, always to begin at the *right* hand; and in *reading* them, to begin at the *left* hand.

2. Since the figures in the first or right hand period always denote units, its name is not pronounced. Hence, in reading figures, when no period is mentioned, it is always understood to be the right hand, or units' period.

EXERCISES IN NUMERATION.

Note.—In numerating large numbers, it is advisable for the pupil first to apply to each figure the name of the order which it occupies. Thus, beginning at the right hand, he should say, " Units, tens, hundreds," &c., and point at the same time to the figures standing in the order which he mentions.

Read the following numbers:

Ex. 1.	3506	11.	706305	21.	967058713
2.	6034	12.	1640030	22.	32100040
3.	5060	13.	830006	23.	106320000
4.	90621	14.	70900038	24.	780507031
5.	73040	15.	3067300	25.	4063107
6.	450302	16.	12604321	26.	29038450
7.	603260	17.	70003000	27.	1046347025
8.	130070	18.	161010602	28.	20380720000
9.	2021305	19.	80367830	29.	8503467030
10.	4506580	20.	400031256	30.	450670412463

QUEST.—40. How do you read numbers expressed by figures? *Obs.* Where begin to point them off? Where to read them? Do you pronounce the name of the right hand period? When no period is named, what is understood?

31.	430812000641	36.	120340078910356
32.	5200240301000	37.	43601000345000
33.	98760000216	38.	506302870045380
34.	82600381000000	39.	4200812053706203 5
35.	403070003462000	40.	653107843604893048

41. 210 256 031 402 385 290 845 381 467.

42. 361 438 201 219 763 281 578 829 318 278.

41. The method of dividing numbers into *periods* of *three* figures, was invented by the *French*, and is therefore called the *French Numeration.*

The *English* divide numbers into periods of *six* figures, in the following manner:

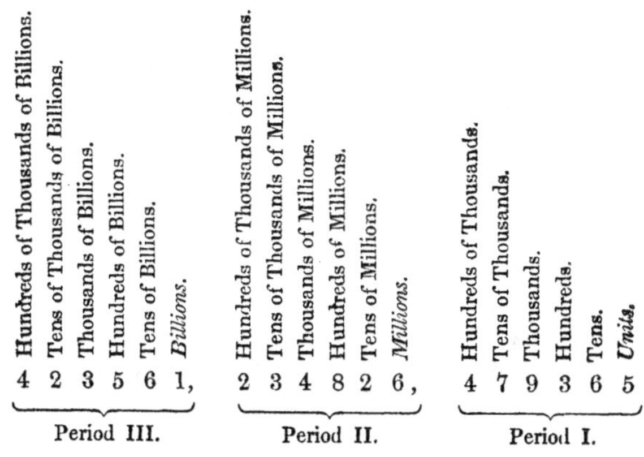

According to this method, the preceding figures are read thus: 423561 billions, 234826 millions, and 479365.

OBS. 1. It will be perceived that the two methods agree as far as hundreds of millions; the former then begins a new period, while the latter continues on through thousands of millions, &c.

2. The French method is generally used throughout the continent of Europe, as well as in America, and has been recently adopted by some English authors. It is very generally admitted to be more simple and convenient than the English method.

QUEST.—41. What is the French method of numeration? What the English method? *Obs.* Which is the more simple and convenient?

EXERCISES IN NOTATION.

42. To express numbers by figures.

Begin at the left hand, and write in each order the figure which denotes the given number in that order.

If any intervening orders are omitted in the proposed number, write ciphers in their places. (Art. 38.)

Write the following numbers in figures:

1. Two thousand, one hundred and nine.
2. Twenty thousand and fifty-seven.
3. Fifty-five thousand and three.
4. One hundred and five thousand, and ten.
5. Seven hundred and ten thousand, three hundred and one
6. Two millions, sixty-three thousand, and eight.
7. Fourteen millions, and fifty-six.
8. Four hundred and forty millions, and seventy-two.
9. Six billions, six millions, six thousand, and six.
10. Forty-five billions, three hundred and forty thousand, and seventy-six.
11. Five hundred and fifty-six millions, three thousand, two hundred and sixty-four.
12. Eight hundred and ten billions, ten millions, and seventy-five thousand.
13. Ninety-six trillions, seven hundred billions, and fifty-four.
14. Three hundred and forty-nine quadrillions, five trillions, seven billions, four millions, and twenty.
15. Nineteen quintillions.
16. Six hundred and thirty sextillions.
17. Two hundred and ninety-eight septillions.
18. Seventy-four octillions.
19. Four hundred and ten decillions.
20. Eight hundred and sixty-three duodecillions.
21. Nine hundred and thirty-five tredecillions.
22. Six hundred and seventy-three quintillions, seventeen quadrillions, and forty-five.
23. Twenty trillions, six hundred and forty-eight billions, and twenty-five thousand.

OBS. The *great facility* with which large numbers may be expressed both in language and by figures, is calculated to give an *imperfect idea* of their real *magnitude*. It may assist the learner in forming a just conception of a *million*, a *billion*, a *trillion*, &c., to reflect, that *to count* a million, at the rate of a hundred a minute, would require nearly *seventeen days* of ten hours each; *to count* a billion, at the same rate, would require more than *forty-five years;* and *to count* a trillion, more than 45,662 years.

43. From the preceding illustrations, the learner will perceive that a variety of other systems of notation may be formed upon the same principle, having different numbers for their *radices*. Thus, if we wished to form a *quinary* system; that is, a system in which the numbers should increase in a *five-fold ratio,* or has *five* for its *radix,* it would require *four significant* figures and a *cipher*. Let the figures 1, 2, 3, 4, and 0, be the characters employed; then five would be expressed by 1 and 0, and would be written thus 10; six by 1 and 1, thus 11; seven by 1 and 2, thus 12; eight by 1 and 3, thus 13; nine by 1 and 4, thus 14; ten by 2 and 0, thus 20; eleven by 2 and 1, thus 21, &c.

44. In the *binary* or *diadic* system of notation developed by Leibnitz, there are two characters employed, 1 and 0. The cipher when placed at the right hand of a number, in this system, multiplies it by *two*. Thus the number one is expressed by 1; two by 10; three by 11; four by 100; five by 101; six by 110; seven by 111; eight by 1000; nine by 1001; ten by 1010; eleven by 1011, &c.

OBS. 1. In like manner other systems of notation may be formed, having *three*, *four*, *six*, *eight*, *twelve*, or *any given number* for their *radix*.

When the radix is two, the system is called *binary* or *diadic;* when three it is called *ternary;* when four, *quaternary;* when five, *quinary;* when six, *senary;* when seven, *septenary;* when eight, *octary;* when nine, *nonary*, &c.

2. It should be observed that every system of notation, formed upon the foregoing principles, will require as many distinct characters, as there are units in the *radix*, and that one of them must be a *cipher*, and another a *unit*.

For the method of changing numbers from the *decimal* to *other* scales of notation, and the converse, see Arts. 162, 163.

QUEST.—43. Is the decimal notation the only system that can be formed on the same principles? How would you form a quinary system of notation? Write six in the quinary scale on the black-board. Write seven, nine, ten, eleven, twelve. *Obs.* How many characters will any system formed upon this principle require?

45. About the commencement of the second century, Ptolemy introduced the *sexagesimal* notation, which has *sixty* for its *radix.*

OBS. 1. It is said that the Chinese and some other eastern nations now employ this system in measuring time, using periods of *sixties*, instead of *centuries.* Relics of the sexagesimal notation may also be seen in our division of the *circle*, and of *time*, where the degree and hour are each divided into 60 minutes, the minute into 60 seconds, &c.

2. The Roman notation seems to have been commenced with V or *five* for its radix, which was afterwards changed to X or *ten*. It may therefore be regarded as a kind of combination of the *quinary* and *decimal* systems.

46. Since the number *eight* may be divided and sub-divided so many times without a remainder, some contend that a system of notation having *eight* for its *radix*, would be preferable to the *decimal* system.

Others claim that the *duodecimal* notation; that is, a system with *twelve* for its *radix*, would be more convenient than either.* However this may be, the decimal system is so firmly rooted, it were hopeless to attempt a change.

OBS. It may be doubted whether any other *ratio* of increase would, on the whole, be more convenient, than that of the present system. If the ratio were *less*, it would require more places of figures to express large numbers; if the ratio were *larger*, it would not indeed require so many figures, but the operations would manifestly be more difficult than at present, on account of the numbers in each order being larger. Besides, the decimal system is sufficiently comprehensive to express with all desirable facility, every conceivable number, the largest as well as the smallest; and yet it is so simple, that a child may understand and apply it. In a word, it is every way adapted to the practical operations of business, as well as the most abstruse mathematical investigations. In whatever light, therefore, it is viewed, the decimal notation must be regarded as one of the most striking monuments of human ingenuity, and its beneficial influence on the progress of science and the arts, on commerce and civilization, must win for its unknown author the everlasting admiration and gratitude of mankind.

* Barlow's Theory of Numbers, Leslie's Philosophy of Arithmetic, Edinburgh Encyclopedia.

SECTION II.

ADDITION.

ART. **49.** Ex. 1. A man bought three lots of land; the first contained 23 acres, the second 9 acres, and the third 15 acres: how many acres did he buy?

Solution.—23 acres and 9 acres are 32 acres, and 15 are 47 acres. *Ans.* 47 acres.

OBS. It will be seen, that the solution of this example consists in finding a *single number*, which will exactly express the value of the *several given numbers united together*.

50. *The process of uniting two or more numbers together, so as to form a single number, is called* ADDITION.

The *answer*, or the number thus found, is called the *Sum* or *Amount*.

OBS. When the numbers to be added are all of the *same denomination*, as all dollars, all pounds, &c., the operation is called *Simple Addition*.

Ex. 2. A miller bought 7864 bushels of wheat of one man, 4952 bushels of another, and 3273 bushels of another: how many bushels did he buy of all?

Operation.
7864
4952
3273
Ans. 16089 *bu.*

Write the numbers under each other, so that units may stand under units, tens under tens, &c., and draw a line beneath them. Then beginning at the right hand or units, add each column separately. Thus, 3 units and 2 units are 5 units, and 4 are 9 units. Write the 9 in units' place under the column added. Next 7 and 5 are 12, and 6 are 18 tens. But 18 requires two figures to express it; (Art. 34;) consequently it cannot all be written under its own column. We therefore write the 8 or right hand figure in tens' place under the column added, and reserving the 1 or left hand figure, add it with the hundreds. Thus, 1 which was reserved, and 2 are 3, and 9 are 12, and 8 are 20 hundreds. Set the 0 or right hand figure under the column

---

QUEST.—50. What is Addition? What is the answer called? *Obs.* When the numbers to be added are all of the same denomination, what is the operation called? 51 What orders of figures do you add together?

added, and reserving the 2 or left hand figure, add it to the next column as before. Thus, 2 which were reserved and 3 are 5, and 4 are 9, and 7 are 16 thousands. Set the 6 under the column added; and since there is no other column to be added, write the 1 in the next place on the left.

51. It will be perceived in this example, that *units* are added to *units*, *tens* to *tens*, &c.; that is, figures of the *same order* are added to each other. All numbers must be added in the same manner. For, figures standing in *different* orders or columns express *different values;* (Art. 35;) consequently, they *cannot be united* together directly in a single sum. Thus, 3 units and 5 tens will neither make *eight units,* nor *eight tens,* any more than 3 oranges and 5 apples will make 8 apples, or 8 oranges. In like manner it is plain that 7 tens and 2 hundreds will neither make 9 tens, nor 9 hundreds.

OBS. The object of writing *units* under *units*, *tens* under *tens*, &c., is to prevent mistakes which might occur from adding *different* orders to each other.

52. When the *sum* of a column does not exceed 9, it will be noticed, we set it under the column added but if it exceeds 9, we set the *units* or *right hand* figure under the column added, and reserving the *tens* or *left hand* figure, *add* it to the *next* column. In adding the *last* column on the left, we set down the *whole sum.*

OBS. The process of *reserving the tens, or left hand figure*, and *adding* it to the next column, is called *carrying tens.*

53. The *principle* of *carrying* may be illustrated in the following manner.

Take, for instance, the last example, and adding as before, write the sum of each column in a separate line. Thus, the sum of the units' column is 9 units; the sum of the tens' column is 18 tens, or 1 hundred and 8 tens; the sum of the hundreds' column is 19 hundred, or 1 thousand 9 hundred; the sum of the

```
 7864
 4952
 3273
 ----
    9 sum of units
  18*  "  "  tens,
 19**  "  "  hund,
14***  "  "  thou,
 ----
16089 Amount.
```

QUEST.—Why not add figures of different orders together?

thousands' column is 14 thousand. Now, adding these results together as they stand, units to units, tens to tens, &c., the amount is 16089 bushels, which is the same as in the solution above.

Thus, it is evident, when the sum of a column exceeds 9, the right hand figure denotes units of the same order as the column added, and the tens or left hand figure denotes units of the next higher order. Hence,

The reason we carry the tens or left hand figure to the next column, is because it is of the same order as the next column, and figures of the same order must always be added together. (Art. 51.)

OBS. 1. The reason for setting down the *whole sum* of the last or left hand column, is because there are no figures in the next order to which the left hand figure can be added. It is, in fact, carrying it to the next column.

2. From the preceding illustration it will also be seen, that the object of beginning to add at the right hand is, that we may *carry the tens*, as we proceed in the operation.

54. From the preceding illustrations and principles we derive the following

GENERAL RULE FOR ADDITION.

I. *Write the numbers to be added, under each other; so that units may stand under units, tens under tens, &c.* (Art. 51. Obs.)

II. *Begin at the right hand, and add each column separately. When the sum of a column does not exceed 9, write it under the column; but if the sum of a column exceeds 9, write the units' figure under the column added, and carry the tens to the next column.* (Arts. 52, 53.)

III. *Proceed in this manner through all the orders, and set down the whole sum of the last or left hand column.* (Art. 53. Obs.)

55. PROOF.—*Beginning at the top, add each column downwards, and if the second result is the same as the first, the work is supposed to be right.*

QUEST.—54 How do you write numbers to be added? Why place units under units, &c.? Where do you begin to add? When the sum of a column does not exceed 9, what do you do with it? When it exceeds 9, how proceed? What is meant by carrying the tens? Why carry the tens to the next column? Why begin to add at the right hand? What do you do with the sum of the last column? 55. How is addition proved?

Note.—The object of beginning at the top and adding downwards, is that the figures may be taken in a different order from that in which they were added before. The order being reversed, the presumption is, that any mistake which may have been made will thus be detected; for it can hardly be supposed that two mistakes exactly equal will occur.

56. *Second Method.*—Cut off the bottom line, and find the sum of the rest of the numbers; then add this sum and the bottom line together, and if the second result is the same as the first, the work is supposed to be right.

Note.—1. This method of proof depends on the axiom, that the *whole* of a quantity is equal to the sum of all its parts. (Ax. 11.)

2. The method of cutting off the *top* line, and afterwards adding it to the sum of the others, is objectionable on account of adding the numbers in the same order as they were added in the solution. (Art. 55. Note.)

57. *Third Method.*—From the *amount*, subtract all the given numbers but one, and if the remainder is equal to the number not subtracted, the work may be supposed to be right.

Note.—This method supposes the pupil to be acquainted with subtraction, before he commences this work. It is placed here on account of the convenience of having all the methods of proving the rule together.

58. *Fourth Method.**—Cast the 9s out of each of the given numbers separately, and place each excess at the right of the number. Then cast the 9s out of the sum of these excesses; also cast the 9s out of the amount; and if these two excesses are equal, the work may be supposed to be right.

Note.—1. This mode of proof is based on a peculiar property of the number 9. For its illustration and demonstration, see Art. 161. Prop. 14.

2. To cast the 9s out of a number, begin at the left hand, add the digits together, and, as soon as the sum is 9 or over, drop the 9, and add the remainder to the next digit, and so on. For example, to cast the 9s out of 4626357, we proceed thus: 4 and 6 are 10; drop the 9 and add the 1 to the next figure. 1 and 2 are 3, and 6 are 9; drop the 9 as above. 3 and 5 are 8, and 7 are 15; dropping the 9, we have 6 remainder.

EXAMPLES FOR PRACTICE.

59. Ex. 1. A man paid 2468 dollars for his farm, 1645 dollars for a house, 865 dollars for stock, and 467 dollars for tools: how much did he pay for the whole?

2. A produce merchant bought 5 cargoes of corn; the first con-

---

QUEST.—*Note.* Why add the columns downwards, instead of upwards? Can addition be proved by any other methods?

* Wallis' Arithmetic, Oxford, 1657.

tained 6725 bushels, the second 7208, the third 5047, the fourth 12586, and the fifth 10391 bushels: how many bushels did he buy?

3. A tavern-keeper bought six loads of hay which weighed as follows: 1725 pounds, 2163 pounds, 1581 pounds, 1908 pounds, 2340 pounds, and 1879 pounds: what was the weight of the whole?

4. A man gave 5460 dollars to his oldest son, to the next 4065, to the next 6750, to the next 8000, and to the youngest 7276 dollars: how much did he give to all?

5. A merchant, on settling up his business, found he owed one creditor 176 dollars, another 841 dollars, another 1356 dollars, another 2370 dollars, another 840 dollars: what was the amount of his debts?

6. The state of Maine contains 32400 square miles; New Hampshire, 9500; Vermont, 9700; Massachusetts, 7800; Rhode Island, 1251; and Connecticut, 4789: how many square miles are there in the New England States?

7. The state of New York contains 46220 square miles; New Jersey, 7948; Pennsylvania, 46215; and Delaware, 2068: how many square miles are there in the Middle States?

8. The state of Maryland contains 10755 square miles; Virginia, 65700; North Carolina, 51632; South Carolina, 31565; Georgia, 61683; Florida, 56336; Alabama, 54084; Mississippi, 49356; Louisiana, 47413; and Texas, 100000: how many square miles are there in the Southern States?

9. The state of Tennessee contains 41752; Kentucky, 40023; Ohio, 40500; Michigan, 60537; Indiana, 35626; Illinois, 56506; Missouri, 70050; Arkansas, 54617; Iowa, 173786; and Wisconsin, 92930: how many square miles are there in the Western States?

10. What is the whole number of square miles in the United States?

11. What is the sum of 75234+41015+19075+176+88350+10040?

12. What is the sum of 250120+30402+7850+465000+10046+65045?

13. What is the sum of 85046+90045+412260+125781+4060+273048?

14. What is the sum of 1500267+45085+4652+4780400+90276+89760841?

15. What is the sum of 45702125+67070420+670856+4230825+750642+8790845?

16. What is the sum of 825760842+35620476+7800490+467243+98371+6425+740?

17. What is the sum of 2503+37621+475290+1223729+10671840+275600312?

18. What is the sum of 463270+2500+7200342+10271+426345+6200705?

19. What is the sum of 80429+7562345+700100+85261798+4000101+3007002?

20. What is the sum of 756+849+934+680+720+843+657689+989876498+8045685+807266780?

21. What is the sum of 6457+29301+82406+7589+63489+101364+46745?

22. Add together 786, 340, 910, 403, 783, 650, 809, 670, 408, 310, and 652.

23. Add together 16075, 250763, 7561, 830654, 293106, 2537104, and 316725.

24. Add together 256, 40, 751, 302, 75, 831, 26, 43, 621, 340, and 510.

25. Add together 493742, 56710607, 23461, 400072, 6811004, 8999003, and 26501.

26. Add together 629405, 7629, 31000401, 263012, 1300512, 390217, and 13268.

27. Add together 286013, 4016702, 1971342, 6894680, 28945, and 2624302.

28. Add together 460167, 296345, 84634123, 64205, 9673108 and 1931456.

29. Add together 432678902, 310046734, 2167005, 327861 and 293000428.

COUNTING-ROOM EXERCISES.

60. To the accountant as well as the mathematician, ***accuracy*** and *expertness* in adding, are *indispensable.* These attainments can be acquired only by frequent exercises in footing up long columns of figures.

Note.—1. Instead of saying 4 and 8 are 12, and 2 are 14, and 7 are 21, and 4 are 25, &c., a skilful accountant, performing the addition at a glance, simply pronounces the results. Thus, four, twelve, twenty-one, thirty-one, (4+6 =10,) thirty-seven, forty-seven, (7+3=10,) fifty-two.

2. When two or three figures taken together make 10, as 6 and 4, or 2, 3, and 5, &c., it accelerates the process to add their sum at once. A little practice will enable the student to run up a long column of figures with as much facility almost as he can count.

3. When the columns are long, accountants sometimes set the figure to be carried below the other figure under the column added. Thus, the sum of the first column in the example above being 52, set the 5 (the figure carried) below the 2. The sum of the second column being 48, set the 4 below the 8. &c. This method saves much time in reviewing an operation, and also enables us, when interrupted, to resume the process where we left off.

30.	
86015	
25163	
85057	
12236	
43026	
67084	
21167	
54042	
42158	
24034	
459982	*Ans.*
3045	*Car.*

Required the amount of each of the following examples:

	31.	32.	33.	34.
	Dollars.	Dollars.	Yards.	Pounds.
	2425	46,519	607,253	421,536
	3282	32,271	232,012	310,101
	2793	17,436	211,849	797,019
	2354	81,587	380,436	233,680
	4262	28,333	578,551	124,402
	9158	52,745	231,349	255,353
	2653	23,052	145,763	852,057
	3424	20,158	605,037	618,041
	1266	71,232	760,155	100,266
	8742	39,464	357,676	971,134
	2126	18,643	544,844	536,920
	5387	42,027	276,232	703,352
Ans.	47872	73,235	803,383	420,503
Car.	465	24,103	725,918	312,675

35.	36.	37.	38.
348,037	460,375	963,172	849,652
272,465	841,681	300,725	361,728
530,634	239,724	463,248	412,381
100,871	763,256	721,003	635,403
693,036	437,891	387,356	872,545
764,543	825,432	241,653	406,223
323,638	285,678	603,280	294,867
428,432	310,720	532,176	811,236
389,763	403,521	278,321	576,037
210,045	687,489	829,248	213,744
760,806	324,061	171,320	764,368
636,215	530,724	206,782	305,216
253,734	623,452	461,027	436,720
251,600	487,638	589,203	823,284
575,453	290,731	248,639	217,436
807,720	803,256	730,461	592,301
930,046	731,463	672,398	243,762
174,173	379,574	246,175	731,445
626,245	823,156	928,340	429,374
342,734	928,348	731,629	684,569

61. Accountants often acquire the habit of adding two columns of figures at a time. The power of rapid addition is easily acquired, and is well worthy the attention of the student. The following examples will illustrate the principle.

39. What is the sum of 312817+527236+141625+462415+251818+234112?

Operation.

Taking the two right hand columns, we say, 12 and 18 are 30, and 15 are 45, and 25 are 70, and 36 are 106, and 17 are 123. Set down the 23 under the columns added, and carry the 1 or left hand figure to the column of hundreds. Proceed in the same manner with the other columns.

```
      312817
      527236
      141625
      462415
      251818
      234112
Ans. 1930023
```

(41.)	(42.)	(43.)	(44.)	(45.)	(46.)
21	22	44	1325	2610	344235
30	13	20	1510	1511	402321
11	40	25	1314	1021	141511
13	25	17	3141	1115	201250
20	14	50	1016	1513	154036
15	11	14	2233	4020	132212
34	33	16	1224	1316	181714
18	45	28	2415	1233	213025
12	12	11	1830	2515	111817
17	20	14	1814	1718	161518
23	18	37	1621	2142	432733

What was the amount of exports and imports of the United States in 1840, and of shipping in 1842?

States.	(47.) Exports.	(48.) Imports.	(49.) Shipping.
Maine, .	Dolls. 1,018,269	Dolls. 628,762	T. 281,930
N. Hampshire,	20,979	114,647	23,921
Vermont, .	305,150	404,617	4,343
Massachusetts,	10,186,261	16,513,858	494,895
Rhode Island,	206,989	274,534	47,243
Connecticut, .	518,210	277,072	67,749
New York, .	34,264,080	60,440,750	518,133
New Jersey, .	16,076	19,209	60,742
Pennsylvania,	6,820,145	8,464,882	113,569
Delaware, .	37,001	802	10,396
Maryland, .	5,768,768	4,910,746	106,856
Dist. of Columbia,	753,923	119,852	17,711
Virginia, .	4,778,220	545,085	47,536
North Carolina,	387,484	252,532	31,682
South Carolina,	10,036,769	2,058,870	23,469
Georgia, .	6,862,959	491,428	16,536
Alabama, .	12,854,694	574,651	14,577
Louisiana, .	34,236,936	10,673,190	144,128
Ohio, . .	991,954	4,915	24,830
Michigan, .	162,229	148,610	12,323
Florida,	1,858,850	190,728	7,288

50 The appropriations of the Government of the United States, for 1847, were as follows: for the Civil and Diplomatic expenses 4,442,790 dolls.; for the Army and Volunteers 32,178,461 dolls.; for the Navy 9,307,958 dolls.; for the Post Office Department 4,145,400 dolls.; for the Indian Department 1,364,204 dolls.; for the Military Academy 124,906 dolls.; for building Steam Ships 1,000,000 dolls.; for Revolutionary and other Pensions 1,358,700 dolls.; for concluding Peace with Mexico 3,000,000 dolls.; for Light Houses 518,830 dolls.; Miscellaneous 540,243 dolls. What was the amount of all the appropriations?

62. It may sometimes be convenient for the learner, as well as gratifying to his curiosity, to be able to add numbers expressed by the Roman Notation.

51. A man paid MDCCCLXXXIII dollars for a farm, DCCXXIIII dollars for stock, and CCCLXVIIII dollars for tools: how much did he pay for all?

Operation.

MDCCCLXXXIII	dolls.
DCCXXIIII	dolls.
CCCLXVIIII	dolls.
MMDCCCCLXXVI	dolls.

Beginning at the right hand, we proceed thus: four Is and four Is are eight, and three Is make eleven, which is equal to two Vs and I. We set down the I, and adding the two Vs to one V makes fifteen, which is equal to X and V. Setting down the V, we count in the X with the other Xs, and find they make seven Xs or seventy, which is expressed by L and XX. We set down the two Xs, and adding the L to the other Ls, it makes three Ls, or one hundred and fifty, which is expressed by C and L. Setting down the L, and counting the C with the other Cs, we have nine Cs or nine hundred, which is expressed by D and CCCC. We set down the four Cs, and counting the D with the other Ds, it makes three Ds or fifteen hundred, which is expressed by M and D. We set down the D, and adding the M to the other M, we have two Ms, which we set down on the left of the other letters. Hence,

63. To add numbers expressed by the Roman Notation.

Beginning at the right hand, count all the letters of each kind together; set down the result, and carry on the principle that five Is *make one* V; *two* Vs, *one* X; *five* Xs, *one* L, &c.

OBS. The teacher can extend the exercises in the Roman Notation as far as he may deem it expedient. A single example is sufficient to illustrate the principle, and to show that the Roman is *greatly inferior* to the Arabic method in its adaptation to business calculations.

SECTION III.

SUBTRACTION.

ART. **65.** Ex. 1. A merchant bought 37 barrels of flour, and afterwards sold 12 of them: how many barrels had he left?

Solution.—12 barrels from 37 barrels leave 25 barrels.

Ans. 25 barrels.

OBS. It will be perceived, that the object in this example, is to *find the difference* between two numbers.

66. *The process of finding the difference between two numbers is called* SUBTRACTION.

The *difference,* or the *answer* to the question, is called the *Remainder.*

OBS. 1. The number to be subtracted is sometimes called the *subtrahend,* and the number from which it is subtracted, the *minuend.*

2. Subtraction, it will be perceived, is the *reverse* of addition. Addition *unites* two or more numbers into *one single* number; subtraction, on the other hand, *separates* a number into *two parts.*

3. When the given numbers are of the *same denomination,* the operation is called *Simple Subtraction.* (Art. 50. Obs.)

Ex. 2. What is the difference between 5364 and 9387?

Write the less number under the greater, units under units, tens under tens, &c. Then, beginning at the right hand, proceed thus: 4 units from 7 units leave 3 units. Write the 3 in the units' place, under the figure subtracted. 6 tens from 8 tens leave 2 tens; set the 2 in tens' place. 3 hundred from 3 hundred leave 0 hundred; we therefore write a cipher in hundreds' place. 5 thousand from 9 thousand leave 4 thousand; set the 4 in the thousands' place. The answer is 4023.

Operation.

```
9387
5364
----
4023 Rem.
```

QUEST.—66. What is subtraction? What is the difference or answer called? *Obs.* What is the number to be subtracted sometimes called? The number from which it is subtracted? Of what is subtraction the reverse? When the given numbers are of the same denomination, what is the operation called?

67. It will be observed, that we subtract *units* from *units*, *tens* from *tens*, &c.; that is, we subtract figures of the *same order* from each other. This is done for the same reason that we *add* figures of the same order to each other. (Art. 51.)

OBS. The *less* number is written *under* the *greater*, simply for convenience in subtracting; and *units* are placed under *units*, *tens* under *tens*, &c., to avoid mistakes which might occur from taking *different* orders from each other.

68. It often happens that a figure in the lower number is *larger* than that above it, and consequently cannot be taken from it.

Ex. 3. What is the difference between 94 and 56?

Analytic solution.

$$94=80+14$$
$$56=50+\ 6$$
Rem. $38=30+\ 8$

It is manifest that we cannot take 6 units from 4 units, for 6 is larger than 4. To obviate this difficulty, we may take 1 ten from the 9 tens, and uniting it with the 4 units, the upper number will become 8 tens and 14 units, or 80+14. Separating the lower number into the parts of which it is composed, it becomes 5 tens and 6 units, or 50+6. Now, subtracting as in the last example, 6 from 14 leaves 8, 50 from 80 leaves 30. The answer is 30+8, or 38. Or, we may simply take 1 ten from the 9 tens, and adding it, mentally, to the 4 units, say 6 from 14 leaves 8; set the 8 under the figure subtracted. Then, having taken 1 from the 9 tens, we have but 8 left, and 5 from 8 leaves 3. The answer is 38.

PROOF.—38+56=94; that is, the sum of the remainder and smaller number being equal to the larger, the answer is right. Hence,

69. When a figure in the lower number is larger than that above it; take 1 from the next higher order in the upper number, and add it to the upper figure; from the sum subtract the lower figure, and diminishing the next upper figure by 1, proceed as before.

OBS. 1. The process of taking *one* from the next higher order and *adding* it to the figure from which the subtraction is to be made, is called *borrowing ten*. It is the *reverse* of carrying.

QUEST.—67. What orders of figures do you subtract from each other? Why not subtract different orders from each other?

2. This method of *borrowing*, it will be seen, does not *affect* the *difference* between the two given numbers; for, it is simply transposing a part of one order to another order in the same number, which, it is obvious, will neither *increase* nor *diminish* its *value*.

3. It may be asked, how can we take *one* from the figure in the next higher order, when that figure is a *cipher?* How can *nothing lend* anything, and how can *nothing* be *diminished* by *one?* The explanation of this apparent contradiction is this: when the next figure is a cipher, we go to the next higher column still, and take *one*, which, added to the figure in the next lower order, makes *ten;* we then take *one* from the *ten* and add it to the upper figure, and proceed as before.

70. There is another method of *borrowing*, or rather of *paying*, which, though perhaps less philosophical than the preceding, is more convenient in practice, especially when the figures in the next higher orders are ciphers. Thus, in the last example, adding 10 to the upper figure, it becomes 14, and 6 from 14 leaves 8. Set down the 8 as before. Now, instead of diminishing the next upper figure by 1, if we add 1 to the next figure in the lower number it becomes 6 tens; and 6 from 9 leaves 3, which is the same as 5 from 8. The answer is 38, the same as before. Hence,

71. When a figure in the lower number is larger than that above it, add 10 to the upper figure, and to compensate this, add 1 to the next left hand figure in the lower number.

OBS. 1. This method of *borrowing* depends on the self-evident principle, that if any two numbers are *equally increased*, their difference will not be *altered*. That the two given numbers are equally increased by this process, is evident from the fact that the 1 added to the lower number is of the next superior order to the 10 added to the upper number, and is therefore equal to it. (Art. 35.)

2. The reason that we borrow 10, instead of 8, or 12, or any other number, is because the radix of the system of Arabic notation, is 10. (Art. 36. Note. 1.) If the radix of the system were 8, it would be necessary to borrow 8; if 12, it would be necessary to borrow 12, &c.

3. On account of borrowing, the learner will perceive it is always necessary to begin to subtract at the right hand.

Ex. 4. A man bought a house for 23006 dollars, and sold it for 21128 dollars: how much did he lose by his bargain?

Operation.		*Proof.*
Cost	23006 dolls.	21128 Less number.
Rec'd.	21128 dolls.	1878 Remainder.
Ans.	1878 dolls.	23006 Larger number.

72. From the preceding illustrations and principles we derive the following

GENERAL RULE FOR SUBTRACTION.

I. *Write the less number under the greater, so that units may stand under units, tens under tens, &c.* (Art. 67. Obs.)

II. *Beginning at the right hand, subtract each figure in the lower number from the figure above it, and set the remainder directly under the figure subtracted.* (Art. 71. Obs. 3.)

III. *When a figure in the lower number is larger than that above it, add* 10 *to the upper figure; then subtract as before, and add* 1 *to the next figure in the lower number, or consider the next upper figure* 1 *less than it is.* (Arts. 69, 71. Obs. 1, 2.)

73. PROOF.—*Add the remainder to the smaller number; and if the sum is equal to the larger number, the work is right.*

OBS. This method of proof depends upon the principle, that the *difference* between two numbers being added to the *less*, the *sum* must be *equal* to the *greater*. For, the difference and the less number are the two parts into which the *greater* is separated, and the *whole* a quantity is equal to the sum of *all its parts.* (Ax. 11.)

74. *Second Method.*—Subtract the *remainder* from the greater of the two given numbers; and if the difference is equal to the *less* number, the work is right.

75. *Third Method.*—Cast the 9s out of the larger number, and place the excess at the right. Next, cast the 9s out of the smaller number, and also out of the remainder; then cast the 9s out of the sum of these two excesses, and if this last excess is the same as the excess of the larger number, the work may be supposed to be right. Thus,

Ex. 5. From 7843 Excess of 9s in the greater number is 4
Take 5675 " " " less " is 5 }
Rem. 2168 " " " remainder is 8 } Now, 8+5=13,
and the excess of 9s in 13 is 4, the same as that of the greater number.

QUEST.—72. How do you write numbers for subtraction? Why write the less number under the greater? Why place units under units, &c.? Where do you begin to subtract? When a figure in the lower line is larger than that above it, how do you proceed? What is meant by borrowing ten? How many methods of borrowing are mentioned? Illustrate the first method upon the black-board. How does it appear that this method of borrowing does not affect the difference between the two given numbers? Explain the second method. Upon what principle does this method depend? Why do you borrow 10, instead of 8, or 12, or any other number? Why do you begin to subtract at the right hand 73. How is subtraction proved? *Obs.* Upon what principle does this method of proof depend? Can subtraction be proved by any other methods?

Note.—This method of proof depends on the same property of the number 9, as that in addition. (Art. 58. Note.) For, since the sum of the smaller number and remainder is equal to the larger number, it follows that the excess of 9s in the larger number must be equal to the excess of 9s in the remainder and smaller number together.

EXAMPLES FOR PRACTICE.

76. Ex. 1. A merchant bought a ship for 35270 dollars, and sold it for 42365 dollars: how much did he make by his bargain?

2. A miller bought 46235 bushels of wheat, and ground 17251 bushels of it: how many bushels had he left?

3. A speculator laid out 50000 dollars in wild land, and afterwards sold it at a loss of 19046 dollars: how much did he get for his land?

4. A man owning a block of buildings worth 155265 dollars, keeps it insured for 109240 dollars: how much would he lose in case the buildings should be destroyed by fire?

5. The distance from the Earth to the Sun is 95000000 of miles; the distance of Mercury is only 37000000: how far is Mercury from the Earth?

6. The imports of Massachusetts in 1840, were 16,513,858 dollars, the exports were 10,186,261 dollars: what was the excess of her imports over her exports?

7. The imports of New York in 1840, were 60,440,750 dollars, the exports were 34,264,080 dollars: what was the excess of her imports over her exports?

8. The imports of Pennsylvania in 1840, were 8,464,882 dollars, the exports were 6,820,145 dollars: what was the excess of her imports over her exports?

9. The imports of South Carolina in 1840, were 2,058,870 dollars, the exports were 10,036,769 dollars: what was the excess of her exports over her imports?

10. The imports of Alabama in 1840, were 574,651 dollars, the exports were 12,854,694 dollars: what was the excess of her exports over her imports?

11. The imports of Louisiana in 1840, were 10,673,190 dollars, the exports were 34,236,936 dollars: what was the excess of her exports over her imports?

12. The tonnage of the United States in 1842, was 2069857, in 1846 it was 2500000 : what was the increase in 4 years?

	13.	14.	15.
From	253760	3856031	54903670
Take	104523	462702	504089

16. 9876102—1050671.
17. 4006723—5001.
18. 3601900—1000000.
19. 5317004—3565.
20. 1000000—456321.
21. 2035024—27040.
22. 45563075—460001.
23. 67030001—300452.
24. 73256300—436020.
25. 56037431—735671.
26. 80200430—250.
27. 96531768—873625.
28. 10000000—999999.
29. 99999999—100000.
30. 83567000—438567.
31. 40600056—7632.
32. 56409250—1057245.
33. 20030000—72534.
34. 83175621—5256360.
35. 70301604—250041.
36. 60050376—6849005.
37. 34200591—8888888.
38. 87035762—753017.
39. 95246300—9438675.
40. From 6764+3764 take 6500+2430.
41. From 2890+8407 take 4251+3042.
42. From 7395+4036 take 8297+1750.
43. From 8404+7296 take 3201—1562.
44. From 6008+9270 take 5136—2352.
45. From 9234+6850 take 9320—4783.
46. From 8564—2573 take 4431—1735.
47. From 7284—5362 take 6045—5729.
48. From 9561—4680 take 7352—6178.
49. From 8630—1763 take 2460+1743.
50. From 7561—2846 take 1734+2056.
51. From 9687—3401 take 3021+1754.

52. A man having 55000 dollars, paid 7520 dollars for a house, 3260 dollars for furniture, 2375 dollars for a library, and invested the balance in bank stock: how much stock did he buy?

53. A gentleman worth 163250 dollars, bequeathed 15200 dollars apiece to his two sons, 16500 dollars to his daughter, and to his wife as much as to his three children, and the remainder to a hospital: how much did his wife receive, and how much the hospital?

54. A man bought three farms; for the first he paid 5260 dollars, for the second 3585, and for the third as much as for the first two. He afterwards sold them all for 15280 dollars: did he make or lose by the operation; and how much?

55. What number is that, to which 3425 being added, the sum will be 175250?

56. A man being asked how much he was worth, replied, if you will give me 325263 dollars, I shall have two millions of dollars: how much was he worth?

57. A jockey gave 150 dollars for a horse, and meeting an acquaintance swapped with him, giving 37 dollars to boot; meeting another, he swapped and received 28 dollars to boot; he finally swapped again and gave 78 dollars to boot, and then sold his last horse for 140 dollars: how much did he lose by all his bargains?

58. A speculator gained 3560 dollars, and afterwards lost 2500 dollars; at another time he gained 6283 dollars, and then lost 3450 dollars: how much more did he gain than lose?

59. A man bought a house for MDCCCCXXXVII dollars, and sold it for DCXVIIII dollars less than he gave: how much did he sell it for?

Operation.

	MDCCCCXXXVII dolls
	DCXVIIII dolls
Ans.	MCCCXVIII dolls

We perceive that the IIII in the lower number cannot be taken from II in the upper number; we therefore borrow a V, which added to the II, makes IIIIIII; then IIII from IIIIIII, leaves III, which we set down. Now since we have borrowed the V in the upper number, there are no Vs left from which we can take the V in the lower number. We must therefore borrow an X; but X is equal to VV; and V from VV leaves V, which we set down. Having borrowed an X from the upper number, there are but XX left, and X from XX leaves X. C from CCCC leaves CCC. D from D leaves nothing. And nothing from M leaves M. Hence,

77. To subtract numbers expressed by the *Roman Notation.*

Write the less number under the greater; then, beginning at the right hand, take the number in the lower line from that expressed by the same letters in the upper line, and set the remainder below. If the number in the lower line is larger than that expressed by the same letters in the upper line, borrow a letter next higher and add it to the number in the upper line; then subtract as before, observing to pay when you borrow as in subtraction of figures. (Art. 72.)

OBS. Other examples expressed by the Roman Notation, can be added by the teacher, if deemed expedient.

SECTION IV.

MULTIPLICATION.

ART. **79.** Ex. 1. What will 3 melons cost, at 15 cents apiece?

Analysis.—If 1 melon costs 15 cents, 3 melons will cost 3 times 15 cents; and 3 times 15 cents are 45 cents. *Ans.* 45 cents.

2. What will 4 sleighs cost, at 21 dollars apiece?

Analysis.—Reasoning as before, if 1 sleigh costs 21 dollars, 4 sleighs will cost 4 times as much; and 4 times 21 dollars are 84 dollars. *Ans.* 84 dollars.

OBS. It is obvious that 3 times 15 cents is the same as 15 cents+15 cents +15 cents, or 15 cents added to itself 3 times; and 4 times 21 dollars is the same as 21 dolls.+21 dolls.+21 dolls.+21 dolls., or 21 dollars added to itself 4 times.

80. *This repeated addition of a number or quantity to itself, is called* MULTIPLICATION.

The number to be *repeated*, or *multiplied*, is called the *Multiplicand.*

The number by which we *multiply*, is called the *multiplier;* and shows *how many times* the multiplicand is to be *repeated.*

The number *produced*, or the *answer* to the question, is called the *product.* Thus, when we say, 8 times 12 are 96, 8 is the multiplier, 12 the multiplicand, and 96 the product.

81. The multiplier and multiplicand together are often called *factors*, because they *make* or *produce* the product.

OBS. 1. The term *factor* is derived from a Latin word which signifies an *agent*, a *doer*, or *producer.*

2. When the multiplicand denotes things of *one denomination only*, the operation is called *Simple Multiplication.*

QUEST.—80. What is multiplication? What is the number to be repeated called? What the number by which we multiply? What does the multiplier show? What is the number produced called? 81. What are the multiplicand and multiplier together called? Why? *Obs.* What does the term factor signify?

MULTIPLICATION TABLE.

1	2	3	4	5	6	7	8	9	10	11	12	13	14	15	16	17	18	19	20
2	4	6	8	10	12	14	16	18	20	22	24	26	28	30	32	34	36	38	40
3	6	9	12	15	18	21	24	27	30	33	36	39	42	45	48	51	54	57	60
4	8	12	16	20	24	28	32	36	40	44	48	52	56	60	64	68	72	76	80
5	10	15	20	25	30	35	40	45	50	55	60	65	70	75	80	85	90	95	100
6	12	18	24	30	36	42	48	54	60	66	72	78	84	90	96	102	108	114	120
7	14	21	28	35	42	49	56	63	70	77	84	91	98	105	112	119	126	133	140
8	16	24	32	40	48	56	64	72	80	88	96	104	112	120	128	136	144	152	160
9	18	27	36	45	54	63	72	81	90	99	108	117	126	135	144	153	162	171	180
10	20	30	40	50	60	70	80	90	100	110	120	130	140	150	160	170	180	190	200
11	22	33	44	55	66	77	88	99	110	121	132	143	154	165	176	187	198	209	220
12	24	36	48	60	72	84	96	108	120	132	144	156	168	180	192	204	216	228	240
13	26	39	52	65	78	91	104	117	130	143	156	169	182	195	208	221	234	247	260
14	28	42	56	70	84	98	112	126	140	154	168	182	196	210	224	238	252	266	280
15	30	45	60	75	90	105	120	135	150	165	180	195	210	225	240	255	270	285	300
16	32	48	64	80	96	112	128	144	160	176	192	208	224	240	256	272	288	304	320
17	34	51	68	85	102	119	136	153	170	187	204	221	238	255	272	289	306	323	340
18	36	54	72	90	108	126	144	162	180	198	216	234	252	270	288	306	324	342	360
19	38	57	76	95	114	133	152	171	190	209	228	247	266	285	304	323	342	361	380
20	40	60	80	100	120	140	160	180	200	220	240	260	280	300	320	340	360	380	400

Note.—This Table was invented by *Pythagoras*, and is therefore sometimes called the *Pythagorean* Table.

The pupil will find assistance in learning the Multiplication Table by observing the following particulars.

1. The several results of multiplying by 10 are formed by simply adding a cipher to the figure that is to be multiplied. Thus, 10 times 2 are 20, 10 times 3 are 30, &c.

2. The results of multiplying by 5 terminate in 5 and 0, alternately. Thus, 5 times 1 are 5, 5 times 2 are 10, 5 times 3 are 15, &c.

3. The first nine results of multiplying by 11 are formed by repeating the figure to be multiplied. Thus, 11 times 2 are 22; 11 times 3 are 33, &c.

4. In the successive results of multiplying by 9, the right hand figure regularly decreases by 1, and the left hand figure regularly increases by 1. Thus, 9 times 2 are 18; 9 times 3 are 27; 9 times 4 are 36, &c.

82. Multiplying by 1, is taking the multiplicand *once:* thus, 4 multiplied by 1=4.

Multiplying by 2, is taking the multiplicand *twice:* thus, 2 times 4, or 4+4=8.

Multiplying by 3, is taking the multiplicand *three times:* thus, 3 times 4, or 4+4+4=12, &c. Hence,

QUEST.—82. What is it to multiply by 1? By 2? By 3?

Multiplying by any whole number, is taking the multiplicand as many times, as there are units in the multiplier.

The application of this principle to *fractional* multipliers will be illustrated under fractions.

OBS. 1. From the definition of multiplication, it is manifest that the *product* is of the *same kind* or *denomination* as the multiplicand: for, *repeating* a number or quantity does not alter its nature. Thus, if we repeat *dollars*, they are *still dollars;* if we repeat *yards*, they are *still yards*, &c. Consequently, if the multiplicand is an *abstract number*, the product will be an *abstract number;* if *money*, the product will be money; if *barrels*, barrels, &c.

2. Every *multiplier* is to be considered an *abstract number*. In familiar language it is sometimes said, that the price multiplied by the *weight* will give the value of an article; and it is often asked how much 25 cents multiplied by 25 cents, &c., will produce. But these are abbreviated expressions, and are liable to convey an erroneous idea, or rather no idea at all. If taken literally, they are absurd; for multiplication is *repeating* a number or quantity a certain *number of times*. Now to say that the price is repeated as many times as the given quantity is *heavy*, or that 25 cents are repeated 25 *cents times*, is nonsense. But we can multiply the price of 1 pound by a *number* equal to the number of pounds in the *weight* of the given article, and the product will be the value of the article. We can also multiply 25 cents by the *number* 25; that is, repeat 25 cents 25 times, and the product is 625 cents. Construed in this manner, the multiplier becomes an *abstract number*, and the expressions have a consistent meaning.

Ex. 3. What will 6 houses cost, at 2341 dollars apiece?

Operation.

	2341	Multiplicand
	6	Multiplier.
Ans.	14046	Dollars.

Write the numbers on the slate as in the margin, and beginning at the right hand, proceed thus: 6 times 1 unit are 6 units; write the 6 under the figure multiplied. 6 times 4 tens are 24 tens; set the 4 or right hand figure under the figure multiplied, and carry the 2 or left hand figure to the next product figure, as in addition. (Art. 52.) 6 times 3 hundreds, are 18 hundreds, and 2 to carry make 20 hundreds; set the 0 under the figure multiplied, and carry the 2 to the next product as before. 6 times 2 thousands are 12 thousands, and 2 to carry make 14 thousands. Since there are no

---

QUEST.—What is it to multiply by any whole number? *Obs.* Of what denomination is the product? How does this appear? What must every multiplier be considered? Can you multiply by a given weight, a measure, or a sum of money?

more figures to be multiplied, set down the 14 in full as in addition. (Art. 53. Obs. 1.) The product is 14046 dollars.

83. *The product of any two numbers will be the same, whichever factor is taken for the multiplier.* Thus,

If an orchard contains 5 rows of trees, and each row has 7 trees, as represented by the stars in the margin, it is evident the whole number of trees is equal either to the number of stars in a horizontal row repeated *five times*, or to the number of stars in a perpendicular row repeated *seven times*, viz: 35. For, $7\times5=35$, also $5\times7=35$.

```
* * * * * * *
* * * * * * *
* * * * * * *
* * * * * * *
* * * * * * *
```

OBS. 1. It is more convenient and therefore customary to place the *larger* number for the multiplicand, and the *smaller* for the multiplier. Thus, it is easier to multiply 8468946 by 3, than it is to multiply 3 by 8468946, but the product would be the same.

Ex. 4. What will 237 coaches cost, at 675 dollars apiece?

Since it is not convenient to multiply by 237 at once, we multiply first by the 7 units, next by the 3 tens, then by the 2 hundreds, and place each result in a separate line, with the first figure of each line directly under that by which we multiply. Finally, adding these results together, units to units, &c., we have 159975 dollars, which is the whole product required. (Ax. 11.)

Operation.

```
   675     Multiplicand.
   237     Multiplier.
  ----
  4725     cost   7 coaches.
 2025*     cost  30    "
1350**     cost 200    "
------
159975     cost 237    "
```

Note.—When the multiplier contains *more* than *one* figure, the several products of the multiplicand into the separate figures of the multiplier, are called *partial* products.

OBS. 2. The reason for placing the first figure of the several partial products under the figure by which we multiply, is to bring the *same orders* under each other, and thus prevent mistakes in adding them together. (Art. 51.)

3. The several *partial* products are added together for the obvious purpose of finding the *whole* product or answer required. (Ax. 11.)

QUEST.—83. Does it make any difference with the result, which of the given numbers is taken for the multiplier? *Obs.* Which is usually taken? Why?

84. The principle of *carrying the tens* in multiplication is the same as in addition, and may be illustrated in a similar manner. (Art. 53.) Thus,

Ex. 5.
```
   9382 Mult'd.
      7 Mult'r.
     14=units,
    56*=tens,
   21**=hunds.
  63***=thou.
  65674 Product.
```

Or, separating the multiplicand into the orders of which it is composed,

$$9382=9000+300+80+2,$$

and
```
9000×7=63000
 300×7= 2100
  80×7=  560
   2×7=   14
```
Adding these results together, we have 65674 *Ans.*

OBS. The reason for always beginning to multiply at the right hand of the multiplicand, is that we may carry the tens as we proceed in the operation.

85. From this illustration it will be observed that units multiplied into units produce units; tens into units, or units into tens, produce tens; (Art. 83;) hundreds into units, or units into hundreds, produce hundreds, &c. Hence,

86. *When units are multiplied into any order whatever, the product will always be of the same order as the other figure.*

And universally, the product of any two integers is of the order next less than that denoted by the sum of the orders of the two given figures. Thus, *hundreds* into *tens* produce *thousands*, or the 4th order, which is *one less* than the *sum* of the two given orders.

OBS. When the multiplier contains *more* than one figure, it is customary to begin to multiply with its units' figure. The result however will be the same, if we begin with its hundreds or any other order of the multiplier, and place the first figure of the partial products, so that the same orders shall stand under each other.

First Operation.
```
     1357
     3574
  4071
   6785
    9499
     5428
  4849918. Prod.
```

Second Operation.
```
  1357
  3574
  4071
   6785
    9499
     5428
  4849918. Prod.
```

QUEST.—85. What do units into units produce? Units into tens, or tens into units?

Ex. 6. What is the product of 5690 into 3008?

After multiplying by the 8 units, we next multiply by the 3 thousands, since there are no tens nor hundreds in the multiplier, and place the first figure of this partial product under the figure 3 by which we are multiplying.

Operation

```
     5690
     3008
    45520
 17070
 17115520 Ans.
```

87. From the preceding illustrations and principles we derive the following

GENERAL RULE FOR MULTIPLICATION.

I. When the multiplier contains but *one* figure.

Write the multiplier under the multiplicand, units under units, tens under tens, &c. (Art. 83. Obs. 1.)

Begin at the right hand and multiply each figure of the multiplicand by the multiplier, setting down the result and carrying as in addition. (Art. 84. Obs.)

II. When the multiplier contains *more* than one figure.

Multiply each figure of the multiplicand by each figure of the multiplier separately, beginning with the units, and write the partial products in separate lines, placing the first figure of each line directly under the figure by which you multiply. (Art. 83. Obs. 2.)

Finally, add the several partial products together, and the sum will be the whole product. (Art. 83. Obs. 3.)

88. PROOF.—*Multiply the multiplier by the multiplicand, and if the product thus obtained is the same as the other product, the work is supposed to be right.*

OBS. This method of proof depends upon the principle, that the product of any two numbers is the same, whichever is taken for the multiplier. (Art. 83.)

89. *Second Method.*—Add the multiplicand to itself as many

QUEST.—86. When units are multiplied into any order, what order is the product? When any two integers are multiplied together, of what order is the product? 87. How do you write the numbers for multiplication? When the multiplier contains but one figure, how proceed? Why begin at the right hand of the multiplicand? When the multiplier contains more than one figure, how proceed? What is meant by partial products? Why place the first figure of each partial product under the figure by which you multiply? What is to be done with the partial products? Why add the several partial products together? Why should this give the whole product? 88. How is multiplication proved? *Obs.* On what principle does this proof depend?

times as there are units in the multiplier, and if the *amount* obtained is equal to the *product*, the work is right.

Note.—When the multiplier is *small*, this is a very convenient mode of proof.

90. *Third Method.*—Cast the 9s out of the multiplicand and multiplier; multiply their remainders together, and casting the 9s out of their product, set down the excess; then cast the 9s out of the answer obtained, and if this excess be the same as that obtained from the multiplier and multiplicand, the work may be considered right.

Ex. 7. Multiply 565 by 356.

Operation.	*Proof.*
565	The excess of 9s in the multiplicand is 7.
356	" " 9s " multiplier is 5.
3390	7×5=35; and the excess of 9s is 8.
2825	
1695	
Prod. 201140.	The excess of 9s in the Ans. is also 8.

91. *Fourth Method.*—Divide the product by one of the factors, and if the quotient thus arising is equal to the other factor, the work is right.

Note.—This method of proof supposes the learner to be acquainted with division before he commences this work. (Art. 57. Note.) It is simply reversing the operation, and must obviously lead us back to the number with which we started: for, if a number is both multiplied and divided by the same number, its value will not be altered. (Ax. 9.)

92. *Fifth Method.**—First, cast the 11s out of the multiplicand and multiplier; multiply their remainders together, cast the 11s out of the product, and set down the excess; then cast the 11s out of the answer obtained, and if the excess is the same as that obtained from the multiplier and multiplicand, the work is right.

Note.—1. This method depends on a peculiar property of the number 11 For its further development and illustration, see Art. 161. Prop. 18.

2. To cast the 11s out of a number, begin at the right hand, mark the alternate figures; then from the sum of the figures marked, increased by 11 if necessary, take the sum of those not marked, and the remainder will be the excess required. Thus to cast the 11s out of 39475025, mark the alternate figures, beginning at the right hand, 3̇94̇75̇02̇5, then the sum of

QUEST.—Can multiplication be proved by any other methods?

* Leslie's Philosophy of Arithmetic.

5+0+7+9=21. Again, the sum of the others, viz: 2+5+4+3=14. Now 21—14=7, the excess of 11s.

Or, as soon as the sum is 11 or over, we may drop the 11, and add the remainder to the next digit. Thus, 5 and 7 are 12; dropping the 11, 1 and 9 are 10. Again, 2 and 5 are 7, and 4 are 11; drop the 11, and there are 3 left. Now, 10—3=7, the same excess as before.

Ex. 8. Multiply 237956 by 3728.

Operation.		*Proof.*
237956	Excess of 11s is 4.	Now, 4×10=40; the excess of 11s
3728	" " 10.	in 40 is 7.
Ans. 887099968	Excess of 11s in the answer is also 7.	

EXAMPLES FOR PRACTICE.

93. Ex. 1. What will 435 acres of land cost, at 57 dollars per acre?

2. What cost 573 oxen, at 63 dollars per head?

3. What cost 1260 tons of iron, at 45 dollars per ton?

4. If a man can travel 248 miles in a day, how far can he travel in 365 days?

5. If an army consume 645 pounds of meat in a day, how much will they consume in 115 days?

6. If 1250 men can build a fort in 298 days, how long would it take 1 man to do it?

7. How many rods is it across the Atlantic Ocean, allowing 320 rods to a mile, and the distance to be 3000 miles?

8. What is the product of 463×45?

9. What is the product of 348×62?

10. What is the product of 793×86?

11. What is the product of 75×42×56?

12. What is the product of 7198×216?

13. 31416×175.

14. 8862×189.

15. 7071×556.

16. 93186×4455.

17. 40930×779.

18. 12345×686.

19. 46481×936.

20. 16734×708.

21. 7575×7575.

22. 8320900×1328.

23. 17500×732.

24. 15607×3094.

25. 7422153×468.

26. 9264397×9584.

27. 4687319×1987.

28. 9507340×7071.

29. 39948123×6007.

30. 73885246×6079.

31. 57902468×5008.
32. 57902468×5080.
33. 57902468×5800.
34. 12481632×1509.
35. 79068025×1386.
36. 92948789×7043.
37. 58763718×6754.
38. 73084163×7584.
39. 144×144×144.
40. 3851×3851×3851.
41. 79094451×764094
42. 89548050×972800

CONTRACTIONS IN MULTIPLICATION.

94. The general rule is adequate to the solution of all examples that occur in multiplication. In many instances, however by the exercise of judgment in applying the preceding principles, the operation may be very much *abridged*.

95. Any number which may be produced by multiplying two or more numbers together, is called a *Composite Number*.

Thus, 4, 15, 21, are composite numbers; for 4=2×2; 15=5×3; 21=7×3.

OBS. 1. The *factors* which, being multiplied together, produce a composite number, are sometimes called the *component parts* of the number.

2. The process of finding the factors of which a given number is composed is called *resolving the number into factors*.

Ex. 1. Resolve 9, 10, 14, 22, into their factors.

2. What are the factors of 35, 54, 56, 63?

3. What are the factors of 45, 72, 64, 81, 96?

96. Some numbers may be resolved into *more* than *two* factors; and also into *different sets of factors*. Thus, 12=2×2×3; also 12=4×3=6×2.

4. What are the different factors and sets of factors of 8, 16, 18, 20, 24?

5. What are the different factors and sets of factors of 27, 32, 36, 40, 48?

96. *a.* We have seen that the product of any two numbers is the same, whichever factor is taken for the multiplier. (Art. 83.) In like manner, it may be shown that the product of any *three* or

QUEST.—95. What is a composite number? *Obs.* What are the factors which produce it sometimes called? What is meant by resolving a number into factors? 96. Are numbers ever composed of more than two factors? 96. *a.* When three or more factors are to be multiplied together, does it make any difference in what order they are taken?

more factors will be the same, in whatever order they are multiplied. For, the *product* of two factors may be considered as *one number*, and this may be taken either for the multiplicand, or the multiplier. Again, the product of three factors may be considered as one number, and be taken for the multiplicand, or the multiplier, &c. Thus, $24=3\times2\times2\times2=6\times2\times2=12\times2=6\times4=4\times2\times3=8\times3$.

Case 1.—*When the multiplier is a composite number.*

6. What will 27 bureaus cost, at 31 dollars apiece?

Analysis.—Since 27 is three times as much as 9; that is, $27=9\times3$, it is manifest that 27 bureaus will cost *three* times as much as 9 bureaus.

Operation.

Dolls. 31 cost of 1 B.
9
Dolls. 279 cost of 9 B.
3
Dolls. 837 cost of 27 B.

Having resolved 27 into the factors 9 and 3, we find the cost of 9 bureaus, then multiplying that by 3, we have the cost of 27 bureaus.

7. What will 36 oxen cost, at 43 dollars per head?

Solution.—$36=9\times4$; and $43\times9\times4=1548$ dolls. *Ans.*

Or, $36=3\times3\times4$; and $43\times3\times3\times4=1548$ dolls. *Ans.* Hence,

97. To multiply by a composite number.

Resolve the multiplier into two or more factors; multiply the multiplicand by one of these factors, and this product by another factor, and so on till you have multiplied by all the factors. The last product will be the answer required.

Obs. The *factors* into which a number may be *resolved*, must not be confounded with the *parts* into which it may be *separated*. (Art. 53.) The former have reference to multiplication, the latter to addition; that is, *factors* must be *multiplied* together, but *parts* must be *added* together to produce the given number. Thus, 56 may be resolved into two *factors*, 8 and 7; it may be separated into two *parts*, 5 tens or 50, and 6. Now, $8\times7=56$, and $50+6=56$.

8. What will 24 horses cost, at 74 dollars a head?

Quest.—97. When the multiplier is a composite number, how do you proceed? *Obs.* What is the difference between the factors into which a number may be resolved, and the parts into which it may be separated?

9. What cost 45 hogsheads of tobacco, at 128 dollars a hogshead?

10. What cost 54 acres of land, at 150 dollars per acre?

11. At 118 shillings per week, how much will it cost a family to board 49 weeks?

12. If a man travels at the rate of 372 miles a day, how far will he travel in 64 days?

13. At 163 dollars per ton, how much will 72 tons of lead cost?

14. What cost 81 pieces of broadcloth, at 245 shillings apiece?

15. What cost 84 carriages, at 384 dollars apiece?

CASE II.—*When the multiplier is* 1 *with ciphers annexed to it.*

98. It is a fundamental principle of notation, that each removal of a figure one place towards the left, increases its value *ten times;* (Art. 36;) consequently, annexing a *cipher* to a number will increase its value *ten times*, or *multiply* it by 10; annexing *two* ciphers will increase its value a *hundred times*, or multiply it by 100; annexing *three* ciphers will increase it a *thousand times*, or multiply it by 1000, &c. Thus, 15 with a cipher annexed, becomes 150, and is the same as 15×10; 15 with *two* ciphers annexed, becomes 1500, and is the same as 15×100; 15 with *three* ciphers annexed, becomes 15000, and is the same as 15×1000, &c. Hence,

99. To multiply by 10, 100, 1000, &c.

Annex as many ciphers to the multiplicand as there are ciphers in the multiplier, and the number thus formed will be the product required.

Note.—To *annex* means to place *after*, or at the *right hand.*

16. What will 10 boxes of lemons cost, at 63 shillings per box? *Ans.* 630 shillings.

17. How many bushels of corn will 465 acres of land produce, at 100 bushels per acre?

QUEST.—98. What is the effect of annexing a cipher to a number? Two ciphers? Three? Four? 99. How do you proceed when the multiplier is 10, 100, 1000, &c.? *Note.* What is the meaning of the term annex?

18. Allowing 365 days for a year, how many days are there in 1000 years?

19. Multiply 153486 by 10000.

20. Multiply 3120467 by 100000.

21. Multiply 52690078 by 1000000.

22. Multiply 689063457 by 10000000.

23. Multiply 4946030506 by 100000000.

24. Multiply 87831206507 by 1000000000.

25. Multiply 67856005109 by 10000000000.

CASE III.—*When the multiplier has ciphers on the right hand.*

26. What will 30 wagons cost, at 45 dollars apiece?

Note.—Any number with ciphers on its right hand, is obviously a composite number; the significant figure or figures being one factor, and 1, with the given ciphers annexed to it, the other factor. Thus, 30 may be resolved into the factors 3 and 10. We may therefore first multiply by 3 and then by 10, by annexing a cipher as above.

Solution.—$45\times3=135$, and $135\times10=1350$ dolls. *Ans.*

27. How many acres of land are there in 3000 farms, if each farm contains 475 acres?

Analysis.—$3000=3\times1000$. Now $475\times3=1425$; and adding three ciphers to this product, multiplies it by 1000. (Art. 99.) Hence,

Operation.

```
      475
        3
     ----
Ans. 1425000 acres.
```

100. When there are ciphers on the right of the multiplier.

Multiply the multiplicand by the significant figures of the multiplier, and to this product annex as many ciphers, as are found on the right of the multiplier.

OBS. It will be perceived that this case combines the principles of the two preceding cases; for, the multiplier is a composite number, and one of its factors is 1 with *ciphers annexed* to it.

28. How much will 50 hogs weigh, at 375 pounds apiece?

29. If 1 barrel of flour weighs 192 pounds, how much will 500 barrels weigh?

30. Multiply 14376 by 25000.

QUEST.—100. When there are ciphers on the right of the multiplier, how do you proceed? *Obs.* What principles does this case combine?

31. Multiply 350634 by 410000.
32. Multiply 4630425 by 6200000.

CASE IV.—*When the multiplicand has ciphers on the right hand.*

33. What will 37 ships cost, at 29000 dollars apiece?

Analysis.—29000=29×1000. But the product of two or more factors is the same in whatever order they are multiplied. (Art. 96. *a.*) We therefore multiply 29 by 37, and this product by 1000 by adding three ciphers to it.

Operation.
```
    29000
    37
    203
    87
Ans. 1073000 dolls.
```

PROOF.—29000×37=1073000, the same as before. Hence,

101. When there are ciphers on the right of the multiplicand.

Multiply the significant figures of the multiplicand by the multiplier, and to the product annex as many ciphers, as are found on the right of the multiplicand.

OBS. When both the multiplier and multiplicand have ciphers on the right multiply the significant figures together as if there were no ciphers, and to their product annex as many ciphers, as are found on the right of both factors.

34. Multiply 2370000 by 52.
35. Multiply 48120000 by 48.
36. Multiply 356300000 by 74.
37. Multiply 1623000000 by 89.
38. Multiply 540000 by 700.

Analysis.—540000=54×10000, and 700=7×100; we therefore multiply the significant figures, or the factors 54 and 7 together, (Art. 96. *a*,) and to this product annex six ciphers. (Art. 99.)

Operation.
```
    540000
    700
Ans. 378000000
```

39. Multiply 1563800 by 20000.
40. Multiply 31230000 by 120000.
41. Multiply 5310200 by 3400000.
42. Multiply 82065000 by 8100000.
43. Multiply 210909000 by 5100000.

QUEST.—101. When there are ciphers on the right of the multiplicand, how proceed? Obs. How, when there are ciphers on the right both of the multiplier and multiplicand?

102. There are *other methods of contracting the operations* in multiplication, which, in certain cases, may be resorted to with advantage. Some of the most useful are the following.

44. How many gallons of water will a hydrant discharge in 13 hours, if it discharges 2325 gallons per hour?

Operation.
2325×13
6975
Ans. 30225 gallons.

Multiplying by the 3 units, we set the first figure of the product one place to the right of the multiplicand. Now, since multiplying by 1 is taking the multiplicand *once*, (Art. 82,) we add together the multiplicand and the partial product already obtained, and the result is the answer.

PROOF.—2325×13=30225 gallons, the same as above. Hence,

103. To multiply by 13, 14, 15, &c., or 1, with either of the other digits *annexed* to it.

Multiply by the units' figure of the multiplier, and write each figure of the partial product one place to the right of that from which it arises; finally, add the partial product to the multiplicand, and the result will be the answer required.

Note—This method is the same, in effect, as if we actually multiplied by the 1 ten, and placed the first figure of the partial product under the figure by which we multiply. (Art. 87. II.)

45. Multiply 3251 by 14.
46. Multiply 4028 by 17.
47. Multiply 25039 by 16.
48. Multiply 50389 by 18.
49. If 21 men can do a job of work in 365 days, how long will it take 1 man to do it?

Operation.
365×21
730
Ans. 7665 days.

We first multiply by the 2 tens, and set the first product figure in tens' place, then adding this partial product to the multiplicand, we have 7665, for the answer.

PROOF.—365×21=7665 days, the same as above. Hence,

104. To multiply by 21, 31, 41, &c., or 1 with either of the other significant figures *prefixed* to it.

Multiply by the tens' figure of the multiplier, and write the first figure of the partial product in tens' place; finally, add this partial product to the multiplicand, and the result will be the answer required.

Note.—The reason of this method of contraction is substantially the same as that of the preceding.

50. Multiply 4275 by 31.
51. Multiply 7504 by 41.
52. Multiply 38256 by 61.
53. Multiply 70267 by 81.
54. How much will 99 carriages cost, at 235 dollars apiece?

Analysis.—Since 1 carriage costs 235 dollars, 100 carriages will cost 100 times as much, which is 23500 dollars. (Art. 99.) But we wished to find the cost of 99 carriages only. Now 99 is 1 less than 100; therefore, if we subtract the price of 1 carriage from the price of 100, it will give the price of 99 carriages. Hence,

Operation.

23500	price	of	100 C.
235	"	of	1 C.
23265	"	of	99 C.

105. To multiply by 9, 99, 999, or any number of 9s.

Annex as many ciphers to the multiplicand as there are 9s in the multiplier; from the result subtract the given multiplicand, and the remainder will be the answer required.

Note.—The *reason* of this method is obvious from the fact that annexing as many ciphers to the multiplicand as there are 9s in the multiplier, multiplies it by 100, or repeats it *once more* than is required; (Art. 99;) consequently, subtracting the multiplicand from the number thus produced, must give the true answer.

55. Multiply 4791 by 99.
56. Multiply 6034 by 999.
57. Multiply 7301 by 999.
58. Multiply 463 by 9999.
59. What is the product of 867 multiplied by 84?

Analysis.—We first multiply by 4 in the usual way. Now, since $8=4\times 2$, it is plain, if the partial product of 4 is multiplied by 2, it will give the partial product of 8. But as 8 denotes tens, the first figure of its product will also be tens. (Art. 86.) The sum of the two partial products will be the answer required.

Operation.

```
 867
  84
 3468×2
6936
72828 Ans.
```

Note.—For the sake of convenience in multiplying, the factor 2 is placed at the right of the partial product of 4, with the sign ×, between them.

60. What is the product of 987 by 486?

Operation.

```
  987
  486
 5922×8
47376
479682 Ans.
```

Since 48=6×8, we multiply the partial product of 6 by 8, and set the first product figure in tens' place as before. (Art. 86.)

PROOF.—987×486=479682, the same as above. Hence,

106. When part of the multiplier is a *composite* number of which the other figure is a *factor*.

First multiply by the figure that is a factor; then multiply this partial product by the other factor, or factors, taking care to write the first figure of each partial product in its proper order, and their sum will be the answer required. (Art. 86.)

OBS. When the figure in thousands, ten thousands, or any other column, is a factor of the other part, or parts of the multiplier, care must be taken to place the first figure of its product under the factor itself, and the first figure of each of the other partial products in its own order. (Art. 86.)

```
   (61.)                    (62.)
   2378                    256841
    936                     85632
 21402   ×4              2054728   ×7×4
  85608                  14383096
2225808 Ans.               8218912
                        21993808512 Ans.
```

63. Multiply 665 by 82. 64. Multiply 783 by 93.

65. Multiply 876 by 396. 66. Multiply 69412 by 95436.

67. 324325×54426. 68. 256721×85632.

69. What is the product of 63 multiplied by 45?

Note.—By multiplying the figures which produce the same order, and adding the results mentally, we may obtain the answer without setting down the partial products.

Operation.

```
  63
  45
2835 Ans.
```

First, multiplying the units into units, we set down the result and carry as usual. Now, since the 6 tens into 5 units, and 3 units into 4 tens will both produce the same order, viz: tens, (Art. 86,) we multiply them and add their products mentally. Thus, 6×5=30, and 3×4=12; now, 30+12=42, and 1 (to carry) makes 43. Finally, 6×4=24, and 4 (to carry) make 28

PROOF.—63×45=2835, the same as before. Hence,

107. To multiply any two numbers together without setting down the partial products.

First multiply the units together; then multiply the figures which produce tens, and adding the products mentally, set down the result and carry as usual. Next multiply the figures which produce hundreds, and add the products, &c., as before. In like manner, perform the multiplications which produce thousands, ten thousands, &c., adding the products of each order as you proceed, and thus continue the operation till all the figures are multiplied.

70. What is the product of 23456789 into 54321?

Analytic Operation.

				2	3	4	5	6	7	8	9
							5	4	3	2	1
				2×1	3×1	4×1	5×1	6×1	7×1	8×1	9×1
			2×2	3×2	4×2	5×2	6×2	7×2	8×2	9×2	
		2×3	3×3	4×3	5×3	6×3	7×3	8×3	9×3		
	2×4	3×4	4×4	5×4	6×4	7×4	8×4	9×4			
2×5	3×5	4×5	5×5	6×5	7×5	8×5	9×5				
12	7	4	1	9	6	2	3	5	2	6	9

Explanation.—Having multiplied by the first two figures of the multiplier, as in the last example, we perceive that there are three multiplications which will produce hundreds, viz: 7×1, 8×2, and 9×3; (Art. 86;) we therefore perform these multiplications, add their products mentally, and proceed to the next order. Again, there are four multiplications which will produce thousands, viz: 6×1, 7×2, 8×3, and 9×4. (Art. 86.) We perform these multiplications as before, and proceed in a similar manner through all the remaining orders. *Ans.* 1274196235269.

Note.—1. In the solution above, the multiplications of the different figures are arranged in separate columns, that the various combinations which produce the same order, may be seen at a glance. In practice it is unnecessary to denote these multiplications. The principle being understood the process of multiplying and adding may easily be carried on in the mind, while the final product only is set down.

2. When the factors contain but two or three figures each, this method is very simple and expeditious. A little practice will enable the student to apply it with facility when the factors contain six or eight figures each, and its application will afford an excellent discipline to the mind. It has sometimes been used when the factors contain twenty-four figures each; but it is doubtful whether the attempt to extend it so far, is profitable.

71. Multiply 25×25.
72. Multiply 54×54.
73. Multiply 81×64.
74. Multiply 45×92.
75. Multiply 194×144.
76. Multiply 1234×125.
77. Multiply 4825×2352.
78. Multiply 6521×5312.

108. By suitable attention, the critical student will discover various other methods of abbreviating the processes of multiplication.

Solve the following examples, contracting the operations when practicable.

79. 42634×63.
80. 50035×56.
81. 72156×1000.
82. 42000×40000.
83. 80000×25000.
84. 2567345×17.
85. 4300450×19.
86. 9803404×41.
87. 6710045×71.
88. 3456710×18.
89. 7000541×91.
90. 4102034×99.
91. 42304×999.
92. 50421×9999.
93. 67243×99999.
94. 78563×93.
95. 34054×639.
96. 52156×756.
97. 41907×54486.
98. 26397×24648.
99. 12900×14000.
100. 64172×42432.
101. 26815678×81.
102. 85×85.
103. 256×256.
104. 322×325.
105. 5234×2435.
106. 48743000×637.
107. 31890420×85672.
108. 80460000×2763.
109. 2364793×8485672.
110. 1256702×999999.
111. 6840005×91×61.
112. 45067034×17×51.
113. 788031245×81×16.
114. 61800000×23000.
115. 12563000×4800000.
116. 91300203×1000000.
117. 680040000×1000000.
118. 4000000000×1000000.

SECTION V.

DIVISION.

ART. **110.** Ex. 1. How many barrels of flour, at 8 dollars per barrel, can you buy for 56 dollars?

Analysis.—Since flour is 8 dollars a barrel, it is obvious you can buy 1 barrel as often as 8 dollars are contained in 56 dollars; and 8 dolls. are contained in 56 dolls. 7 times. *Ans.* 7 barrels.

Ex. 2. A man wished to divide 72 dollars equally among 9 beggars: how many dollars would each receive?

Solution.—Reasoning as before, each beggar would receive as many dollars as 9 is contained times in 72; and 9 is contained in 72, 8 times. *Ans.* 8 dollars.

OBS. The learner will at once perceive that the object in the first example, is to find *how many times* one number is contained in another; and that the object of the second, is to *divide* a given number into equal parts, but its solution consists in finding how many times one number is contained in another, and is the *same* in principle as that of the first.

111. *The Process of finding how many times one number is contained in another, is called* DIVISION.

The number to be *divided,* is called the *dividend.*

The number by which we *divide,* is called the *divisor.*

The number *obtained* by division, or the *answer* to the question, is called the *quotient.* It shows how *many times* the divisor is contained in the dividend. Hence, it may be said,

112. *Division is finding a quotient, which multiplied into the divisor, will produce the dividend.*

Note.—The term *quotient* is derived from the Latin word *quoties,* which signifies *how often,* or *how many times.*

QUEST.—111. What is division? What is the number to be divided called? The number by which we divide? What is the number obtained called? What does the quotient show? 112. What then may division be said to be?

113. The number which is sometimes *left* after division, is called the *remainder*. Thus, when we say 5 is contained in 38, 7 times, and 3 over, 5 is the divisor, 38 the dividend, 7 the quotient, and 3 the remainder.

OBS. 1. The remainder is of the same denomination as the dividend; for, it is a part of it.

2. The remainder is always *less* than the divisor; for, if it were *equal* to, or *greater* than the divisor, the divisor could be contained *once more* in the dividend.

114. It will be perceived that division is *similar in principle* to subtraction, and may be performed by it. For instance, to find how many times 7 is contained in 21, subtract 7 (the divisor) continually from 21 (the dividend), until the latter is exhausted; then counting these repeated subtractions, we shall have the true quotient. Thus, 7 from 21 leaves 14; 7 from 14 leaves 7; and 7 from 7 leaves 0. Now by counting, we find that 7 has been taken from 21, 3 times; consequently, 7 is contained in 21, 3 times. Hence,

Division is sometimes defined to be a short way of performing repeated subtractions of the same number.

OBS. 1. It will be observed that division is the *reverse* of multiplication. Multiplication is the *repeated addition* of the same number; division is the *repeated subtraction* of the same number. The *product* of the one answers to the *dividend* of the other; but the latter is always *given*, while the former is *required*.

2. When the dividend denotes things of *one denomination only*, the operation is called *Simple Division*.

SHORT DIVISION.

Ex. 3. How many hats, at 2 dollars apiece, can be bought for 4862 dollars?

Operation.

Divisor. Divid.

2) 4862

Quot. 2431

We write the divisor on the left of the dividend with a curve line between them; then, beginning at the left hand, proceed thus: 2 is contained in 4, 2 times. Now, since the 4 de-

QUEST.—113. What is the number called which is sometimes left after division? *Obs.* Of what denomination is the remainder? Why? Is the remainder greater or less than the divisor? Why? 114 To what rule is division similar in principle? *Obs.* Of what is division the reverse? When the dividend denotes things of one denomination only, what is the operation called?

notes thousands, the 2 must be thousands; we therefore write it in thousands' place, under the figure divided. 2 is contained in 8, 4 times; and as the 8 is hundreds, the 4 must also be hundreds; hence we write it in hundreds' place, under the figure divided. 2 in 6, 3 times; the 6 being tens, the 3 must also be tens, and should be set in tens' place. 2 in 2, once; and since the 2 is units, the 1 is a unit, and must therefore be written in units' place. The answer is 2431 hats.

115. *When the process of dividing is carried on in the mind, and the quotient only is set down, as in the last example, the operation is called* SHORT DIVISION.

116. The *reason* that each quotient figure is of the *same order* as the *figure divided,* may be shown in the following manner:

Analytic Solution.

4862=4000+800+60+2
2)4000+800+60+2
  2000+400+30+1

Having separated the dividend of the last example into the *orders* of which it is composed, we perceive that 2 is contained in 4000, 2000 times; for 2×2000=4000, Again, 2 is contained in 800, 400 times; for 2×400=800, &c. *Ans.* 2431.

Ex. 4. A man left an estate of 209635 dollars, to be divided equally among 4 children: how much did each receive?

Operation.

4)209635
Ans. 52408¾ dolls.

Since the divisor 4, is not contained in 2, the first figure of the dividend, we find how many times it is contained in the first two figures. Thus, 4 is contained in 20, 5 times; write the 5 under the 0. Again, 4 is contained in 9, 2 times and 1 over; set the 2 under the 9. Now, as we have 1 thousand over, we prefix it mentally to the 6 hundreds, making 16 hundreds; and 4 in 16, 4 times. Write the 4 under the 6. But 4 is not contained in 3, the next figure, we therefore put a cipher in the quotient, and prefix the 3 to the next figure of the dividend, as if it were a remainder. Then 4 in 35, 8 times and 3 over; place the 8 under the 5, and setting the remainder over the divisor thus ¾, place it on the right of the quotient.

Note.—To *prefix* means to place *before*, or at the *left hand.*

117. When the divisor is not contained in any figure of the dividend, a cipher must always be placed in the quotient.

OBS. The *reason* for placing a cipher in the quotient, is to preserve the *true local value* of each figure of the quotient. (Art. 116.)

118. In order to render the division complete, it is obvious that the *whole* of the dividend must be divided. But when there is a remainder after dividing the last figure of the dividend, it must of necessity be smaller than the divisor, and cannot be divided by it. (Art. 113. Obs. 2.) We therefore represent the division by placing the remainder over the divisor, and annex it to the quotient. (Art. 25.)

OBS. 1. The learner will observe that in *dividing* we begin at the *left hand*, instead of the *right*, as in Addition, Subtraction, and Multiplication. The *reason* is, because there is frequently a remainder in dividing a *higher* order, which must necessarily be united with the next *lower* order, before the division can be performed.

2. The divisor is placed on the left of the dividend, and the quotient under it, merely for the sake of convenience. When division is represented by the sign ÷, the divisor is placed on the *right* of the dividend; and when represented in the form of a fraction, the divisor is placed *under* the dividend.

LONG DIVISION.

Ex. 5. At 15 dollars apiece, how many cows can be bought for 3525 dollars?

Operation.

```
Divisor. Divid. Quot.
  15)    3525  (235
         30
         --
          52
          45
          --
           75
           75
           --
```

Having written the divisor on the left of the dividend as before, we find that 15 is contained in 35, 2 times, and place the 2 on the right of the dividend, with a curve line between them. We next multiply the divisor by this quotient figure, place the product under the figures divided, and subtract it therefrom. We now bring down the next figure of the dividend, and placing it on the right of the remainder 5, we perceive that 15 is contained in 52, 3 times. Set the 3 on the right of the last quotient figure, multiply the divisor by it, and subtract the product from the figures divided as before. We then bring down the next, which is the last figure of the dividend, to the right of this remainder, and finding 15 is contained in 75, 5 times, we place the 5 in the quotient, multiply and subtract as before. The answer is 235 cows.

119. *When the result of each step in the operation is written down, as in the last example, the process is called* LONG DIVISION. Long Division is the *same* in *principle* as Short Division. The only difference between them is, that in the *former*, the result of each step in the operation is written down, while in the *latter*, we carry on the process in the mind, and simply write the *quotient*.

OBS. 1. When the divisor contains but *one* figure, the operation by *Short Division* is the most expeditious, and therefore should always be practiced; but when the divisor contains *two* or *more* figures, it will generally be the most convenient to use *Long Division*.

2. To prevent mistakes, it is advisable to put a dot under each figure of the dividend, when it is brought down.

3. The *French* place the divisor on the *right* of the dividend, and the quotient *below* the divisor,* as seen in the following example.

Ex. 6. How many times is 72 contained in 5904?

Operation.

```
5904 (72 divisor.
576   82 quotient.
 144
 144
```

The divisor is contained in 590, the first three figures of the dividend, 8 times. Set the 8 under the divisor, multiply, &c., as before.

Ex. 7. How many times is 435 contained in 262534?

Operation.

```
435)262534(603 229/435 Ans.
    2610..
      1534
      1305
       229 rem.
```

Since the divisor is not contained in the first *three* figures of the dividend, we find how many times it is contained in the first *four*, the fewest that will contain it, and write the 6 in the quotient; then multi-

QUEST.—115. What is short division? 119. What is long division? What is the difference between them?

* Elèments D'Arithmétique, par M. Bourdon. Also, Lacroix's Arithmetic, translated by Professor Farrar.

plying and subtracting as before, the remainder is 15. Bringing down the next figure, we have 153 to be divided by 435. But 435 is not contained in 153; we therefore place a cipher in the quotient, and bring down the next figure. Then 435 in 1534, 3 times. Place the 3 in the quotient, and proceed as before.

Note.—After the first quotient figure is obtained, for *each figure of the dividend which is brought down*, either *a significant figure*, or a *cipher*, **must be put in the** *quotient*. (Art. 117.)

120. From the preceding illustrations and principles we derive the following

GENERAL RULE FOR DIVISION.

I. When the divisor contains but *one* figure.

Write the divisor on the left of the dividend, with a curve line between them. Begin at the left hand, divide successively each figure of the dividend by the divisor, and place each quotient figure directly under the figure divided. (Arts. 116, 118. Obs. 1, 2.)

If there is a remainder after dividing any figure, prefix it to the next figure of the dividend and divide this number as before; and if the divisor is not contained in any figure of the dividend, place a cipher in the quotient and prefix this figure to the next one of the dividend, as if it were a remainder. (Arts. 117, 118.)

II. When the divisor contains *more* than *one* figure.

Beginning on the left of the dividend, find how many times the divisor is contained in the fewest figures that will contain it, and place the quotient figure on the right of the dividend with a curve line between them. Then multiply the divisor by this figure and subtract the product from the figures divided; to the right of the remainder bring down the next figure of the dividend and divide this number as before. Proceed in this manner till all the figures of the dividend are divided.

QUEST.—120. How do you write the numbers for division? When the divisor contains but one figure, how proceed? Why place the divisor on the left of the dividend and the quotient under the figure divided? When there is a remainder after dividing a figure, what is to be done with it? When the divisor is not contained in any figure of the dividend, how proceed? Why? Why begin to divide at the left hand? When the divisor contains more than one figure, how proceed?

Whenever there is a remainder after dividing the last figure, write it over the divisor and annex it to the quotient. (Art. 118.)

Demonstration.—The principle on which the operations in Division depend, is that *a part* of the quotient is found, and the product of this part into the divisor is taken from the dividend, showing how much of the latter remains to be divided; then *another part* of the quotient is found, and its product into the divisor is taken from what remained before. Thus the operation proceeds till the *whole* of the dividend is divided, or till the *remainder* is less than *the divisor.* (Art. 113. Obs. 2.)

OBS. When the divisor is large, the pupil will find assistance in determining the quotient figure, by finding how many times the first figure of the divisor is contained in the first figure, or if necessary, the first *two* figures of the dividend. This will give pretty nearly the right figure. Some allowance must, however, be made for carrying from the product of the other figures of the divisor, to the product of the first into the quotient figure.

121. PROOF.—*Multiply the divisor by the quotient, to the product add the remainder, and if the sum is equal to the dividend, the work is right.*

OBS. Since the quotient shows how many times the divisor is contained in the dividend, (Art. 111,) it follows, that if the divisor is repeated as many times as there are units in the quotient, it must produce the dividend.

Ex. 8. Divide 256329 by 723.

Operation.	*Proof.*
723)256329(354$\frac{387}{723}$ *Ans.*	723 divisor.
2169	354 quotient.
3942	2892
3615	3615
3279	2169
2892	387 rem.
387 rem.	256329 dividend.

122. *Second Method.*—Subtract the remainder, if any, from the dividend, divide the dividend thus diminished, by the quotient; and if the *result* is *equal* to the given *divisor,* the work is *right.*

QUEST.—When there is a remainder after dividing the last figure of the dividend, what must be done with it? 121. How is division proved? *Obs.* How does it appear that the product of the divisor and quotient will be equal to the dividend, if the work is right? Can division be proved by any other methods?

123. *Third Method.*—First cast the 9s out of the divisor and quotient, and multiply the remainders together; to the product add the remainder, if any, after division; cast the 9s out of this sum, and set down the excess; finally cast the 9s out of the dividend, and if the *excess* is the *same* as that obtained from the divisor and *quotient*, the work may be considered *right*.

Note.—Since the divisor and quotient answer to the multiplier and multiplicand, and the dividend to the product, it is evident that the principle of casting out the 9s will apply to the proof of division, as well as that of multiplication. (Art. 90.)

124. *Fourth Method.*—Add the remainder and the respective products of the divisor into each quotient figure together, and if the *sum* is equal to the *dividend*, the work is *right*.

Note.—This mode of proof depends upon the principle that the *whole of a quantity is equal to the sum of all its parts.* (Ax. 11.)

125. *Fifth Method.*—First cast the 11s out of the divisor and quotient, and multiply the remainders together; to the product add the remainder, if any, after division, and casting the 11s out of this sum, set down the excess; finally, cast the 11s out of the dividend, and if the *excess* is the *same* as that obtained from the divisor and quotient, the work is *right*. (Art. 92. Note 2.)

EXAMPLES FOR PRACTICE.

127. Ex. 1. A farmer raised 2970 bushels of wheat on 66 acres of land: how many bushels did he raise per acre?

2. A garrison consumed 8925 barrels of flour in 105 days: how much was that per day?

3. The President of the United States receives a salary of 25000 dollars a year: how much is that per day?

4. A drover paid 2685 dollars for 895 head of cattle: how much did he pay per head?

5. If a man's expenses are 3560 dollars a year, how much are they per week?

6. If the annual expenses of the government are 27 millions of dollars, how much will they be per day?

7. How long will it take a ship to sail from New York to Liverpool, allowing the distance to be 3000 miles, and the ship to sail 144 miles per day?

8. Sailing at the same rate, how long would it take the same ship to sail round the globe, a distance of 25000 miles?

10. 47839÷42.
11. 75043÷52.
12. 93840÷63.
13. 421645÷74.
14. 325000÷85.
15. 400000÷96.
16. 999999÷47.
17. 352417÷29.
18. 47981÷251.
19. 423405÷485.
20. 16512÷344.
21. 304916÷6274.
22. 12689÷145.
23. 145260÷1345.
24. 147735÷3283.
25. 1203033÷327.
26. 1912500÷425.
27. 5184673÷102.
28. 301140÷478.
29. 8893810÷37846.
30. 9302688÷14356.
31. 9749320÷365.
32. 3228242÷5734.
33. 75843639426÷8593.
34. 65358547823÷2789.
35. 102030405060÷123456.
36. 908070605040÷654321.
37. 1000000000000000÷111.
38. 1000000000000000÷1111.
39. 1000000000000000÷11111.

CONTRACTIONS IN DIVISION.

128. The operations in division, as well as those in multiplication, may often be shortened by a careful attention tc the application of the preceding principles.

CASE 1.—*When the divisor is a composite number.*

Ex. 1. A man divided 837 dollars equally among 27 persons, who belonged to 3 families, each family containing 9 persons: how many dollars did each person receive?

Analysis.—Since 27 persons received 837 dollars, each one must have received as many dollars, as 27 is contained times in 837. But as 27 (the number of persons), is a composite number whose factors are 3 (the number of families), and 9 (the number of persons in each family), it is obvious we may first find how many dollars each family received, and then how many each person received.

Operation.

3)837 whole sum divided.

9)279 portion of each Fam.

Ans. 31 " " " person.

If 3 families received 837 dollars, 1 family must have received as many dollars, as 3 is contained times in 837; and 3 in 837, 279 times. That is, each family received 279 dollars.

Again, if 9 persons, (the number in each family,) received 279 dollars, 1 person must have received as many dollars, as 9 is contained times in 279; and 9 in 279, 31 times. *Ans.* 31 dollars.

PROOF.—31×27=837, the same as the dividend. Hence,

129. To divide by a *composite* number.

I. *Divide the dividend by one of the factors of the divisor, then divide the quotient thus obtained by another factor; and so on till all the factors are employed. The last quotient will be the answer.*

II. To find the *true* remainder.

If the divisor is resolved into but two factors, multiply the last remainder by the first divisor, to the product add the first remainder, if any, and the result will be the true remainder.

When more than two factors are employed, multiply each remainder by all the preceding divisors, to the sum of their products, add the first remainder, and the result will be the true remainder.

OBS. 1. The *true* remainder may also be found by multiplying the quotient by the divisor, and subtracting the product from the dividend.

2. This contraction is exactly the *reverse* of that in multiplication. (Art. 97.) The result will evidently be the same, in whatever order the factors are taken.

2. A man bought a quantity of clover seed amounting to 507 pints, which he wished to divide into parcels containing 64 pints each: how many parcels can he make?

Note.—Since 64=2×8×4, we divide by the factors respectively.

Operation.

```
2)507
 8)253—1 rem.   . . .        =  1 pt.
   4)31—5 rem.  Now 5×2      = 10 pts.
      7—3 rem.  and 3×8×2    = 48 pts.
Ans. 7 parcels, and 59 pts. over.   59 pts. True Rem.
```

Demonstration.—1. Dividing 507 the number of pints, by 2, gives 253 for the quotient, or distributes the seed into 253 equal parcels, leaving 1 pint over. Now the units of this quotient are evidently of a *different value* from those of the given dividend; for since there are but *half* as many parcels as at first, it

QUEST.—129. How proceed when the divisor is a composite number? How find the true remainder?

is plain that each parcel must contain 2 pints, or 1 quart; that is, every unit of the first quotient contains 2 of the units of the given dividend; consequently, every unit of it that remains will contain the same; (Art. 113. Obs. 2;) therefore this remainder must be multiplied by 2, in order to find the units of the given dividend which it contains.

2. Dividing the quotient 253 parcels, by 8, will distribute them into 31 other equal parcels, each of which will evidently contain 8 times the quantity of the preceding, viz: 8 times 1 quart =8 quarts, or 1 peck; that is, every unit of the second quotient contains 8 of the units in the first quotient, or 8 times 2 of the units in the given dividend; therefore what remains of it, must be multiplied by 8×2, or 16, to find the units of the given dividend which it contains.

3. In like manner, it may be shown, that dividing by *each successive factor* reduces each quotient to a class of units of a higher value than the preceding; that every unit which remains of any quotient, is of the same value as that quotient, and must therefore be multiplied by all the preceding divisors, in order to find the units of the given dividend which it contains.

4. Finally, the several remainders being reduced to the same units as those of the given dividend according to the rule, their sum must evidently be the *true* remainder. (Ax. 11.)

3. How many acres of land, at 35 dollars an acre, can you buy for 4650 dollars?

4. Divide 16128 by 24.

5. Divide 25760 by 56.

6. Divide 17220 by 84.

7. Divide 91080 by 72.

CASE II.—*When the divisor is* 1 *with ciphers annexed to it.*

130. It has been shown that *annexing* a cipher to a number *increases* its value *ten times*, or *multiplies* it by 10. (Art. 98.) Reversing this process; that is, *removing* a cipher from the right hand of a number, will evidently *diminish* its value *ten times*, or *divide* it by 10; for, each figure in the number is thus restored to its original place, and consequently to its original value. Thus, annexing a cipher to 15, it becomes 150, which is the same as 15×10. On the other hand, removing the cipher from 150, it becomes 15, which is the same as 150÷10.

In the same manner it may be shown, that removing *two* ciphers from the right of a number, divides it by 100; removing *three*, divides it by 1000; removing *four*, divides it by 10000, &c. Hence,

QUEST.—130. What is the effect of annexing a cipher to a number? What is the effect of removing a cipher from the right of a number? How does this appear?

131. To divide by 10, 100, 1000, &c.

Cut off as many figures from the right hand of the dividend as there are ciphers in the divisor. The remaining figures of the dividend will be the quotient, and those cut off the remainder.

8. In one dime there are 10 cents: how many dimes are there in 200 cents? In 340 cents? In 560 cents?

9. In one dollar there are 100 cents: how many dollars are there in 65000 cents? In 765000 cents? In 4320000 cents?

10. Divide 26750000 by 100000.

11. Divide 144360791 by 1000000.

12. Divide 582367180309 by 100000000.

CASE III.—*When the divisor has ciphers on the right hand.*

13. How many hogsheads of molasses, at 30 dollars apiece, can you buy for 9643 dollars?

OBS. The divisor 30, is a composite number, the factors of which are 3 and 10. (Arts. 95, 96.) We may, therefore, divide first by one factor and the quotient thence arising by the other. (Art. 129.) Now cutting off the right hand figure of the dividend, divides it by ten; (Art. 131;) consequently dividing the remaining figures of the dividend by 3, the other factor of the divisor, will give the quotient.

Operation.

3|0)964|3

321 $\frac{13}{30}$ *Ans.*

We first cut off the cipher on the right of the divisor, and also cut off the right hand figure of the dividend; then dividing 964 by 3, we have 1 remainder. Now as the 3 cut off, is part of the remainder, we therefore annex it to the 1. *Ans.* 321$\frac{13}{30}$ hogsheads. Hence,

132. When there are ciphers on the right hand of the divisor.

Cut off the ciphers, also cut off as many figures from the right of the dividend. Then divide the other figures of the dividend by the remaining figures of the divisor, and annex the figures cut off from the dividend to the remainder.

14. How many buggies, at 70 dollars apiece, can you buy for 7350 dollars?

QUEST.—131. How proceed when the divisor is 10, 100, 1000, &c.? 132. When there are ciphers on the right hand of the divisor, how proceed? What is to be done with figures cut off from the dividend?

15. How many barrels will it take to pack 36800 pounds of pork, allowing 200 pounds to a barrel?

16. Divide 3360000 by 17000.

133. Operations in Long Division may be shortened by subtracting the product of the respective figures in the divisor into each quotient figure as we proceed in the operation, setting down the remainders only. This is called the *Italian Method.*

17. How many times is 21 contained in 4998?

Operation.

```
21)4998(238
    79
    168
```

This method, it will be seen, requires a much smaller number of figures than the ordinary process.

18. Divide 1188 by 33.

19. Divide 2516 by 37.

20. Divide 3128 by 86.

21. Divide 7125 by 95.

22. A merchant laid out 873 dollars in flour, at 5 dollars a barrel: how many barrels did he get?

Operation.

```
      873
        2
1|0)174|6
    174 3/5 Ans.
```

We first double the dividend, and then divide the product by 10, which is done by cutting off the right hand figure. (Art. 131.) But since we multiplied the dividend by 2, it is plain that the 6 cut off, is 2 times too large for the remainder; we therefore divide it by 2, and we have 3 for the true remainder. Hence,

134. When the divisor is 5.

Multiply the dividend by 2, *and divide the product by* 10. (Art. 131.)

Note.—1. When the figure cut off is a significant figure, it must be divided by 2 for the *true* remainder.

2. This contraction depends upon the principle that any given divisor is contained in any given dividend, just as many times as *twice* that divisor is contained in *twice* that dividend, *three times* that divisor in *three times* that dividend, &c. For a further illustration of this principle see General Principles in Division.

23. Divide 6035 by 5.

24. Divide 8450 by 5.

25. Divide 32561 by 5.

26. Divide 43270 by 5.

135. When the divisor is 15, 35, 45, or 55.

Double the dividend, and divide the product by 30, 70, 90, *or* 110, *as the case may be.* (Art. 132.)

Note.—This method is simply doubling both the divisor and dividend. We must therefore divide the remainder, if any, by 2, for the *true* remainder.

27. Divide 1256 by 15.
28. Divide 2673 by 35.
29. Divide 3507 by 45.
30. Divide 7853 by 55.

136. When the divisor is 25.

Multiply the dividend by 4, *and divide the product by* 100. (Art. 131.)

Note.—This is obviously the same as multiplying both the dividend and divisor by 4. (Art. 134. Note 2.) Hence, we must divide the remainder, if any thus found, by 4, for the *true* remainder.

31. Divide 2350 by 25.
32. Divide 4860 by 25.
33. Divide 42340 by 25.
34. Divide 94880 by 25.

137. To divide by 125.

Multiply the dividend by 8, *and divide the product by* 1000. (Art. 131.)

Note.—This contraction is multiplying both the dividend and divisor by 8. For the *true* remainder, therefore, we must divide the remainder, if any, by 8.

35. Divide 8375 by 125.
36. Divide 25426 by 125.

138. To divide by 75, 175, 225, or 275.

Multiply the dividend by 4, *and divide the product by* 300, 700, 900, *or* 1100, *as the case may be.* (Art. 132.)

Note.—For the *true* remainder, divide the remainder, if any thus found, by 4.

37. Divide 1125 by 75.
38. Divide 2876 by 175.
39. Divide 3825 by 225.
40. Divide 8250 by 275.

139. The preceding are among the most frequent and useful modes of contracting operations in division. Various other methods might be added, but they will naturally suggest themselves to the inventive student, as opportunities occur for their application.

41. How long would it take a vessel sailing 100 miles per day to circumnavigate the earth, whose circumference is 25000 miles?

42. The distance of the Earth from the Sun is 95,000,000 of miles: how long would it take a balloon going at the rate of 100,000 miles a year, to reach the sun?

43. The debts of the several States of the Union, in 1840, amounted to 171,000,000 of dollars, and the number of inhabitants was 17,000,000: how much must each individual have been taxed to pay the debt?

44. The national debt of Holland is 800,000,000 of dollars, and the number of inhabitants 2,800,000: what is the amount of indebtedness of each individual?

45. The national debt of Spain is 467,000,000 of dollars, and the number of inhabitants 11,900,000: what is the amount of indebtedness of each individual?

46. The national debt of Russia is 150,000,000 of dollars, and the number of inhabitants 51,100,000: what is the amount of indebtedness of each individual?

47. The national debt of Austria is 380,000,000 of dollars, and the number of inhabitants 34,100,000: what is the amount of indebtedness of each individual?

48. The national debt of France is 1,800,000,000 of dollars, and the number of inhabitants 33,300,000: what is the amount of indebtedness of each individual?

49. The national debt of Great Britain is 5,556,000,000 of dollars, and the number of inhabitants 25,300,000: what is the amount of indebtedness of each individual?

50. Divide 467000000000 by 25000000000.

51. 568240÷42.
52. 785372÷63.
53. 896736÷72.
54. 67234568÷5.
55. 34256726÷15.
56. 42367581÷45.
57. 16753672÷35.
58. 3256385÷55.
59. 45672400÷25.
60. 6245634÷45.
61. 8245623÷125.
62. 462156÷75.
63. 3562189÷225.
64. 685726÷32000
65. 723564÷175.
66. 892565÷225.
67. 456212÷275.
68. 925673÷125.
69. 763421÷175.
70. 876240÷275.
71. 7825600÷80000.
72. 92004578÷100000.

GENERAL PRINCIPLES IN DIVISION.

140. From the nature of division, it is evident, that the *value* of the *quotient* depends both on the *divisor* and the *dividend.*

141. If a given divisor is contained in a given dividend a certain number of times, the same divisor will obviously be contained,

In *double* that dividend, *twice* as many times.

In *three times* that dividend, *thrice* as many times, &c. Hence,

If the divisor remains the same, multiplying the dividend by any number, is in effect multiplying the quotient by that number.

Thus, 6 is contained in 12, 2 times; in 2 times 12 or 24, 6 is contained 4 times; (i. e. twice 2 times;) in 3 times 12 or 36, 6 is contained 6 times; (i. e. thrice 2 times;) &c.

142. Again, if a given divisor is contained in a given dividend a certain number of times, the same divisor is contained,

In *half* that dividend, *half* as many times;

In a *third* of that dividend, a *third* as many times, &c. Hence,

If the divisor remains the same, dividing the dividend by any number, is in effect dividing the quotient by that number.

Thus, 8 is contained in 48, 6 times; in 48÷2 or 24, (half of 48,) 8 is contained 3 times; (i. e. half of 6 times;) in 48÷3 or 16, (a third of 48,) 8 is contained 2 times; (i. e. a third of 6 times;) &c.

143. If a given divisor is contained in a given dividend a certain number of times, then, in the same dividend,

Twice that divisor is contained only *half* as many times;

Three times that divisor, a *third* as many times, &c. Hence,

If the dividend remains the same, multiplying the divisor by any number, is in effect dividing the quotient by that number.

Thus, 4 is contained in 24, 6 times; 2 times 4 or 8 is con-

QUEST.—140. Upon what does the value of the quotient depend? 141. If the divisor remains the same, what effect has it on the quotient to multiply the dividend? 142. What is the effect of dividing the dividend by any given number? 143. If the dividend remains the same, what is the effect of multiplying the divisor by any given number?

tained in 24, 3 times; (i. e. half of 6 times;) 3 times 4 or 12 is contained in 24, 2 times; (i. e. a third of 6 times;) &c.

144. If a given divisor is contained in a given dividend a certain number of times, then, in the same dividend,

Half that divisor is contained *twice* as many times;

A *third* of that divisor, *three times* as many times, &c. Hence,

If the dividend remains the same, dividing the divisor by any number, is in effect multiplying the quotient by that number.

Thus, 6 is contained in 36, 6 times; 6÷2 or 3, (half of 6,) is contained in 36, 12 times; (i. e. twice 6 times;) 6÷3 or 2, (a third of 6,) is contained in 36, 18 times; (i. e. thrice 6 times;) &c.

145. From the preceding articles, it is evident that any given divisor is contained in any given dividend, just as many times as *twice* that divisor is contained in *twice* that dividend; *three times* that divisor in *three times* that dividend, &c.

Conversely, any given divisor is contained in any given dividend just as many times, as *half* that divisor is contained in *half* that dividend; a *third* of that divisor, in a *third* of that dividend, &c. Hence,

146. *If the divisor and dividend are both multiplied, or both divided by the same number, the quotient will not be altered.*

Thus, 6 is contained in 12, 2 times;

2 times 6 is contained in 2 times 12, 2 times;

3 times 6 is contained in 3 times 12, 2 times, &c.

Again, 12 is contained in 48, 4 times;

12÷2 is contained in 48÷2, 4 times;

12÷3 is contained in 48÷3, 4 times, &c.

147. If the *sum* of two or more numbers is divided by any number, the *quotient* will be equal to the *sum* of the quotients which will arise from dividing the given numbers separately.

Thus, the sum of 12+18=30; and 30÷6=5.

Now, 12÷6=2; and 18÷6=3; but the sum of 2+3=5.

Again, the sum of 32+24+40=96; and 96÷8=12.

Now, 32÷8=4; 24÷8=3; and 40÷8=5; but 4+3+5=12.

QUEST.—144. What of dividing the divisor? 146. What is the effect upon the quotient if the divisor and dividend are both multiplied, or both divided by the same number?

CANCELATION.*

148. We have seen that division is finding a quotient, which, multiplied into the divisor, will produce the dividend. (Art. 112.) If, therefore, the dividend is resolved into two such factors that one of them is the divisor, the other factor will, of course, be the quotient. Suppose, for example, 42 is to be divided by 6. Now the factors of 42 are 6 and 7, the first of which being the divisor, the other must be the quotient. Therefore,

Canceling a factor of any number, divides the number by that factor. Hence,

149. When the dividend is the product of two factors, one of which is the same as the divisor.

Cancel the factor common to the dividend and divisor; the other factor of the dividend will be the answer. (Ax. 9.)

Note.—The term *cancel*, signifies to *erase* or *reject.*

1. Divide the product of 34 into 28 by 34.

Common Method.

```
    34
    28
   ---
   272
   68
34)952(28 Ans.
   68
   ---
   272
   272
```

By Cancelation.

$\not{3}\not{4})\not{3}\not{4}\times 28$

28 *Ans.*

Canceling the factor 34, which is common both to the divisor and dividend, we have 28 for the quotient, the same as before.

150. *The method of contracting arithmetical operations, by rejecting equal factors, is called* CANCELATION.

OBS. It applies with *great advantage* to that class of examples and problems, which involve both multiplication and division; that is, which require the *product* of two or more numbers to be divided by *another number*, or by the *product* of two or more numbers.

2. Divide 76×45 by 76.
3. Divide 63×81 by 81.
4. Divide 65×82 by 82.
5. Divide 95×73 by 95.
6. Divide the product of 45 times 84 by 9.

* Birk's Arithmetical Collections: London, 1764.

Analysis.—The factor $45=5\times9$; hence the dividend is composed of the factors $84\times5\times9$. We may therefore cancel 9 which is common both to the divisor and dividend, and 84×5, the other factors of the dividend, will be the answer required.

Operation.	*Proof.*
$\cancel{9})84\times5\times\cancel{9}$	$84\times5\times9=3780$
420 *Ans.*	And $3780\div9=420$.

7. Divide the product of $45\times6\times3$ by 18×5.

Operation.	*Proof.*
$\cancel{18}\times5)45\times\cancel{6}\times\cancel{3}$	$45\times6\times3=810$; and $18\times5=90$
9 *Ans.*	Now, $810\div90=9$

Note.—We cancel the factors 6 and 3 in the dividend and 18 in the divisor; for $6\times3=18$. Canceling the same or equal factors in the divisor and dividend, is dividing them both by the same number, and therefore does not affect the quotient. (Arts. 146, 148.) Hence,

151. When the divisor and dividend have common factors.

Cancel the factors common to both; then divide the product of those remaining in the dividend by the product of those remaining in the divisor.

8. Divide $15\times7\times12$ by $5\times3\times7\times2$.
9. Divide $27\times3\times4\times7$ by $9\times12\times6$.
10. Divide $75\times15\times24$ by $25\times3\times6\times4\times5$.

Note.—The further *development* and *application* of the principles of Cancelation, may be seen in reduction of compound fractions to simple ones; in multiplication and division of fractions; in simple and compound proportion, &c.

151. *a.* The four preceding rules, viz: *Addition, Subtraction, Multiplication,* and *Division,* are usually called the FUNDAMENTAL RULES of Arithmetic, because they are the *foundation* or *basis* of all arithmetical calculations.

OBS. Every *change* that can be made upon the value of a number, must necessarily either *increase* or *diminish* it. Hence, the fundamental operations in arithmetic are, strictly speaking, but two, *addition* and *subtraction*; that is, *increase* and *decrease*. Multiplication, we have seen, is an abbreviated form of addition; division of subtraction. (Arts. 80, 114.)

QUEST.—151. *a.* Name the fundamental rules of Arithmetic. Why are these rules called fundamental?

APPLICATIONS OF THE FUNDAMENTAL RULES.

152. When the *sum* of two numbers and *one* of the numbers are given, to find the *other* number.

From the given sum, subtract the given number, and the remainder will be the other number.

Ex. 1. The sum of two numbers is 87, one of which is 25: what is the other number?

Solution.—87—25=62, the other number. (Art. 72.)

PROOF.—62+25=87, the given sum. (Ax. 11.)

2. A and B together own 350 acres of land, 95 of which belong to A: how many does B own?

3. Two merchants bought 1785 bushels of barley together, one of them took 860 bushels: how many bushels did the other have?

153. When the *difference* and the *greater* of two numbers are given, to find the *less.*

Subtract the difference from the greater, and the remainder will be the less number.

4. The greater of two numbers is 72, and the difference between them is 28: what is the less number?

Solution.—72—28=44, the less number. (Art. 72.)

PROOF.—44+28=72, the greater number. (Art. 73. Obs.)

5. A man bought a horse and chaise; for the chaise he gave 265 dollars, which was 75 dollars more than he paid for the horse: how much did he give for the horse?

6. A traveler met two droves of sheep; the first contained 1250, which was 125 more than the second had: how many sheep were there in the second drove?

154. When the *difference* and the *less* of two numbers are given, to find the *greater.*

QUEST.—152. When the sum of two numbers and one of them are given, how is the other found? 153. When the difference and the greater of two numbers are given, how is the less found? 154. When the difference and the less of two numbers are given, how is the greater found?

Add the difference and the less number together, and the sum will be the greater number. (Art. 73. Obs.)

7. The difference between two numbers is 12, and the less number is 45 : what is the greater number ?

Solution.—45+12=57, the greater number.

PROOF.—57—45=12, the given difference. (Art. 72.)

8. A is worth 1890 dollars, and B is worth 350 dollars more than A : how much is B worth ?

9. A man's expenses are 2561 dollars a year, and his income exceeds his expenses 875 dollars : how much is his income ?

155. When the *sum* and *difference* of two numbers are given, to find the *two numbers.*

From the sum subtract the difference, divide the remainder by 2, and the quotient will be the smaller number.

To the smaller number thus found, add the given difference, and the sum will be the larger number.

10. The sum of two numbers is 48, and their difference is 18 : what are the numbers ?

Solution.—48—18=30, and 30÷2=15, the smaller number. And 15+18=33, the greater number.

PROOF.—33+15=48, the given sum. (Ax. 11.)

11. The sum of the ages of two men is 173 years, and the difference between them is 15 years : what are their ages ?

12. A man bought a span of horses and a carriage for 856 dollars ; the carriage was worth 165 dollars more than the horses : what was the price of each ?

156. When the *product* of two numbers and *one* of the numbers are given, to find the *other* number.

Divide the given product by the given number, and the quotient will be the number required. (Art. 91.)

QUEST.—155. When the sum and difference of two numbers are given, how are the numbers found ? 156. When the product of two numbers and one of them are given, how is the other found ?

13. The product of two numbers is 144, and one of the numbers is 8: what is the other number?

Solution.—144÷8=18, the required number. (Art. 120.)

PROOF.—18×8=144, the given product. (Art. 88.)

14. The product of A and B's ages is 3250 years, and B's age is 50 years: what is the age of A?

15. The product of the length of a field multiplied by its breadth is 15925 rods, and its breadth is 91 rods: what is its length?

157. When the *divisor* and *quotient* are given, to find the *dividend.*

Multiply the given divisor and quotient together, and the product will be the dividend. (Art. 121.)

16. If a certain divisor is 12, and the quotient is 30, what is the dividend?

Solution.—30×12=360, the dividend required.

PROOF.—360÷12=30, the given quotient. (Art. 120.)

17. If the quotient is 275 and the divisor 683, what must be the dividend?

18. If the divisor is 1031 and the quotient 1002, what must be the dividend?

158. When the *dividend* and *quotient* are given, to find the *divisor.*

Divide the given dividend by the given quotient, and the quotient thus obtained will be the number required. (Art. 122.)

19. A certain dividend is 864, and the quotient is 12: what is the divisor?

Solution.—864÷12=72, the divisor required. (Art. 120.)

PROOF.—72×12=864, the given dividend. (Art. 121.)

20. A gentleman handed a purse containing 1152 shillings, to

QUEST.—157. When the divisor and quotient are given, how is the dividend found? 158. When the dividend and quotient are given, how is the divisor found?

a company of beggars, which was sufficient to give them 24 shillings apiece: how many beggars were there?

21. A farmer having 2500 sheep, divided them into flocks of 125 each: how many flocks did they make?

159. When the *product* of three numbers and *two* of the numbers are given, to find the *other* number.

Divide the given product by the product of the two given numbers, and the quotient will be the other number.

22. There are three numbers whose product is 288; one of them is 8, and another 9: it is required to find the other number.

Solution.—$9\times8=72$; and $288\div72=4$, the number required.

PROOF.—$9\times8\times4=288$, the given product.

23. The product of three persons' ages is 14880 years; the age of the oldest is 31 years, and that of the second is 24 years: what is the age of the youngest?

24. If a garrison of 75 men have 18750 pounds of meat, how long will it last them, allowing 25 pounds to each man per month?

25. The sum of two numbers is 3471, and the less is 1629: what is the greater?

26. The sum of two numbers is 4136, and the greater is 3074: what is the less?

27. The difference between two numbers is 128, and the greater is 760: what is the less?

28. The difference between two numbers is 340, and the less is 634: what is the greater?

29. The sum of two numbers is 12640, and their difference is 1608: what are the numbers?

30. The sum of two numbers is 25264, and their difference is 736: what are the numbers?

31. The sum of two numbers is 42126, and their difference is 176: what are the numbers?

32. The product of two numbers is 246018, and one of them is 313: what is the other number?

SECTION VI.

PROPERTIES OF NUMBERS.*

ART. **160.** The *progress* as well as the *pleasure* of the student in Arithmetic, depends very much upon the *accuracy* of his *knowledge* of the terms, which are employed in mathematical reasoning. Particular pains should therefore be taken to understand their true import.

DEF. 1. An integer signifies a *whole* number. (Art. 28. Obs. 2.)

2. Whole numbers or integers are divided into *prime* and *composite* numbers.

3. A *composite* number, we have seen, is one which may be produced by multiplying two or more numbers together; as, 4, 10, 15. (Art. 95.)

4. A *prime* number is one which *cannot* be produced by multiplying any two or more numbers together; or which *cannot* be exactly divided by any *whole* number, except a *unit* and *itself*. Thus, 1, 2, 3, 5, 7, 11, 13, &c., are prime numbers.

OBS. 1. One number is said to be *prime to another*, when a *unit* is the only number by which both can be divided without a remainder.

2. The learner must be careful not to confound numbers which are *prime to each other* with *prime* numbers; for numbers that are prime to each other, may themselves be *composite* numbers. Thus 4 and 9 are prime to each other, while they are composite numbers.

3. The number of prime numbers is unlimited. For those under 3413, see Table, page 94.

5. An *even* number is one which can be divided by 2 without a remainder; as, 4, 6, 8, 10.

QUEST.—160. Upon what does the progress and pleasure of the student in Arithmetic very much depend? What is an integer? What is a composite number? What is a prime number? Are prime numbers divisible by other numbers? *Obs.* When is one number said to be prime to another? How many prime numbers are there? What is an even number? An odd number? *Obs.* Are even numbers prime or composite? What is true of odd numbers in this respect?

* Barlow on the Theory of Numbers; also, Bonnycastle's Arithmetic

6. An *odd* number is one which cannot be divided by 2 without a remainder; as, 1, 3, 5, 7, 9, 15.

OBS. All even numbers except 2, are *composite* numbers; an odd number is sometimes a *composite*, and sometimes a *prime* number.

7. One number is a *measure* of another, when the former is *contained* in the latter, any number of times without a remainder. Thus, 3 is a measure of 15; 7 is a measure of 28, &c.

8. One number is a *multiple* of another, when the former can be *divided* by the latter without a remainder. Thus, 6 is a multiple of 3; 20 is a multiple of 5, &c.

OBS. A *multiple* is therefore a *composite* number, and the number thus contained in it, is always one of its factors.

9. The *aliquot parts* of a number, are the parts by which it can be *measured* or *divided* without a remainder. Thus, 5 and 7 are the aliquot parts of 35.

10. The *reciprocal* of a number is the quotient arising from dividing a *unit* by that number. Thus, the reciprocal of 2 is $\frac{1}{2}$; the reciprocal of 3 is $\frac{1}{3}$; that of $\frac{4}{5}$ is $\frac{5}{4}$, &c.

11. The *difference* between a given number and 10, 100, 1000, &c., that is, between the given number and the next *higher order*, is called the ARITHMETICAL COMPLEMENT *of that number*. Thus, 3 is the complement of 7; 15 is the complement of 85.

OBS. The arithmetical complement of a number consisting of *one* integral figure, either with or without decimals, is found by *subtracting* the number from 10. If there are *two* integral figures, they are subtracted from 100; if *three*, from 1000, &c.

12. A *perfect* number is one which is equal to the *sum* of all its *aliquot* parts. Thus, $6=1+2+3$, the sum of its aliquot parts, and is a perfect number.

OBS. 1. All the numbers known, to which this property really belongs, are the following: 6; 28; 496; 8128; 33,550,336; 8,589,869,056; 137,438,691,328; and 2,305,843,008,139,952,128.*

2. All perfect numbers terminate with 6, or 28.

QUEST.—When is one number a measure of another? What is a multiple? What are aliquot parts? What is the reciprocal of a number?

* Hutton's Mathematical Recreations.

161. By the term *properties* of numbers, is meant those qualities or elements which are inherent and inseparable from them. Some of the more prominent are the following:

1. The sum of any *two* or *more even* numbers, is an even number.

2. The difference of any *two even* numbers, is an even number.

3. The sum or difference of *two odd* numbers, is *even;* but the sum of *three odd* numbers, is *odd.*

4. The sum of any *even* number of odd numbers, is even; but the sum of any *odd* number of odd numbers, is odd.

5. The sum, or difference, of an *even* and an *odd* number, is an odd number.

6. The product of an *even* and an *odd* number, or of *two even* numbers, is even.

7. If an even number be divisible by an odd number, the *quotient* is an even number.

8. The product of any number of factors, is *even,* if any one of them be even.

9. An odd number cannot be *divided* by an even number without a remainder.

10. The product of any *two* or *more odd* numbers, is an odd number.

11. If an odd number divides an even number, it will also divide the *half* of it.

12. If an even number be divisible by an odd number, it will also be divisible by *double* that number.

13. Any number that *measures* two others, must likewise measure their *sum,* their *difference,* and their *product.*

14. A number that *measures* another, must also measure its *multiple,* or its *product* by any *whole* number.

15. Any number expressed by the decimal notation, divided by 9, will leave the *same remainder,* as the sum of its figures or digits divided by 9.

Demonstration.—Take any number, as 6357; now separating it into its several parts, it becomes $6000+300+50+7$. But $6000=6\times1000=6\times(999+1)=6\times999+6$. In like manner $300=3\times99+3$, and $50=5\times9+5$. Hence $6357=6\times999+3\times99+5\times9+6+3+5+7$; and $6357\div9=(6\times999+3\times99+$

QUEST.—161. What is meant by properties of numbers?

$5\times9+6+3+5+7)\div9$. But $6\times999+3\times99+5\times9$ is evidently divisible by 9 therefore if 6357 be divided by 9, it will leave the same remainder as $6+3+5+7\div9$. The same will be found true of any other number whatever.

OBS. 1. This property of the number 9 affords an ingenious method of proving each of the fundamental rules. (Arts. 90, 123.) The same property belongs to the number 3; for, 3 is a measure of 9, and will therefore be contained an exact number of times in any number of 9s. But it belongs to no other digit.

2. The preceding is not a *necessary* but an *incidental* property of the number 9. It arises from the *law* of *increase* in the decimal notation. If the *radix* of the system were 8, it would belong to 7; if the radix were 12, it would belong to 11; and universally, it belongs to the number that is *one less* than the *radix* of the system of notation.

16. If the number 9 is multiplied by any *single* figure or digit, the *sum* of the figures composing the product, will make 9. Thus, $9\times4=36$, and $3+6=9$.

17. If we take any two numbers whatever; then *one* of them, or their *sum*, or their *difference*, is divisible by 3. Thus, take 11 and 17; though neither of the numbers themselves, nor their sum is divisible by 3, yet their difference is, for it is 6.

18. Any number divided by 11, will leave the *same remainder*, as the sum of its *alternate* digits in the *even* places reckoning from the right, taken from the sum of its alternate digits in the *odd* places, increased by 11 if necessary.

Take any number, as 38405603, and mark the alternate figures. Now the sum of those marked, viz: $8+0+6+3=17$. The sum of the others, viz: $3+4+5+0=12$. And $17-12=5$, the remainder sought. That is, 38405603 divided by 11, will leave 5 remainder.

Again, take 5847362, the sum of the marked figures is 14; the sum of those not marked is 21. Now 21 taken from 25, $(=14+11,)$ leaves 4, the remainder sought.

19. Every *composite number* may be resolved into *prime factors*. For, since a composite number is produced by multiplying two or more factors together, (Art. 160. Def. 3,) it may evidently be resolved into those factors; and if these factors themselves are *composite*, they also may be resolved into *other* factors, and thus the analysis may be continued, until all the factors are *prime* numbers.

20. The *least* divisor of every number is a *prime* number. For, every whole number is either *prime*, or *composite*; (Art. 160.

Def. 2;) but a composite number, we have just seen, can be resolved into prime factors; consequently, the *least* divisor of every number must be a *prime* number.

21. Every prime number except 2, if increased or diminished by 1, is divisible by 4. See table of prime numbers, next page.

22. Every prime number except 2 and 3, if increased or diminished by 1, is divisible by 6.

23. Every *prime* number, except 2 and 5, is contained without a remainder, in the number expressed in the common notation by as many 9s as there are units, less one, in the prime number itself.* Thus, 3 is a measure of 99; 7 of 999,999; and 13 of 999,999, 999,999.

24. Every prime number, except 2, 3, and 5, is a measure of the number expressed in common notation, by as many 1s as there are units, less one, in the prime number. Thus, 7 is a measure of 111,111; and 13 of 111,111,111,111.

25. All prime numbers except 2, are *odd;* and consequently terminate with an odd digit. (Art. 160. Def. 4.)

Note.—1. It must not be inferred from this that all *odd* numbers are *prime.* (Art. 160. Def. 6. Obs.)

2. It is plain that any number terminating with 5, can be divided by 5 without a remainder. Hence,

26. All *prime* numbers, except 2 and 5, must terminate with 1, 3, 7, or 9; all other numbers are *composite.*

161. *a.* To find the *prime* numbers in any series of numbers.

Write in their proper order all the odd numbers contained in the series. Then reckoning from 3, place a point over every third number in the series; reckoning from 5, place a point over every fifth number; reckoning from 7, place a point over every seventh number, and so on. The numbers remaining without points, together with the number 2, are the primes required.

Take the series of numbers up to 40, thus, 1, 3, 5, 7, $\dot{9}$, 11, 13, $\ddot{15}$, 17, 19, $\dot{21}$, 23, $\dot{25}$, $\dot{27}$, 29, 31, $\ddot{33}$, $\dot{35}$, 37, $\dot{39}$; then adding the number 2, the primes are 1, 2, 3, 5, 7, 11, 13, &c.

Note.—This method of excluding the numbers which are not prime from a series, was invented by Eratosthenes, and is therefore called *Eratosthenes' Sieve.*

* Théorie des Nombres, par M. Legendre.

TABLE OF PRIME NUMBERS FROM 1 TO 3413.

1	173	409	659	941	1223	1511	1811	2129	2423	2741	3079
2	179	419	661	947	1229	1523	1823	2131	2437	2749	3083
3	181	421	673	953	1231	1531	1831	2137	2441	2753	3089
5	191	431	677	967	1237	1543	1847	2141	2447	2767	3109
7	193	433	683	971	1249	1549	1861	2143	2459	2777	3119
11	197	439	691	977	1259	1553	1867	2153	2467	2789	3121
13	199	443	701	983	1277	1559	1871	2161	2473	2791	3137
17	211	449	709	991	1279	1567	1873	2179	2477	2797	3163
19	223	457	719	997	1283	1571	1877	2203	2503	2801	3167
23	227	461	727	1009	1289	1579	1879	2207	2521	2803	3169
29	229	463	733	1013	1291	1583	1889	2213	2531	2819	3181
31	233	467	739	1019	1297	1597	1901	2221	2539	2833	3187
37	239	479	743	1021	1301	1601	1907	2237	2543	2837	3191
41	241	487	751	1031	1303	1607	1913	2239	2549	2843	3203
43	251	491	757	1033	1307	1609	1931	2243	2551	2851	3209
47	257	499	761	1039	1319	1613	1933	2251	2557	2857	3217
53	263	503	769	1049	1321	1619	1949	2267	2579	2861	3221
59	269	509	773	1051	1327	1621	1951	2269	2591	2879	3229
61	271	521	787	1061	1361	1627	1973	2273	2593	2887	3251
67	277	523	797	1063	1367	1637	1979	2281	2609	2897	3253
71	281	541	809	1069	1373	1657	1987	2287	2617	2903	3257
73	283	547	811	1087	1381	1663	1993	2293	2621	2909	3259
79	293	557	821	1091	1399	1667	1997	2297	2633	2917	3271
83	307	563	823	1093	1409	1669	1999	2309	2647	2927	3299
89	311	569	827	1097	1423	1693	2003	2311	2657	2939	3301
97	313	571	829	1103	1427	1697	2011	2333	2659	2953	3307
101	317	577	839	1109	1429	1699	2017	2339	2663	2957	3313
103	331	587	853	1117	1433	1709	2027	2341	2671	2963	3319
107	337	593	857	1123	1439	1721	2029	2347	2677	2969	3323
109	347	599	859	1129	1447	1723	2039	2351	2683	2971	3329
113	349	601	863	1151	1451	1733	2053	2357	2687	2999	3331
127	353	607	877	1153	1453	1741	2063	2371	2689	3001	3343
131	359	613	881	1163	1459	1747	2069	2377	2693	3011	3347
137	367	617	883	1171	1471	1753	2081	2381	2699	3019	3359
139	373	619	887	1181	1481	1759	2083	2383	2707	3023	3361
149	379	631	907	1187	1483	1777	2087	2389	2711	3037	3371
151	383	641	911	1193	1487	1783	2089	2393	2713	3041	3373
157	389	643	919	1201	1489	1787	2099	2399	2719	3049	3389
163	397	647	929	1213	1493	1789	2111	2411	2729	3061	3391
167	401	653	937	1217	1499	1801	2113	2417	2731	3067	3407

DIFFERENT SCALES OF NOTATION.

162. A number expressed in the *decimal* notation, may be changed to *any required scale* of notation in the following manner.

Divide the given number by the radix of the required scale continually, till the quotient is less than the radix; then annex to the last quotient the several remainders in a retrograde order, placing ciphers where there is no remainder, and the result will be the number in the scale required. (Arts. 43, 44.)

Ex. 1. Express 429 in the quinary scale of notation.

```
5)429
5) 85—4
5) 17—0
    3—2
Ans. 3204
```

Explanation.—By Dividing the given number by 5, it is evidently distributed into 85 parts, each of which is equal to 5, with 4 remainder. Dividing again by 5, these parts are distributed into 17 other parts, each of which is equal to 5 times 5, and the remainder is nothing. Dividing by 5 the third time, the parts last found are again distributed into 3 other parts, each of which is equal to 5 times 5 into 5, with 2 remainder. Thus, the given number is resolved into $3\times5\times5\times5+2\times5\times5+0\times5+4$, or 3204, which is the answer required.

2. Change 7854 from the decimal to the binary scale. *Ans.* 1111010101110.

3. Change 7854 from the decimal to the ternary scale. *Ans.* 101202220.

4. Change 7854 from the decimal to the quaternary scale. *Ans.* 1322232.

5. Change 7854 from the decimal to the quinary scale. *Ans.* 222404

6. Change 7854 from the decimal to the senary scale. *Ans.* 100210.

7. Change 7854 from the decimal to the octary scale. *Ans.* 17256.

8. Change 7854 from the decimal to the nonary scale. *Ans.* 11686.

9. Change 7854 from the decimal to the duodecimal scale. *Ans.* 4666.

10. Change 35261 from the decimal to the quaternary scale.
11. Change 643175 from the decimal to the octary scale.
12. Change 175683 from the decimal to the septenary scale.
13. Change 534610 from the decimal to the octary scale.
14. Change 841568 from the decimal to the nonary scale.
15. Change 592835 from the decimal to the duodecimal scale.

Note.—Since every scale requires as many characters as there are units in the radix, we will denote 10 by *t*, and 11 by *e*. *Ans.* 2470 *t e*.

163. To change a number expressed in *any given scale* of notation, to the *decimal* scale.

Multiply the left hand figure by the given radix, and to the product add the next figure; then multiply this sum by the radix again, and to this product add the next figure; thus continue the operation till all the figures in the given number have been employed, and the last product will be the number in the decimal scale.

16. Change 3204 from the quinary to the decimal scale.

Explanation.—Multiplying the left hand figure by 5, the given radix, evidently reduces it to the next lower order; for in the quinary scale, 5 in an *inferior* order make one in the next superior order. For the same reason, multiplying this sum by 5 again, reduces it to the next lower order, &c.

Operation.
3204
5
17
5
85
5
429 *Ans.*

OBS. This and the preceding operations are the same in principle, as reducing compound numbers from one denomination to another.

17. Change 1322232 from the quaternary to the decimal scale. *Ans.* 7854.
18. Change 2546571 from the octary to the decimal scale.
19. Change 34120521 from the senary to the decimal scale.
20. Change 145620314 from the septenary to the decimal scale.
21. Change 834107621 from the nonary to the decimal scale.
22. Change 403130021 from the quinary to the decimal scale.
23. Change 704400316 from the octary to the decimal scale.
24. Change 903124106 from the duodecimal to the decimal scale.

ANALYSIS OF COMPOSITE NUMBERS.

164. Every *composite* number, it has been shown, may be resolved into *prime* factors. (Art. 161. Prop. 19.)

Ex. 1. Resolve 210 into its prime factors.

Operation.

```
2)210
3)105
5)35
   7
```

Ans. 2, 3, 5, and 7.

We first divide the given number by 2, which is the least number that will divide it without a remainder, and which is also a prime number. (Prop. 20.) We next divide by 3, then by 5. The several divisors and the last quotient are the prime factors required.

PROOF.—$2 \times 3 \times 5 \times 7 = 210$. Hence,

165. To resolve a *composite number* into its prime factors.

Divide the given number by the smallest number which will divide it without a remainder; then divide the quotient in the same way, and thus continue the operation till a quotient is obtained which can be divided by no number greater than 1. *The several divisors with the last quotient, will be the prime factors required.* (Art. 161. Prop. 19.)

Demonstration.—Every *division* of a number, it is plain, resolves it into *two factors*, viz.: the divisor and dividend. (Art. 112.) But according to the rule, the divisors, in every case, are the *smallest* numbers that will divide the given number and the successive quotients without a remainder; consequently they are all *prime* numbers. (Art. 161. Prop. 20.) And since the division is continued till a quotient is obtained, which cannot be divided by any number greater than 1, it follows that the *last* quotient must also be a *prime* number; for, a prime number is one which cannot be exactly divided by any whole number except a *unit* and *itself.* (Art. 160. Def. 4.)

OBS. 1. Since the *least divisor* of every number is a *prime* number, it is evident that a composite number may be resolved into its prime factors, by dividing it continually by *any prime number* that will divide the given number and the quotients without a remainder. Hence,

2. A composite number can be divided by any of its *prime factors* without a remainder, and by the product of any two or more of them, but by *no other* number. Thus, the prime factors of 42 are 2, 3, and 7. Now 42 can be di-

QUEST.—165. How do you resolve a composite number into its prime factors? *Obs.* Will the same result be obtained, if we divide by any of its prime factors?

vided by 2, 3, and 7; also by 2×3, 2×7, 3×7, and 2×3×7; but it can be divided by no other number.

2. Resolve 4 and 6 into their prime factors.

Solution.—4=2×2; and 6=2×3.

3. Resolve 8 into its prime factors. *Ans.* 8=2×2×2.

Resolve the following composite numbers into their prime factors:

4. 9.	22. 34.	40. 57.	58. 81.
5. 10.	23. 35.	41. 58.	59. 82.
6. 12.	24. 36.	42. 60.	60. 84.
7. 14.	25. 38.	43. 62.	61. 85.
8. 15.	26. 39.	44. 63.	62. 86.
9. 16.	27. 40.	45. 64.	63. 87.
10. 18.	28. 42.	46. 65.	64. 88.
11. 20.	29. 44.	47. 66.	65. 90.
12. 21.	30. 45.	48. 68.	66. 91.
13. 22.	31. 46.	49. 69.	67. 92.
14. 24.	32. 48.	50. 70.	68. 93.
15. 25.	33. 49.	51. 72.	69. 94.
16. 26.	34. 50.	52. 74.	70. 95.
17. 27.	35. 51.	53. 75.	71. 96.
18. 28.	36. 52.	54. 76.	72. 98.
19. 30.	37. 54.	55. 77.	73. 99.
20. 32.	38. 55.	56. 78.	74. 100.
21. 33.	39. 56.	57. 80.	75. 102.

76. Resolve 120 and 144 into their prime factors.
77. Resolve 180 and 420 into their prime factors.
78. Resolve 714 and 836 into their prime factors.
79. Resolve 574 and 2898 into their prime factors.
80. Resolve 11492 and 980 into their prime factors.
81. What are the prime factors of 650 and 1728?
82. What are the prime factors of 1492 and 8032?
83. What are the prime factors of 4604 and 16806?
84. What are the prime factors of 71640 and 20780?
85. What are the prime factors of 84570 and 65480?
86. What are the prime factors of 92352 and 81660?

GREATEST COMMON DIVISOR.

166. A *common divisor* of two or more numbers, is a number which will divide each of them without a remainder. Thus 2 is a common divisor of 6, 8, 12, 16, 18, &c.

167. The *greatest common divisor* of two or more numbers, is the *greatest* number which will divide them without a remainder. Thus 6 is the greatest common divisor of 12, 18, 24, and 30.

OBS. A common divisor is sometimes called a *common measure.* It will be seen that a *common divisor* of two or more numbers, is simply a factor which is *common* to those numbers, and the *greatest* common divisor is the *greatest* factor common to them. Hence,

168. To find a *common divisor* of two or more numbers.

Resolve each number into two or more factors, one of which shall be common to all the given numbers.

Or, resolve the given numbers into their prime factors, then if the same factor is found in each, it will be a common divisor. (Art. 165. Obs. 2.)

OBS. If the given numbers have not a *common factor*, they cannot have a common divisor greater than a unit; consequently they are either *prime numbers, or are prime to each other.* (Art. 160. Def. 4. Obs. 2.)

Note.—The following facts may assist the learner in finding common divisors:

1. Any number ending in 0, or an even number, as 2, 4, 6, &c., may be divided by 2.
2. Any number ending in 5 or 0, may be divided by 5.
3. Any number ending in 0, may be divided by 10.
4. When the two right hand figures are divisible by 4, the whole number may be divided by 4.
5. If the three right hand figures of any number are divisible by 8, the whole is divisible by 8.

Ex. 1. Find a common divisor of 6, 15, and 21.

Solution.—$6=3\times 2$; $15=3\times 5$; and $21=3\times 7$. The factor 3 is common to each of the given numbers, and is therefore a common divisor of them.

QUEST.—166. What is a common divisor of two or more numbers? 167. What is the greatest common divisor of two or more numbers? *Obs.* What is a common divisor sometimes called? 168. How do you find a common divisor of two or more numbers? *Obs.* If the given numbers have not a common factor, what is true as to a common divisor?

2. Find a common divisor of 15, 18, 24, and 36.
3. Find a common divisor of 14, 28, 42, and 35.
4. Find a common divisor of 10, 35, 50, 75, and 60.
5. Find a common divisor of 82, 118, and 146.
6. Find a common divisor of 42 and 66. *Ans.* 2, 3, or 6.

169. It will be seen from the last example that two numbers may have more than *one* common divisor. In many cases it is highly important to find the *greatest divisor* that will divide two or more given numbers without a remainder.

7. What is the greatest common divisor of 35 and 50?

Operation.

```
35)50(1
   35
   15)35(2
      30
       5)15(3
         15
```

Dividing 50 by 35, the remainder is 15, then dividing 35 (the preceding divisor) by 15 (the last remainder) the remainder is 5; finally, dividing 15 (the preceding divisor) by 5 (the last remainder) nothing remains; consequently 5, the last divisor, is the greatest common divisor. Hence,

170. To find the *greatest common divisor* of two numbers.

Divide the greater number by the less; then divide the preceding divisor by the last remainder, and so on, till nothing remains The last divisor will be the greatest common divisor.

When there are *more* than *two* numbers given.

First find the greatest common divisor of any two of them; then, that of the common divisor thus obtained and of another given number, and so on through all the given numbers. The last common divisor found, will be the one required.

Demonstration.—Since 5 is a measure of the last dividend 15, in the preceding solution, it must therefore be a measure of the preceding dividend 35; because $35=2\times15+5$; and 35 is one of the given numbers. Now, since 5 measures 15 and 35, it must also measure their sum, viz: $35+15$, or 50, which is the other given number. (Art. 161. Prop. 13.) In a similar manner it may be shown that the last divisor will, in all cases, be the *greatest common divisor.*

Note.—Numbers which have no *common measure* greater than 1, are said to be *incommensurable.* Thus 17 and 29 are incommensurable.

QUEST.—170. How find the greatest common divisor of two numbers? Of more than two

8. What is the greatest common divisor of 285 and 465?
9. What is the greatest common divisor of 532 and 1274?
10. What is the greatest common divisor of 888 and 2775?
11. What is the greatest common divisor of 2145 and 3471?
12. What is the greatest common divisor of 1879 and 2425?
13. What is the greatest common divisor of 75, 125, and 160?

Suggestion.—Find the greatest common divisor of 75 and 125, which is 25. Then that of 25 and 160. *Ans.* 5.

14. What is the greatest common divisor of 183, 3996, 108?
15. What is the greatest common divisor of 672, 1440, and 3472?
16. What is the greatest common divisor of 30, 42, and 66?

Operation.

$30=2\times3\times5$
$42=2\times3\times7$
$66=2\times3\times11$
Now $2\times3=6$ *Ans.*

Analysis.—By resolving the given numbers into their prime factors, (Art. 165,) we find that the factors 2 and 3 are both common divisors of them. But we have seen that a composite number can be divided by the product of any two or more of its prime factors; (Art. 165. Obs. 2;) consequently 30, 42, and 66 can all be divided by 2×3; for 2×3 is the product of two prime factors common to each. And since they are the only factors common to the given numbers, their product must be the greatest common divisor of them. Hence, we deduce a

171. *Second Method* of finding the greatest common divisor of two or more numbers.

Resolve the given numbers into their prime factors, and the continued product of those factors which are common to each, will be the greatest common divisor.

OBS. If the given numbers have but *one common* factor, that factor itself is the *greatest common divisor.*

17. What is the greatest common divisor of 105 and 165?
18. What is the greatest common divisor of 36, 60, and 108?
19. What is the greatest common divisor of 108, 126, and 162?
20. What is the greatest common divisor of 140, 210, and 315?
21. What is the greatest common divisor of 24, 42, 54, and 60?
22. What is the greatest common divisor of 56, 84, 140, and 168?

LEAST COMMON MULTIPLE.

172. One number is said to be a *multiple* of another, when the former can be divided by the latter without a remainder. (Art. 160. Def. 8.) Hence,

173. A *common multiple* of two or more numbers, is a number which can be *divided* by *each* of them without a remainder. Thus, 12 is a common multiple of 2, 3, and 4; 15 is a common multiple of 3 and 5, &c.

OBS. A *common* multiple is always a composite number, of which each of the given numbers must be a *factor;* otherwise it could not be divided by them. (Art. 165. Obs. 2.)

174. The *continued product* of two or more given numbers will always form a common multiple of those numbers. The same numbers may have an *unlimited number* of common multiples; for, multiplying their continued product by any number, will form a new common multiple. (Art. 161. Prop. 14.)

175. The *least common multiple* of two or more numbers, is the *least* number which can be divided by each of them without a remainder. Thus, 12 is the least common multiple of 4 and 6, for it is the least number which can be exactly divided by them.

OBS. The least common multiple of two or more numbers, is evidently composed of all the prime factors of each of the given numbers repeated *once*, and only *once.* For, if it did not contain all the prime factors of any one of the given numbers, it could not be divided by that number. (Art. 165. Obs. 2.) On the other hand, if any prime factor is employed *more times* than it is repeated as a factor in some one of the given numbers, then it would not be the *least* common multiple.

Ex. 1. What is the least common multiple of 10 and 15?

Analysis.—$10=2\times5$, and $15=3\times5$. The prime factors of the given numbers are 2, 5, 3, and 5. Now since the factor 5 occurs *once* in each number, we may therefore cancel it in one

QUEST.—172. When is one number said to be a multiple of another? 173. What is a common multiple? 174. How may a common multiple of two or more numbers be formed? How many common multiples may there be of any given numbers? 175. What is the least common multiple of two or more numbers?

instance, and the continued product of the remaining factors $2\times3\times5$, or 30, will be the least common multiple.

Operation.
5)10 ″ 15
2 ″ 3
$5\times2\times3=30$ *Ans.*

We first divide both the numbers by 5 in order to resolve them into prime factors. (Art. 175. Obs.) Thus, all the different factors of which the given numbers are composed, are found in the divisor and quotients *once*, and only *once*. Therefore the product of the divisor and quotients $5\times2\times3$, is the least common multiple required. Hence,

176. To find the least common multiple of two or more numbers.

Write the given numbers in a line with two points between them. Divide by the smallest number which will divide any two or more of them without a remainder, and set the quotients and the undivided numbers in a line below. Divide this line and set down the results as before; thus continue the operation till there are no two numbers which can be divided by any number greater than 1. *The continued product of the divisors into the numbers in the last line, will be the least common multiple required.*

OBS. 1. We have seen that the *least divisor* of every number is a *prime* number; hence, dividing by the *smallest* number which will divide two or more of the given numbers, is dividing them by a *prime* number. (Art. 161. Prop. 20.)

The result will evidently be the same, if, instead of dividing by the *smallest* number, we divide the given numbers by *any prime* number, that will divide two or more of them, without a remainder.

2. The preceding operation, it will be seen, resolves the given numbers into their *prime factors*, (Art. 165,) then multiplies all the different factors together, taking each factor as many times in the product, as are equal to the *greatest number of times* it is found in either of the given numbers.

3. If the given numbers are prime numbers, or are prime to each other, the continued product of the numbers themselves will be their least common multiple. (Art. 168. Obs.) Thus, the least common multiple of 5 and 7 is 35; of 8 and 9 is 72.

QUEST.—176. How is the least common multiple of two or more numbers found? *Obs.* If the given numbers are prime, or are prime to each other, what is the least common multiple of them? 176. *a.* Upon what principle does this rule depend? *Obs.* Why do you divide by the smallest number that will divide two or more of the given numbers without a remainder?

Ex. 2. What is the least common multiple of 6, 8, and 12?

Analysis.—By resolving the given numbers into their prime factors, it will be seen that 2 is found *once* as a factor in 6; *twice* in 12; and *three times* in 8. It must therefore be taken *three times* in the *product.* Again, 3 is a factor of 6, and 12, consequently it must be taken only *once* in the product. (Art. 176. Obs. 2.) Thus, 2×2×2×3=24 *Ans.*

Operation.
6=2×3
8=2×2×2
12=2×2×3
2×2×2×3=24

Ex. 3. What is the least common multiple of 12, 18, and 36?

First Operation.	*Second Operation.*	*Third Operation.*
2)12 " 18 " 36	9)12 " 18 " 36	12)12 " 18 " 36
2) 6 " 9 " 18	2)12 " 2 " 4	3) 1 " 18 " 3
3) 3 " 9 " 9	2) 6 " 1 " 2	1 " 6 " 1
3) 1 " 3 " 3	3 " 1 ' 1	And 12×3×6=216.
1 " 1 " 1	Now 9×2×2×3=108.	
2×2×3×3=36 *Ans.*		

Explanation.—In the first operation, we divide by the *smallest* numbers which will divide any two or more of the given numbers without a remainder, and the product of the divisors, &c., is 36, which is the answer required.

In the second and third operations, we divide by numbers that will divide two or more of the given numbers without a remainder, and in both cases, obtain erroneous answers.

Note.—It will be seen from the second and third operations above, that "dividing by any number, which will divide two or more of the given numbers without a remainder," according to the rule given by some authors, does not always give the *least* common multiple of the numbers.

176. *a.* The *reason* of the preceding rule depends upon the principle that the least common multiple of any two or more numbers, is composed of all the *prime factors* of the given numbers, each taken as many times, as are equal to the *greatest number* of times it is found in either of the given numbers. (Art. 175. Obs.)

Note.—1. The reason for dividing by the *smallest* number, is because the divisor may otherwise be a *composite* number, (Art. 161. Prop. 20,) and have a factor *common* to it and one of the quotients, or undivided numbers in the last line; consequently the continued product of them would be too *large* for the *least* common multiple. (Art. 175. Obs.) Thus, in the second operation the divisor 9, is a composite number, containing the factor 3 common to the 3 in the quotient; consequently the product is *three times too large.* In the third operation the divisor 12, is a composite number, and contains the factor 6 common to the 6 in the quotient; therefore the product is *six times too large.*

2. The object of arranging the given numbers in a line, is that all of them may be resolved into their prime factors at the same time; and also to present at a glance the factors which compose the least common multiple required.

4. Find the least common multiple of 6, 9, and 15.
5. Find the least common multiple of 8, 16, 18, and 24.
6. Find the least common multiple of 9, 15, 12, 6, and 5.
7. Find the least common multiple of 5, 10, 8, 18, and 15.
8. Find the least common multiple of 24, 16, 18, and 20.
9. Find the least common multiple of 36, 25, 60, 72, and 35.
10. Find the least common multiple of 42, 12, 84, and 72.
11. Find the least common multiple of 27, 54, 81, 14, and 63.
12. Find the least common multiple of 7, 11, 13, 3, and 5.

177. The process of finding the least common multiple may often be shortened, by *canceling every number which will divide any other given number, without a remainder, and also those which will divide any other number in the same line. The least common multiple of the numbers that remain, will be the answer required.*

OBS. By attention and practice, the student will be able to discover, by inspection, the *least common multiple* of numbers, when they are not large.

13. Find the least common multiple of 4, 6, 10, 8, 12, and 15.

Operation.

```
2)4 " 6 " 10 " 8 " 12 " 15
   2) 5 " 4 "  6 " 15
        3) 2 "  3 " 15
           2 "  1 "  5
```

Now, $2 \times 2 \times 3 \times 2 \times 5 = 120$ *Ans.*

Since 4 and 6, will exactly divide 8, and 12, we cancel them. Again, since 5 in the second line will exactly divide 15 in the same line, we therefore cancel it, and proceed with the remaining numbers as before.

14. Find the least common multiple of 9, 12, 72, 36, and 144.
15. Find the least common multiple of 8, 12, 20, 24, and 25.
16. Find the least common multiple of 1, 2, 3, 4, 5, 6, 7, 8, 9.

17. Find the least common multiple of 63, 12, 84, and 7.
18. Find the least common multiple of 54, 81, 63, and 14.
19. Find the least common multiple of 72, 120, 180, 24, and 36.

177. *a.* The least common multiple of two or more numbers may also be found in the following manner.

First find the greatest common divisor of two of the given numbers; by this divide one of these two numbers, and multiply the quotient by the other. Then perform a similar operation on the product and another of the given numbers; thus continue the process until all of the given numbers have been employed, and the final result will be the least common multiple required.

20. What is the least common multiple of 24, 16, and 12?

Solution.—By inspection, we find the greatest common divisor of 24 and 16, is 8. Now 24÷8=3; and 3×16=48. Again, the greatest common divisor of 48 and 12, is 12. Now 48÷12=4; and 4×12=48. *Ans.*

PROOF.—Resolving the given numbers into their prime factors, 24=2×2×2×3; 16=2×2×2×2; and 12=2×2×3; (Art. 165;) consequently, 2×2×2×2×3=48, the least common multiple. (Art. 175. Obs.)

OBS. The *reason* of this rule depends upon the principle, that if the *product* of any two numbers be divided by *any* factor which is common to both, the *quotient* will be a *common multiple* of the two numbers. Thus, if 48, the product of 6 and 8, be divided by 2, a factor of both, the quotient 24, will be a multiple of each, since it may be regarded either as 8 multiplied by the quotient of 6 by the factor 2, or as 6 multiplied by the quotient of 8 by the same factor. Hence, it is obvious, that the *greater* the common measure is, the *less* will be the multiple; and, consequently, the greatest common measure will produce the least common multiple.

When the common multiple of the first two numbers is found, it is evident, that any number which is a common multiple of it and the third number, will be a multiple of the first, second, and third numbers.

21. What is the least common multiple of 75, 120, and 300?
22. What is the least common multiple of 96, 144, and 720?
23. What is the least common multiple of 256, 512, and 1728?
24. What is the least common multiple of 375, 850, and 3400?

SECTION VII.

FRACTIONS.

ART. **178.** When a number or thing is divided into *two equal* parts, one of those parts is called *one half.* If the number or thing is divided into *three equal* parts, one of the parts is called *one third;* if it is divided into *four equal* parts, one of the parts is called *one fourth,* or *one quarter;* and, universally,

When a number or thing is divided into equal parts, the parts take their name from the number of parts into which the thing or number is divided.

179. The *value* of one of these equal parts manifestly depends upon the number of parts into which the given number or thing is divided. Thus, if an orange is successively divided into 2, 3, 4, 5, 6, &c., equal parts, the thirds will be less than the halves; the fourths, than the thirds; the fifths, than the fourths, &c.

OBS. A *half* of any number is equal to as many units, as 2 is contained times in that number; a *third* of a number is equal to as many, as 3 is contained times in the given number; a *fourth* is equal to as many, as 4 is contained in the number, &c.

180. *When a number or thing is divided into equal parts, these parts are called* FRACTIONS.

OBS. Fractions are used to express parts of a *collection* of things, as well as of a *single* thing; or parts of any *number* of units, as well as of *one* unit. Thus, we speak of $\frac{1}{3}$ of *six* oranges; $\frac{3}{5}$ of 75, &c. In this case the *collection,* or *number* to be divided into equal parts, is regarded as a *whole.*

181. Fractions are divided into two classes, *Common and Decimal.* For the illustration of Decimal Fractions, see Section IX.

QUEST.—178. What is meant by one half? What is meant by one third? What is meant by a fourth? What is meant by fifths? By sixths? How many sevenths make a whole one? How many tenths? What is meant by twentieths? By hundredths? When number or thing is divided into equal parts, from what do the parts take their name? 79. Upon what does the value of one of these equal parts depend? 180. What are fractions? 181. Into how many classes are fractions divided?

182. *Common Fractions* are expressed by two numbers, one placed over the other, with a line between them. One half is written thus $\frac{1}{2}$; one third, $\frac{1}{3}$; one fourth, $\frac{1}{4}$; nine tenths, $\frac{9}{10}$; thirteen forty-fifths, $\frac{13}{45}$, &c.

The number below the line is called the *denominator*, and shows into *how many parts* the number or thing is *divided*.

The number above the line is called the *numerator*, and shows *how many* parts are *expressed* by the fraction. Thus, in the fraction $\frac{2}{3}$, the denominator 3, shows that the number is divided into *three* equal parts; the numerator 2, shows that *two* of those parts are expressed by the fraction.

The denominator and numerator together are called the *terms* of the fraction.

OBS. 1. The term *fraction*, is of Latin origin, and signifies *broken*, or *separated* into parts. Hence, fractions are sometimes called *broken numbers*.

2. *Common* fractions are often called *vulgar* fractions. This term, however, is very properly falling into disuse.

3. The number below the line is called the *denominator*, because it gives the *name* or *denomination* to the fraction; as, halves, thirds, fifths, &c.

The number above the line is called the *numerator*, because it *numbers* the parts, or shows how many parts are expressed by the fraction.

183. A *proper* fraction is a fraction whose numerator is *less* than its denominator; as, $\frac{1}{2}$, $\frac{2}{3}$, $\frac{4}{5}$.

An *improper* fraction is one whose numerator is *equal to*, or *greater* than its denominator; as, $\frac{3}{3}$, $\frac{7}{2}$.

A *mixed* number is a whole number and a fraction expressed together; as, $4\frac{2}{3}$, $25\frac{11}{12}$.

A *simple* fraction is a fraction which has but *one* numerator and *one* denominator, and may be proper, or improper; as, $\frac{3}{5}$, $\frac{9}{4}$.

A *compound* fraction is a fraction of a fraction; as, $\frac{2}{3}$ of $\frac{4}{5}$ of $\frac{7}{8}$, $\frac{1}{5}$ of $\frac{9}{12}$ of $\frac{6}{13}$ of $\frac{27}{35}$.

QUEST.—182. How are common fractions expressed? What is the number below the line called? What does it show? What is the number above the line called? What does it show? What are the denominator and numerator, taken together, called? *Obs.* What is the meaning of the term fraction? What are common fractions sometimes called? Why is the lower number called the denominator? Why is the upper one called the numerator? 183. What is a proper fraction? An improper fraction? A mixed number? A simple fraction? A compound fraction?

A *complex* fraction is one which has a fraction in its numerator or denominator, or in both; as, $\frac{2\frac{1}{2}}{5}$, $\frac{4}{5\frac{1}{3}}$, $\frac{2\frac{1}{3}}{8\frac{3}{4}}$.

184. Fractions, it will be seen both from the definition and the mode of expressing them, arise from *division*, and may be treated as expressions of *unexecuted* division. The *numerator* answers to the *dividend*, and the *denominator* to the *divisor*. (Arts. 25, 182.) Hence,

185. The *value* of a fraction is the *quotient* of the numerator divided by the denominator. Thus, the value of $\frac{6}{3}$ is *two*; of $\frac{4}{4}$ is *one*; of $\frac{1}{3}$ is *one third*, &c. Hence,

186. *If the denominator remains the same, multiplying the numerator by any number, multiplies the value of the fraction by that number.* For, since the numerator and denominator answer to the dividend and divisor, multiplying the numerator is the same as multiplying the dividend. But multiplying the dividend, we have seen, multiplies the quotient, (Art. 141,) which is the same as the value of the fraction. (Art. 185.) Thus, the value of $\frac{6}{3}=2$; now, multiplying the numerator by 3, the fraction becomes $\frac{18}{3}$, whose value is 6, and is the same as 2×3.

187. *Dividing the numerator by any number, divides the value of the fraction by that number.* For, dividing the dividend, divides the quotient. (Art. 142.) Thus, $\frac{6}{3}=2$; now dividing the numerator by 2, the fraction becomes $\frac{3}{3}$, whose value is 1, and is the same as $2\div2$. Hence,

Obs. With a given denominator, the *greater* the *numerator*, the *greater* will be the value of the fraction.

188. *If the numerator remains the same, multiplying the denominator by any number, divides the value of the fraction by that number.* For, multiplying the divisor, we have seen, divides the

Quest.—What is a complex fraction? 184. From what do fractions arise? 185. What is the value of a fraction? 186. What is the effect of multiplying the numerator, while the denominator remains the same? Explain the reason. 187. What is the effect of dividing the numerator? *Obs.* With a given denominator, what is the effect of increasing the numerator? 188. What is the effect of multiplying the denominator?

quotient. (Art. 143.) Thus, $\frac{24}{6}=4$; now multiplying the denominator by 2, the fraction becomes $\frac{24}{12}$, whose value is 2, and is the same as $4\div 2$.

189. *Dividing the denominator by any number, multiplies the value of the fraction by that number.* For, dividing the divisor multiplies the quotient. (Art. 144.) Thus, $\frac{24}{6}=4$; now dividing the denominator by 2, the fraction becomes $\frac{24}{3}$, whose value is 8, and is the same as 4×2. Hence,

OBS. With a given numerator, the *greater* the *denominator*, the *less* will be the value of the fraction.

190. It is evident from the preceding articles, that *multiplying the numerator* by any number, has the same effect on the value of the fraction, as *dividing the denominator* by that number. (Arts. 186, 189.) And,

Dividing the numerator has the same effect, as *multiplying the denominator.* (Arts. 187, 188.)

OBS. It will be observed, that multiplying or dividing the *numerator* of a fraction, has the same effect upon its value, as the same operation has upon a whole number; but, the effect of multiplying or dividing the *denominator* is *exactly contrary* to that of the same operation upon a whole number.

191. If the numerator and denominator are both *multiplied* or both *divided* by the same number, *the value of the fraction will not be altered.* (Art. 146.) Thus, $\frac{12}{4}=3$; now if the numerator and denominator are both multiplied by 2, the fraction becomes $\frac{24}{8}$, whose value is 3. If both terms are divided by 2, the fraction becomes $\frac{6}{2}$, whose value is 3; that is, $\frac{12}{4}=\frac{24}{8}=\frac{6}{2}=3$.

192. Since the value of a fraction is the quotient of the numerator divided by the denominator, it follows,

If the *numerator* and *denominator* are *equal*, the value is a *unit* or *one.* Thus, $\frac{5}{5}=1$, $\frac{7}{7}=1$, &c.

QUEST.—189. What is the effect of dividing the denominator? Why? *Obs.* With a given numerator, what is the effect of increasing the denominator? 190. What may be done to the denominator to produce the same effect on the value of the fraction, as multiplying the numerator by any given number? What, to produce the same effect as dividing the numerator by any given number? 191. What is the effect if the numerator and denominator are both multiplied, or both divided by the same number? 192. When the numerator and denominator are equal, what is the value of the fraction?

If the numerator is *greater* than the denominator, the value is greater than *one*. Thus, $\frac{4}{2}=2$, $\frac{5}{3}=1\frac{2}{3}$.

If the numerator is *less* than the denominator, the value is less than *one*. Thus, $\frac{1}{3}=1$ third of 1, $\frac{4}{5}=4$ fifths of 1.

193. Fractions may be *added, subtracted, multiplied,* and *divided,* as well as whole numbers. But, in order to perform these operations, it is often necessary to make certain changes in the terms of the fractions.

OBS. It is evident that any changes may be made in the terms of a fraction, which do not alter the quotient of the numerator divided by the denominator; for, if the quotient is not altered, the value remains the same. Thus, the terms of the fraction $\frac{4}{2}$ may be changed into $\frac{2}{1}$, $\frac{8}{4}$, $\frac{16}{8}$, &c., without altering its value; for in each case the quotient of the numerator divided by the denominator is 2. Hence, for any given fraction, we may substitute any other fraction, which will give the *same quotient.*

REDUCTION OF FRACTIONS.

194. The process of changing the *terms* of a fraction into others, without altering its value, is called REDUCTION OF FRACTIONS.

CASE I.

Ex. 1. Reduce $\frac{10}{20}$ to its lowest terms.

First Operation.
$2)\frac{10}{20}=\frac{5}{10}$: again, $5)\frac{5}{10}=\frac{1}{2}$ *Ans.*

Dividing both terms of the fraction by 2, it becomes $\frac{5}{10}$: again, dividing both by 5, we obtain $\frac{1}{2}$, whose terms are the lowest to which the given fraction can be reduced.

Second Operation.
$10)\frac{10}{20}=\frac{1}{2}$ *Ans.*

If we divide both terms by 10, their greatest common divisor, (Art. 170,) the given fraction will be reduced to its lowest terms by a single division. Hence,

QUEST.—When the numerator is larger than the denominator, what? When smaller, what? *Obs.* What changes may be made in the terms of a fraction? 194. What is meant by reduction of fractions? 195. How is a fraction reduced to its lowest terms?

195. To reduce a fraction to its lowest terms.

Divide the numerator and denominator by any number which will divide them both without a remainder; and thus continue the operation, till there is no number greater than 1 *that will divide them exactly.*

Or, divide both the numerator and denominator by their greatest common divisor; the two quotients thence arising will be the lowest terms to which the given fraction can be reduced. (Art. 170.)

OBS. 1. Since *halves* are *larger* than *twentieths*, it may be asked, how the fraction $\frac{1}{2}$, can be said to be in *lower terms* than $\frac{10}{20}$. It should be observed, the expression *lowest term*, has reference to the *number* of parts into which the unit or thing is divided, and not to the *value* or *size* of the parts. Thus, in $\frac{1}{2}$, there are *fewer* parts than in $\frac{10}{20}$; in $\frac{1}{4}$, there are *fewer* parts than in $\frac{3}{12}$, &c. Hence, a fraction is said to be reduced to its *lowest terms*, when its numerator and denominator are expressed in the *smallest* numbers possible.

2. The value of a fraction is not altered by reducing it to its lowest terms for, the numerator and denominator are both divided by the same number.

3. When the terms of the fraction are small, the former method will generally be found to be the shorter and more convenient; but when the terms are large, it is often difficult to determine whether the fraction is in its simplest form, without finding the *greatest common divisor* of its terms.

2. Reduce $\frac{9}{18}$ to its lowest terms. *Ans.* $\frac{1}{2}$.

3. Reduce $\frac{6}{15}$.
4. Reduce $\frac{9}{21}$.
5. Reduce $\frac{16}{20}$.
6. Reduce $\frac{36}{48}$.
7. Reduce $\frac{56}{72}$.
8. Reduce $\frac{75}{87}$.
9. Reduce $\frac{66}{100}$.
10. Reduce $\frac{77}{121}$.
11. Reduce $\frac{299}{713}$.
12. Reduce $\frac{522}{1044}$.
13. Reduce $\frac{360}{468}$.
14. Reduce $\frac{252}{324}$.
15. Reduce $\frac{374}{561}$.
16. Reduce $\frac{1293}{1724}$.
17. Reduce $\frac{1457}{5921}$.
18. Reduce $\frac{6465}{7335}$.

CASE II.

19. Reduce $\frac{23}{7}$ to a whole or mixed number.

Analysis.—The object in this example, is to find a *whole*, or *mixed* number, whose value is *equal* to the given fraction. Now, since 7

Operation.

7)23
$3\frac{2}{7}$ *Ans.*

QUEST.—*Obs.* What is meant by the expression, lowest terms? When is a fraction said to be reduced to its lowest terms? Is the value of a fraction altered by reducing it to its lowest terms? Why not?

sevenths make 1 *whole one,* 23 sevenths will make as many *whole ones* as 7 is contained times in 23. And $23 \div 7 = 3\frac{2}{7}$. But the value of a fraction is the *quotient* of the numerator divided by the denominator. (Art. 185.) Hence,

196. To reduce an improper fraction to a whole, or mixed number,

Divide the numerator by the denominator, and the quotient will be the whole, or mixed number required.

20. Reduce $\frac{34}{5}$ to a whole or mixed number. *Ans.* $6\frac{4}{5}$.

Reduce the following fractions to whole or mixed numbers:

21. Reduce $\frac{27}{3}$.
22. Reduce $\frac{45}{9}$.
23. Reduce $\frac{36}{10}$.
24. Reduce $\frac{64}{7}$.
25. Reduce $\frac{54}{54}$.
26. Reduce $\frac{720}{12}$.
27. Reduce $\frac{756}{36}$.
28. Reduce $\frac{18200}{350}$.
29. Reduce $\frac{2430}{40}$.
30. Reduce $\frac{45672}{425}$.

CASE III.

31. Reduce the mixed number $27\frac{2}{5}$ to an improper fraction.

Analysis.—In 1 there are 5 *fifths,* and in 27 there are 27 times as many. Now $5 \times 27 = 135$, and 2 fifths make 137 fifths. Hence,

Operation.

$27\frac{2}{5}$
5
$\frac{137}{5}$ *Ans.*

197. To reduce a mixed number to an improper fraction.

Multiply the whole number by the denominator of the fraction, and to the product add the given numerator. The sum placed over the given denominator, will form the improper fraction required.

OBS. 1. Any whole number may be expressed in the form of a fraction without altering its value, by *making* 1 *the denominator.*

2. A whole number may also be reduced to a fraction of any denominator, by *multiplying* the given number by the proposed denominator; the product will be the numerator of the fraction required.

QUEST.—196. How is an improper fraction reduced to a whole or mixed number? 197. How reduce a mixed number to an improper fraction? *Obs.* How express a whole number in the form of a fraction? How reduce it to a fraction of a given denominator?

195. To reduce a fraction to its lowest terms.

Divide the numerator and denominator by any number which will divide them both without a remainder; and thus continue the operation, till there is no number greater than 1 *that will divide them exactly.*

Or, divide both the numerator and denominator by their greatest common divisor; the two quotients thence arising will be the lowest terms to which the given fraction can be reduced. (Art. 170.)

OBS. 1. Since *halves* are *larger* than *twentieths*, it may be asked, how the fraction $\frac{1}{2}$, can be said to be in *lower terms* than $\frac{10}{20}$. It should be observed, the expression *lowest term*, has reference to the *number* of parts into which the unit or thing is divided, and not to the *value* or *size* of the parts. Thus, in $\frac{1}{2}$, there are *fewer* parts than in $\frac{10}{20}$; in $\frac{1}{4}$, there are *fewer* parts than in $\frac{3}{12}$, &c. Hence, a fraction is said to be reduced to its *lowest terms*, when its numerator and denominator are expressed in the *smallest* numbers possible.

2. The value of a fraction is not altered by reducing it to its lowest terms for, the numerator and denominator are both divided by the same number.

3. When the terms of the fraction are small, the former method will generally be found to be the shorter and more convenient; but when the terms are large, it is often difficult to determine whether the fraction is in its simplest form, without finding the *greatest common divisor* of its terms.

2. Reduce $\frac{9}{18}$ to its lowest terms. *Ans.* $\frac{1}{2}$.

3. Reduce $\frac{6}{15}$.
4. Reduce $\frac{9}{21}$.
5. Reduce $\frac{16}{20}$.
6. Reduce $\frac{36}{48}$.
7. Reduce $\frac{56}{72}$.
8. Reduce $\frac{75}{87}$.
9. Reduce $\frac{66}{100}$.
10. Reduce $\frac{77}{121}$.
11. Reduce $\frac{209}{713}$.
12. Reduce $\frac{522}{1044}$.
13. Reduce $\frac{360}{468}$.
14. Reduce $\frac{252}{324}$.
15. Reduce $\frac{374}{561}$.
16. Reduce $\frac{1293}{1724}$.
17. Reduce $\frac{1457}{5921}$.
18. Reduce $\frac{6465}{7335}$.

CASE II.

19. Reduce $\frac{23}{7}$ to a whole or mixed number.

Analysis.—The object in this example, is to find a *whole*, or *mixed* number, whose value is *equal* to the given fraction. Now, since 7

Operation.
7)23
$3\frac{2}{7}$ *Ans.*

QUEST.—*Obs.* What is meant by the expression, lowest terms? When is a fraction said to be reduced to its lowest terms? Is the value of a fraction altered by reducing it to its lowest terms? Why not?

49. Reduce $\frac{2}{5}$ of $\frac{3}{8}$ of $\frac{7}{15}$ of $\frac{8}{13}$ to a simple fraction.

50. Reduce $\frac{1}{2}$ of $\frac{2}{3}$ of $\frac{3}{4}$ of $\frac{5}{7}$ of $\frac{4}{9}$ to a simple fraction.

Operation.

$$\frac{1}{\not 2} \text{ of } \frac{\not 2}{\not 3} \text{ of } \frac{\not 3}{\not 4} \text{ of } \frac{5}{7} \text{ of } \frac{\not 4}{9} = \frac{5}{63}$$

Analysis.—Since the product of the numerators is to be divided by the product of the denominators, we may cancel the factors 2, 3, and 4, which are common to both; for, this is dividing the terms of the new fraction by the same number, (Art. 148,) and therefore does not alter its value. (Art. 191.) Multiplying the remaining factors together, we have $\frac{5}{63}$, which is the answer required. Hence,

199. To reduce compound fractions to simple ones by CANCELATION.

Cancel all the factors which are common to the numerators and denominators; then multiply the remaining terms together as before. (Art. 198.)

OBS. 1. The *reason* of this rule depends upon the fact that the numerator and denominator of the new fraction are, in effect, divided by the *same* numbers; for, *canceling* a factor of a number *divides* the number by that factor. (Art. 148.) Consequently the value of the fraction is not altered. (Art. 191.)

2. This method not only *shortens* the operation of multiplying, but at the same time reduces the answer to its *lowest* terms. A little practice will give the student great facility in its application.

51. Reduce $\frac{3}{5}$ of $\frac{15}{24}$ of $\frac{8}{7}$ to a simple fraction.

Operation.

$$\frac{\not 3}{\not 5} \text{ of } \frac{\overset{3}{\not 1\not 5}}{\not 2\not 4} \text{ of } \frac{\not 8}{7} = \frac{3}{7} \text{ Ans.}$$

First we cancel the 3 and 8 in the numerator, then the 24 in the denominator, which is equal to the factors 3 into 8. Finally, we cancel the 5 in the denominator and the factor 5 in the numerator 15, placing the other factor 3 above. We have 3 left in the numerator, and 7 in the denominator. *Ans.* $\frac{3}{7}$.

52. Reduce $\frac{7}{8}$ of $\frac{4}{7}$ of $\frac{2}{3}$ of $\frac{10}{12}$ to a simple fraction.

53. Reduce $\frac{2}{5}$ of $\frac{4}{7}$ of $\frac{1}{4}$ of $\frac{7}{14}$ of $\frac{5}{3}$ to a simple fraction.

QUEST.—199. How by cancelation? How does it appear that this method will give the true answer? *Obs* What advantages does this method posses?

54. Reduce $\frac{3}{4}$ of $\frac{5}{8}$ of $\frac{2}{9}$ of $\frac{4}{5}$ of $\frac{9}{10}$ to a simple fraction.
55. Reduce $\frac{4}{5}$ of $\frac{8}{10}$ of $\frac{5}{9}$ of $\frac{20}{32}$ to a simple fraction.
56. Reduce $\frac{6}{5}$ of $\frac{15}{18}$ of $\frac{4}{7}$ of $\frac{1}{20}$ to a simple fraction.
57. Reduce $\frac{2}{9}$ of $\frac{10}{12}$ of $\frac{13}{17}$ of $\frac{18}{20}$ to a simple fraction.
58. Reduce $\frac{7}{8}$ of $\frac{3}{12}$ of $\frac{6}{17}$ of $\frac{9}{7}$ of $\frac{2}{3}$ to a simple fraction.
59. Reduce $\frac{1}{3}$ of $\frac{3}{4}$ of $4\frac{2}{3}$ of $\frac{6}{7}$ of $\frac{4}{5}$ to a simple fraction.
60. Reduce $\frac{4}{5}$ of $3\frac{1}{3}$ of $\frac{8}{9}$ of $\frac{9}{10}$ of $\frac{1}{8}$ to a simple fraction.

Note.—For reduction of *complex* fractions to simple ones, see Art. 239.

CASE V.

Ex. 61. Reduce $\frac{1}{3}$ and $\frac{1}{4}$ to a common denominator.

Note.—Two or more fractions are said to have a *common denominator*, when they have the *same* denominator.

Solution.—If both terms of the first fraction $\frac{1}{3}$, are multiplied by the denominator of the second, it becomes $\frac{4}{12}$; and if both terms of the second fraction $\frac{1}{4}$, are multiplied by the denominator of the first, it becomes $\frac{3}{12}$. Thus the fractions $\frac{4}{12}$ and $\frac{3}{12}$ have a common denominator, and are respectively equal to the given fractions, viz: $\frac{4}{12}=\frac{1}{3}$, and $\frac{3}{12}=\frac{1}{4}$. (Art. 191.) Hence,

200. To reduce fractions to a common denominator.

Multiply each numerator into all the denominators except its own for a new numerator, and all the denominators together for a common denominator.

62. Reduce $\frac{1}{3}$, $\frac{3}{4}$, and $\frac{5}{6}$ to a common denominator.

Operation.

$$\left.\begin{array}{l} 1\times4\times6=24 \\ 3\times3\times6=54 \\ 5\times3\times4=60 \end{array}\right\}\text{ the three numerators.}$$

$$3\times4\times6=72 \text{ the common denominator.}$$

Ans. $\frac{24}{72}$, $\frac{54}{72}$, and $\frac{60}{72}$.

OBS. The reason that the process of reducing fractions to a common denominator does not *alter* their *value*, is because the numerator and denominator of each of the given fractions, are multiplied by the *same* numbers; and *multiplying*

QUEST.—*Note.* What is meant by a common denominator? 200. How are fractions reduced to a common denominator? *Obs.* Does the process of reducing fractions to a common denominator alter their value? Why not?

both the numerator and denominator of a fraction by the same number. does not alter its value. (Art. 191.)

63. Reduce $\frac{2}{3}$, $\frac{3}{5}$, $\frac{1}{4}$, and $\frac{5}{7}$ to a common denominator.

64. Reduce $\frac{3}{5}$, $\frac{1}{9}$, $\frac{2}{3}$, and $\frac{3}{4}$ to a common denominator.

Reduce the following fractions to a common denominator:

65. Reduce $\frac{4}{9}$, $\frac{1}{2}$, $\frac{4}{5}$, and $\frac{2}{3}$.

66. Reduce $\frac{5}{7}$, $\frac{4}{7}$, $\frac{6}{9}$, and $\frac{2}{5}$.

67. Reduce $\frac{8}{9}$, $\frac{5}{7}$, $\frac{6}{10}$, and $\frac{7}{12}$.

68. Reduce $\frac{8}{11}$, $\frac{6}{7}$, $\frac{12}{15}$, and $\frac{2}{5}$.

69. Reduce $\frac{15}{33}$, $\frac{63}{70}$, and $\frac{27}{40}$.

70. Reduce $\frac{12}{26}$, $\frac{70}{100}$, and $\frac{52}{25}$.

71. Reduce $\frac{20}{45}$, $\frac{35}{50}$, and $\frac{10}{25}$.

72. Reduce $\frac{50}{81}$, $\frac{65}{30}$, and $\frac{128}{250}$.

CASE VI.

73. Reduce $\frac{1}{3}$, $\frac{3}{4}$, and $\frac{5}{8}$ to the least common denominator.

Analysis.—We first find the least common multiple of all the given denominators, which is 24. (Art. 176.) The next step is to reduce the given fractions to *twenty-fourths* without altering their value. This may evidently be done by multiplying both terms of each fraction by such a number as will make its denominator 24. (Art. 191.) Thus 3, the denominator of the first fraction, is contained in 24, 8 times; now, multiplying both terms of the fraction $\frac{1}{3}$ by 8, it becomes $\frac{8}{24}$. The denominator 4, is contained in 24, 6 times; hence, multiplying the second fraction $\frac{3}{4}$ by 6, it becomes $\frac{18}{24}$. The denominator 8, is contained in 24, 3 times; and multiplying the third fraction $\frac{5}{8}$ by 3, it becomes $\frac{15}{24}$. Therefore $\frac{8}{24}$, $\frac{18}{24}$, and $\frac{15}{24}$ are the fractions required. Hence,

Operation.

```
2)3 " 4 " 8
  ---------
2)3 " 2 " 4
  ---------
  3 " 1 " 2
```

Now $2 \times 2 \times 3 \times 2 = 24$, the least common denominator

201. To reduce fractions to their *least* common denominator.

I. *Find the least common multiple of all the denominators of the given fractions, and it will be the least common denominator.* (Art. 176.)

II. *Divide the least common denominator by the denominator of each given fraction, and multiply the quotient by the numerator; the products will be the numerators of the fractions required.*

QUEST.—201. How are fractions reduced to the least common denominator?

Obs. 1. This process, in effect, multiplies both the numerator and denominator of the given fractions by the same number, and consequently does not alter their value. (Art. 191.)

2. The rule supposes each of the given fractions to be reduced to its *lowest* terms; otherwise, the least common multiple of their denominators *may* not be the least common denominator to which the given fractions are capable of being reduced. Thus, the fractions $\frac{1}{4}$, $\frac{3}{6}$, and $\frac{9}{12}$, when reduced to the least common denominator as they stand, become $\frac{3}{12}$, $\frac{6}{12}$, and $\frac{9}{12}$. But it is obvious that these fractions are not reduced to their *least* common denominator; for, they can be reduced to $\frac{1}{4}$, $\frac{2}{4}$, and $\frac{3}{4}$. Now, if the given fractions are reduced to the *lowest terms*, they become $\frac{1}{4}$, $\frac{1}{2}$, and $\frac{3}{4}$, and the *least common multiple* of their denominators, is also 4. (Art. 176.)

3. By a moment's reflection the student will often discover the least common denominator of the given fractions, without going through the ordinary process of finding the least common multiple of their denominators. Take the fractions $\frac{1}{2}$, $\frac{3}{4}$, and $\frac{3}{12}$; the least common denominator, it will be seen at a glance, is 4. Now if we multiply both terms of $\frac{1}{2}$ by 2, it becomes $\frac{2}{4}$; and if we divide both terms of $\frac{3}{12}$ by 3, or reduce it to its lowest terms, it becomes $\frac{1}{4}$. Thus the given fractions are equal to $\frac{2}{4}$, $\frac{3}{4}$, and $\frac{1}{4}$, and are reduced to the *least common denominator.*

74. Reduce $\frac{3}{4}$, $\frac{5}{6}$, and $\frac{7}{8}$ to the least common denominator.

Operation.	
2)4 " 6 " 8	Now $2\times2\times3\times2=24$, the least com. denom.
2)2 " 3 " 4	Then $24\div4=6$, and $6\times3=18$, the 1st num
1 " 3 " 2	$24\div6=4$, and $4\times5=20$, the 2d "
	$24\div8=3$, and $3\times7=21$, the 3d "

Ans. $\frac{18}{24}$, $\frac{20}{24}$, and $\frac{21}{24}$.

75. Reduce $\frac{5}{7}$ and $\frac{5}{14}$ to the least common denominator.

Reduce the following fractions to the least common denominator:

76. $\frac{5}{18}$, $\frac{5}{6}$, $\frac{7}{9}$, and $\frac{7}{12}$.
77. $\frac{3}{4}$, $\frac{5}{6}$, and $\frac{7}{8}$.
78. $\frac{1}{4}$, $\frac{5}{9}$, $\frac{7}{8}$, and $\frac{11}{12}$.
79. $\frac{3}{4}$, $\frac{5}{8}$, $\frac{8}{9}$, and $\frac{1}{12}$.
80. $\frac{2}{3}$, $\frac{1}{4}$, $\frac{5}{9}$, and $\frac{15}{18}$.
81. $\frac{7}{8}$, $\frac{8}{10}$, $\frac{13}{20}$, and $\frac{35}{40}$.
82. $\frac{7}{12}$, $\frac{19}{20}$, $\frac{9}{10}$, and $\frac{28}{30}$.
83. $\frac{5}{7}$, $\frac{4}{6}$, $\frac{18}{21}$, and $\frac{11}{14}$.
84. $\frac{12}{15}$, $\frac{25}{30}$, $\frac{19}{20}$, and $\frac{18}{60}$.
85. $\frac{7}{12}$, $\frac{41}{48}$, $\frac{13}{18}$, and $\frac{7}{42}$.
86. $\frac{5}{21}$, $\frac{17}{43}$, $\frac{12}{17}$, and $\frac{25}{35}$.
87. $\frac{16}{35}$, $\frac{21}{60}$, $\frac{7}{45}$, and $\frac{25}{50}$.
88. $\frac{28}{48}$, $\frac{15}{36}$, $\frac{38}{72}$, and $\frac{16}{18}$.
89. $\frac{31}{45}$, $\frac{17}{60}$, $\frac{14}{35}$, and $\frac{24}{70}$.
90. $\frac{41}{60}$, $\frac{13}{48}$, $\frac{61}{72}$, and $\frac{25}{144}$.
91. $\frac{81}{90}$, $\frac{65}{180}$, $\frac{49}{60}$, and $\frac{25}{270}$.

QUEST.—*Obs.* Does this process alter the value of the given fractions? Why not? What does this rule suppose respecting the given fractions?

ADDITION OF FRACTIONS.

Ex. 1. A beggar meeting four persons, obtained $\frac{1}{6}$ of a dollar from the first, $\frac{3}{6}$ from the second, $\frac{4}{6}$ from the third, and $\frac{5}{6}$ from the fourth: how much did he receive from all?

Solution.—Since the several donations are all in the same parts of a dollar, viz: sixths, it is plain they may be added together in the same manner as whole dollars, whole yards, &c. Thus, 1 sixth and 3 sixths are 4 sixths, and 4 are 8 sixths, and 5 are 13 sixths. *Ans.* $\frac{13}{6}$, or $2\frac{1}{6}$ dollars.

Ex. 2. What is the sum of $\frac{2}{3}$ and $\frac{3}{4}$?

OBS. A difficulty here presents itself to the learner; for, it is evident, that 2 *thirds* and 3 *fourths* neither make 5 *thirds*, nor 5 *fourths*. (Art. 51.) This difficulty may be removed by reducing the given fractions to a common denominator. (Art. 200.) Thus,

Operation.

$$\left.\begin{array}{l} 2\times4=8 \\ 3\times3=9 \end{array}\right\} \text{the new numerators.}$$

$3\times4=12$, the common denominator.

The fractions, when reduced, are $\frac{8}{12}$ and $\frac{9}{12}$; now 8 twelfths + 9 twelfths=17 twelfths. *Ans.* $\frac{17}{12}$, or $1\frac{5}{12}$.

202. From these illustrations we deduce the following general

RULE FOR ADDITION OF FRACTIONS.

Reduce the fractions to a common denominator; add their numerators, and place the sum over the common denominator.

OBS. 1. *Compound fractions* must, of course, be reduced to simple ones, before attempting to reduce them to a common denominator. (Art. 198.)

2. *Mixed numbers* may be reduced to improper fractions, and then be added according to the rule; or, we may add the whole numbers and fractional parts separately, and then unite their sums.

3. In many instances the operation may be shortened by reducing the given fractions to the *least common denominator*. (Art. 201.)

QUEST.—202. How are fractions added? *Obs.* What must be done with compound fractions? How are mixed numbers added? How may the operation frequently be shortened?

EXAMPLES.

3. What is the sum of $\frac{1}{2}$, $\frac{4}{6}$, and $\frac{5}{6}$? *Ans.* $\frac{12}{6}=2$.
4. What is the sum of $\frac{1}{2}$, $\frac{2}{5}$, $\frac{3}{4}$, and $\frac{5}{3}$?
5. What is the sum of $\frac{3}{4}$, $\frac{3}{8}$, $\frac{1}{2}$, and $\frac{2}{3}$?
6. What is the sum of $\frac{2}{3}$, $\frac{4}{9}$, $\frac{11}{15}$, and $\frac{1}{5}$?
7. What is the sum of $\frac{5}{8}$, $\frac{3}{12}$, $\frac{6}{7}$, and $\frac{9}{13}$?
8. What is the sum of $\frac{5}{9}$, $\frac{3}{5}$, $\frac{2}{11}$, and $\frac{6}{18}$?
9. What is the sum of $\frac{4}{5}$, $\frac{1}{10}$, $\frac{2}{7}$, and $\frac{5}{6}$?
10. What is the sum of $\frac{5}{5}$, $\frac{4}{13}$, $\frac{7}{8}$, and $\frac{12}{7}$?
11. What is the sum of $\frac{1}{2}$, $\frac{1}{3}$, $\frac{1}{6}$, $\frac{2}{8}$, and $\frac{9}{2}$?
12. What is the sum of $\frac{2}{3}$ of $\frac{1}{8}$, $\frac{3}{5}$ of $\frac{6}{3}$, and $\frac{8}{7}$?
13. What is the sum of $\frac{1}{5}$ of $\frac{6}{8}$, $\frac{2}{3}$ of $\frac{1}{2}$, and $\frac{7}{12}$?
14. What is the sum of $\frac{3}{5}$ of $\frac{2}{3}$ of $\frac{7}{8}$ of $\frac{1}{4}$, and $\frac{1}{6}$?
15. What is the sum of $\frac{5}{6}$, $\frac{1}{7}$ of 3, $\frac{3}{5}$ of $\frac{1}{3}$, and $\frac{5}{8}$?
16. What is the sum of $4\frac{1}{5}$, $8\frac{1}{4}$, $2\frac{1}{2}$, $6\frac{1}{3}$, and $\frac{2}{3}$?
17. What is the sum of $\frac{1}{4}$ of 6, $\frac{2}{3}$ of 2, $3\frac{1}{2}$, and $5\frac{3}{7}$?
18. What is the sum of $\frac{4}{5}$, $\frac{19}{20}$, $\frac{24}{30}$, $\frac{81}{45}$, and $\frac{19}{10}$?
19. What is the sum of $21\frac{1}{2}$, $35\frac{1}{8}$, $\frac{53}{12}$, and $\frac{2}{3}$ of $\frac{7}{8}$?
20. What is the sum of $\frac{1}{2}$ of $\frac{3}{5}$, $\frac{25}{4}$, $6\frac{1}{2}$, $1\frac{2}{3}$, and $\frac{5}{8}$?
21. What is the sum of $\frac{1}{8}$ and $\frac{1}{12}$?

Note.—It is obvious, if two fractions, each of whose numerators is 1, are reduced to a common denominator, the new numerators will be the same as the given denominators. (Art. 200.) Thus, if $\frac{1}{8}$ and $\frac{1}{12}$ are reduced to a common denominator, the new numerators will be 12 and 8, the same as the given denominators. Now, the sum of the new numerators, placed over the product of the denominators, will be the answer; (Art. 202;) that is $\frac{12+8}{12\times8}=\frac{20}{96}$, or $\frac{5}{24}$, the answer required. Hence,

203. To find the sum of any two fractions whose numerators are one.

Add the denominators together, place this sum over their product, and the result will be the answer required.

OBS. 1. The *reason* of this rule may be seen from the fact that the operation is the same as reducing the given fractions to a common denominator, then adding their numerators.

2. When the *numerators* of two fractions are the *same*, their *sum* may be found

---

QUEST.—203. How is the sum of any two fractions found whose numerators are 1? *Obs.* How find the sum of two fractions whose numerators are the same?

by multiplying the sum of the two denominators by the *common numerator*, and placing the result over the product of the given denominators. Thus, the sum of $\frac{3}{4}$ and $\frac{3}{5}$ is equal to $\frac{(4+5)\times 3}{4\times 5}=\frac{9\times 3}{4\times 5}=\frac{27}{20}$, or $1\frac{7}{20}$.

22. What is the sum of $\frac{1}{20}$ and $\frac{1}{35}$? Of $\frac{1}{40}$ and $\frac{1}{65}$?
23. What is the sum of $\frac{1}{56}$ and $\frac{1}{77}$? Of $\frac{1}{63}$ and $\frac{1}{89}$?
24. What is the sum of $\frac{1}{68}$ and $\frac{1}{52}$? Of $\frac{1}{465}$ and $\frac{1}{925}$?
25. What is the sum of $\frac{5}{8}$ and $\frac{5}{17}$? Of $\frac{7}{18}$ and $\frac{7}{34}$?
26. What is the sum of $\frac{15}{20}$ and $\frac{15}{13}$? Of $\frac{15}{60}$ and $\frac{15}{75}$?
27. What is the sum of $\frac{30}{45}$ and $\frac{30}{67}$? Of $\frac{30}{70}$ and $\frac{30}{120}$?
28. What is the sum of 5 and $\frac{2}{3}$?

Note.—The design in this and the following examples is to *incorporate* the integers with the fractions, and express the answer fractionally.

Solution.—$5=\frac{15}{3}$. (Art. 197. Obs. 2.) Now $\frac{15}{3}+\frac{2}{3}=\frac{17}{3}$ *Ans.*

204. Hence, to add a whole number and a fraction together. *Reduce the whole number to a fraction of the same denominator as that of the given fraction; then add their numerators together.* (Arts. 202, 197, Obs. 1, 2.)

Note.—The process of incorporating a whole number with a fraction, is the same as that of reducing a mixed number to an improper fraction. (Art. 197.)

29. What is the sum of 45 and $\frac{2}{9}$?
30. What is the sum of 320 and $\frac{7}{15}$?
31. What is the sum of 452 and $\frac{30}{100}$?
32. What is the sum of $635\frac{15}{16}+427\frac{12}{48}+1625\frac{7}{8}$?
33. What is the sum of $195\frac{61}{75}+600\frac{19}{20}+5630\frac{31}{40}+160\frac{3}{4}$?
34. What is the sum of $671\frac{29}{45}+483\frac{13}{18}+8421\frac{17}{21}+4325\frac{1}{8}$?
35. What is the sum of $590\frac{11}{12}+100\frac{61}{81}+4005\frac{47}{56}+3020\frac{1}{12}$?
36. What is the sum of $239\frac{12}{17}+644\frac{21}{33}+1650\frac{16}{19}+4500\frac{1}{10}$?
37. What is the sum of $6563\frac{1}{3}+1000\frac{1}{5}+1830\frac{2}{3}+8396\frac{1}{6}$?
38. What is the sum of $356\frac{10}{11}+46\frac{3}{7}+165\frac{1}{2}+600\frac{5}{9}+321\frac{2}{3}$?
39. What is the sum of $41\frac{1}{2}+105\frac{2}{9}+300\frac{3}{4}+241\frac{3}{5}+472\frac{1}{4}$?
40. What is the sum of $8672\frac{1}{4}+163645\frac{1}{5}+1800\frac{3}{8}+66251\frac{3}{10}$?
41. What is the sum of $26003\frac{1}{7}+19352\frac{2}{3}+92831+68693\frac{3}{4}$?
42. What is the sum of $19256\frac{6}{11}+45600\frac{5}{8}+\frac{3}{4}$ of $\frac{2}{3}$ of $\frac{6}{7}$?
43. What is the sum of $\frac{3}{5}$ of $28+6\frac{1}{2}+45\frac{2}{7}+\frac{4}{5}$ of 300?

QUEST.—204. How add a whole number and a fraction?

SUBTRACTION OF FRACTIONS.

205. Ex. 1. A man bought $\frac{9}{10}$ of an acre of land, and afterwards sold $\frac{7}{10}$ of it: how much land had he left?

Solution.—7 tenths from 9 tenths leave 2 tenths.

Ans. $\frac{2}{10}$ of an acre.

2. A laborer having received $\frac{7}{8}$ of a dollar for a day's work, spent $\frac{3}{5}$ of a dollar for liquor: how much money had he left?

Note.—The learner meets with the same difficulty here as in the second example of adding fractions; that is, he can no more subtract *fifths* from *eighths*, than he can add *fifths* to *eighths*; for, $\frac{3}{5}$ of a dollar taken from $\frac{7}{8}$ of a dollar will neither leave 4 *fifths*, nor 4 *eighths*. The fractions must therefore be reduced to a common denominator before the subtraction can be performed.

Operation.

$$\left.\begin{array}{l}7\times5=35\\3\times8=24\end{array}\right\}\text{ the numerators. (Art. 200.)}$$

$8\times5=40$, the common denominator.

The fractions become $\frac{35}{40}$ and $\frac{24}{40}$. Now $\frac{35}{40}-\frac{24}{40}=\frac{11}{40}$ *Ans.*

206. From these illustrations we deduce the following general

RULE FOR SUBTRACTION OF FRACTIONS.

Reduce the given fractions to a common denominator; subtract the less numerator from the greater, and place the remainder over the common denominator.

OBS. *Compound fractions* must be reduced to simple ones, as in addition of fractions. (Art. 198.)

EXAMPLES.

3. From $\frac{3}{5}$ take $\frac{1}{4}$. *Ans.* $\frac{7}{20}$.
4. From $\frac{12}{15}$ take $\frac{8}{15}$.
5. From $\frac{25}{35}$ take $\frac{17}{35}$.
6. From $\frac{18}{27}$ take $\frac{3}{9}$.
7. From $\frac{21}{40}$ take $\frac{16}{47}$.
8. From $\frac{63}{74}$ take $\frac{26}{43}$.
9. From $\frac{48}{55}$ take $\frac{42}{67}$.
10. From $\frac{2}{3}$ of $\frac{5}{6}$ take $\frac{1}{5}$ of $\frac{7}{8}$.
11. From $\frac{3}{5}$ of $\frac{7}{8}$ take $\frac{1}{8}$ of $\frac{5}{9}$.
12. From $\frac{7}{8}$ of 40 take $\frac{3}{5}$ of 20.
13. From $\frac{2}{7}$ of $\frac{8}{9}$ of $\frac{4}{5}$ take $\frac{2}{3}$ of $\frac{1}{3}$.

QUEST.—206. How is one fraction subtracted from another? *Obs.* What is to be done with compound fractions?

207. *Mixed numbers* may be reduced to improper fractions, then to a common denominator, and be subtracted; or, the fractional part of the less number may be taken from the fractional part of the greater, and the less whole number from the greater.

14. From $9\frac{1}{4}$ take $7\frac{3}{4}$.

First Operation.	*Second Operation.*
$9\frac{1}{4}=\frac{37}{4}$	$9\frac{1}{4}$
$7\frac{3}{4}=\frac{31}{4}$	$7\frac{3}{4}$
Ans. $\frac{6}{4}=1\frac{2}{4}$, or $1\frac{1}{2}$.	*Ans.* $1\frac{2}{4}$, or $1\frac{1}{2}$.

Note.—Since we cannot take 3 fourths from 1 fourth, we borrow a *unit* in the second operation and reduce it to *fourths*, which added to the 1 fourth, make 5 fourths. Now 3 fourths from 5 fourths leave 2 fourths: 1 to carry to 7 makes 8, and 8 from 9 leaves 1.

15. From $25\frac{5}{7}$ take $13\frac{2}{3}$.
16. From $230\frac{7}{25}$ take $160\frac{3}{10}$.
17. From $178\frac{12}{40}$ take $56\frac{8}{9}$.
18. From $761\frac{25}{75}$ take $482\frac{60}{100}$.
19. From 5 take $\frac{2}{3}$.

Suggestion.—Since 3 *thirds* make a *whole* one, in 5 whole ones there are 15 thirds; now 2 thirds from 15 thirds leave 13 thirds, *Ans.* $\frac{13}{3}$, or $4\frac{1}{3}$. Hence,

208. To subtract a fraction from a whole number.

Change the whole number to a fraction having the same denominator as the fraction to be subtracted, and proceed as before. (Art. 197. Obs. 2.)

OBS. If the fraction to be subtracted is a proper fraction, we may simply borrow a unit and take the fraction from this, remembering to diminish the whole number by 1. (Art. 69. Obs. 1.)

20. From 20 take $\frac{3}{5}$. *Ans.* $19\frac{2}{5}$.
21. From 135 take $9\frac{7}{8}$.
22. From 263 take $24\frac{11}{12}$.
23. From 168 take $30\frac{5}{8}$.
24. From 567 take $100\frac{24}{40}$.
25. From 634 take $342\frac{2}{9}$.
26. From 729 take $125\frac{33}{40}$.
27. From 1000 take $25\frac{9}{10}$.
28. From 563 take $562\frac{45}{63}$.
29. From 9263 take $999\frac{1}{4}$.
30. From 857 take $785\frac{73}{84}$.

QUEST.—207. How are mixed numbers subtracted? 208. How is a fraction subtracted from a whole number?

MULTIPLICATION OF FRACTIONS.

209. We have seen that multiplying by a *whole number*, is taking the multiplicand as many times as there are *units* in the multiplier. (Art. 82.) On the other hand,

If the multiplier is only a *part* of a unit, it is plain we must take only a *part* of the multiplicand. That is,

Multiplying by $\frac{1}{2}$, is taking 1 *half* of the multiplicand *once*. Thus, $12\times\frac{1}{2}=6$.

Multiplying by $\frac{1}{3}$, is taking 1 *third* of the multiplicand *once*. Thus, $12\times\frac{1}{3}=4$.

Multiplying by $\frac{2}{3}$, is taking 1 *third* of the multiplicand *twice* Thus, $12\times\frac{2}{3}=8$. Hence,

210. *Multiplying by a fraction is taking a certain* PORTION *of the multiplicand as many times, as there are like portions of a unit in the multiplier.*

OBS. If the multiplier is a *unit* or 1, the product is *equal* to the multiplicand; if the multiplier is *greater* than a unit, the product is *greater* than the multiplicand; (Art. 82;) and if the multiplier is *less* than a unit, the product is *less* than the multiplicand.

CASE I.

211. *To multiply a fraction and a whole number together.*

Ex. 1. If 1 man drinks $\frac{2}{3}$ of a barrel of cider in a month, how much will 5 men drink in the same time?

Analysis.—Since 1 man drinks $\frac{2}{3}$ of a barrel, 5 men will drink 5 times as much; and 5 times 2 thirds are 10 thirds; that is, $\frac{2}{3}\times5=\frac{10}{3}$, or $3\frac{1}{3}$. (Art. 196.) *Ans.* $3\frac{1}{3}$ barrels.

Ex. 2. If a pound of tea costs $\frac{5}{8}$ of a dollar, how much will 4 pounds cost?

Solution.—$\frac{5}{8}\times4=\frac{20}{8}$; and $\frac{20}{8}=2\frac{4}{8}$, or $2\frac{1}{2}$ dolls. *Ans.*

Or, since dividing the denominator of a fraction by any number, multiplies the value of the fraction by that number, (Art. 189,)

QUEST.—209. What is meant by multiplying by a whole number? 210. What is meant by multiplying by a fraction? *Obs.* If the multiplier is a unit or 1, what is the product equal to? When the multiplier is greater than 1, how is the product, compared with the multiplicand? When less, how?

if we divide the denominator 8 by 4, the fraction will become $\frac{5}{2}$, which is equal to $2\frac{1}{2}$, the same as before. Hence,

212. To multiply a fraction by a whole number.

Multiply the numerator of the fraction by the whole number, and write the product over the denominator.

Or, divide the denominator by the whole number, when this can be done without a remainder. (Art. 189.)

OBS. 1. A fraction is multiplied into a number *equal* to its *denominator* by *canceling* the denominator. (Ax. 9.) Thus $\frac{4}{7}\times7=4$.

2. On the same principle, a fraction is multiplied into *any factor* in its *denominator*, by *canceling* that factor. (Art. 189.) Thus, $\frac{3}{15}\times3=\frac{3}{5}$.

3. Since multiplication is the *repeated addition* of a number or quantity to *itself*, (Art. 80,) the student sometimes finds it difficult to account for the fact that the product of a number or quantity by a proper fraction, is always *less* than the number multiplied. This difficulty will at once be removed by reflecting that *multiplying* by a *fraction* is *taking* or *repeating* a certain *portion* of the multiplicand as many times, as there are *like portions* of a *unit* in the multiplier. (Art. 210.)

EXAMPLES.

3. Multiply $\frac{21}{30}$ by 15. *Ans.* $\frac{315}{30}$, or $10\frac{1}{2}$.
4. Multiply $\frac{17}{40}$ by 8.
5. Multiply $\frac{25}{60}$ by 30.
6. Multiply $\frac{63}{94}$ by 27.
7. Multiply $\frac{151}{180}$ by 45.
8. Multiply $\frac{21}{43}$ by 100.
9. Multiply $\frac{40}{74}$ by 165.
10. Multiply $\frac{456}{683}$ by 100.
11. Multiply $\frac{75}{238}$ by 530.
12. Multiply $\frac{52}{67}$ by 1000.
13. Multiply $\frac{761}{955}$ by 831.
14. Multiply $12\frac{2}{3}$ by 8.

Operation.

$12\frac{2}{3}$
8
Ans. $101\frac{1}{3}$.

8 times $\frac{2}{3}$ are $\frac{16}{3}$, which are equal to 5 and $\frac{1}{3}$. Set down the $\frac{1}{3}$. 8 times 12 are 96, and 5 (which arose from the fraction) make 101. Hence,

213. To multiply a mixed number by a whole one.

Multiply the fractional part and the whole number separately, and unite the products.

QUEST.—212. How multiply a fraction by a whole number? *Obs.* How is a fraction multiplied by a number equal to its denominator? How by any factor in its denominator? 213. How is a mixed number multiplied by a whole one?

15. Multiply $45\frac{1}{8}$ by 10. *Ans.* $451\frac{2}{8}$.
16. Multiply $81\frac{3}{4}$ by 9.
17. Multiply $31\frac{10}{15}$ by 20.
18. Multiply $148\frac{10}{16}$ by 25.
19. Multiply $127\frac{1}{9}$ by 35.
20. Multiply $48\frac{2}{11}$ by 47.
21. Multiply $250\frac{7}{18}$ by 50.

214. *Multiplying* by a *fraction*, we have seen, is taking a certain portion of the multiplicand as many times, as there are like portions of a unit in the multiplier. Hence,

To multiply by $\frac{1}{2}$: *Divide the multiplicand by* 2.
To multiply by $\frac{1}{3}$: *Divide the multiplicand by* 3.
To multiply by $\frac{1}{4}$: *Divide the multiplicand by* 4, &c.
To multiply by $\frac{2}{3}$: *Divide by* 3, *and multiply the quotient by* 2.
To multiply by $\frac{3}{4}$: *Divide by* 4, *and multiply the quotient by* 3.

215. Hence, to multiply a whole number by a fraction.

Divide the multiplicand by the denominator, and multiply the quotient by the numerator.

Or, multiply the given number by the numerator, and divide the product by the denominator.

OBS. 1. When the given number cannot be divided by the denominator without a remainder, the latter method is generally preferred.

2. Since the product of any two numbers is the same, whichever is taken for the multiplier, (Art. 83,) the fraction may be taken for the multiplicand, and the whole number for the multiplier, when it is more convenient.

22. If 1 ton of hay costs 21 dollars, how much will $\frac{3}{4}$ of a ton cost?

Analysis.—Since 1 ton costs 21 dollars, $\frac{3}{4}$ of a ton will cost $\frac{3}{4}$ as much. Now, 1 fourth of 21 is $\frac{21}{4}$; and $\frac{3}{4}$ of 21 is 3 times as much; but $\frac{21}{4}\times3=\frac{21\times3}{4}=\frac{63}{4}$, or $15\frac{3}{4}$ dollars.

Operation.
$$\begin{array}{r} 4)\underline{21} \\ 5\frac{1}{4} \\ 3 \end{array}$$
Ans. $15\frac{3}{4}$ dolls.

23. Multiply 136 by $\frac{1}{3}$. *Ans.* $45\frac{1}{3}$.
24. Multiply 432 by $\frac{1}{4}$.
25. Multiply 635 by $\frac{1}{5}$.
26. Multiply 360 by $\frac{2}{3}$.
27. Multiply 580 by $\frac{3}{4}$.

QUEST.—215. How is a whole number multiplied by a fraction? 216. How find a fractional part of a number?

28. Multiply 672 by $\frac{5}{6}$.
29. Multiply 710 by $\frac{7}{8}$.
30. Multiply 765 by $\frac{11}{12}$.
31. Multiply 660 by $\frac{3}{26}$.
32. Multiply 840 by $\frac{48}{80}$.
33. Multiply 975 by $\frac{145}{463}$.

216. Since *multiplying* by a *fraction* is taking a certain portion of the multiplicand as many times, as there are like portions of a unit in the multiplier, it is plain, that the process of finding a *fractional part* of a number, is simply *multiplying the number by the given fraction, and is therefore performed by the same rule.*

Thus, $\frac{2}{3}$ of 12 dollars is the same as the product of 12 dollars multiplied by $\frac{2}{3}$; and $12\times\frac{2}{3}=8$ dollars.

OBS. The process of finding a fractional part of a number, is often a source of confusion and perplexity to the learner. The difficulty arises from the erroneous impression that finding a *fractional part*, implies that the given number must be *divided* by the *fraction*, instead of being *multiplied* by it.

34. What is $\frac{7}{12}$ of 457? *Ans.* $266\frac{7}{12}$.
35. What is $\frac{16}{25}$ of 16245?
36. What is $\frac{44}{63}$ of 25000?
37. What is $\frac{11}{128}$ of 4261?
38. What is $\frac{144}{367}$ of 5268?
39. What is $\frac{256}{325}$ of 45260?
40. What is $\frac{521}{1000}$ of 452120?

41. Multiply 64 by $5\frac{1}{2}$.

Operation.

$$\begin{array}{r} 2)64 \\ \underline{5\frac{1}{2}} \\ 320 \\ \underline{32} \\ \textit{Ans. } 352. \end{array}$$

We first multiply 64 by 5, then by $\frac{1}{2}$, and the sum of the products is 352. But multiplying by $\frac{1}{2}$ is taking *one half* of the multiplicand *once.* (Arts. 82, 214.) Hence,

217. To multiply a whole by a mixed number.

Multiply first by the integer, then by the fraction, and add the products together. (Art. 214.)

42. Multiply 83 by $7\frac{1}{5}$. *Ans.* $597\frac{3}{5}$.
43. Multiply 45 by $8\frac{1}{3}$.
44. Multiply 72 by $10\frac{1}{4}$.
45. Multiply 93 by $12\frac{2}{3}$.
46. Multiply 184 by $18\frac{3}{4}$.
47. Multiply 225 by $30\frac{1}{5}$.
48. Multiply 342 by $20\frac{1}{6}$.
49. Multiply 432 by $35\frac{3}{8}$.
50. Multiply 685 by $42\frac{4}{5}$.

QUEST.—217. How is a whole number multiplied by a mixed number?

51. Multiply 125 by $10\frac{6}{25}$.
52. Multiply 26 by $10\frac{19}{26}$.
53. Multiply 256 by $17\frac{9}{16}$.
54. Multiply 196 by $41\frac{11}{28}$.
55. Multiply 341 by $30\frac{7}{13}$.
56. Multiply 457 by $12\frac{31}{75}$.
57. Multiply 107 by $47\frac{21}{39}$.
58. Multiply 510 by $85\frac{13}{65}$.
59. Multiply 834 by $89\frac{1}{13}$.
60. Multiply 963 by $95\frac{81}{98}$.

CASE II.

218. *To multiply a fraction by a fraction.*

Ex. 1. A man bought $\frac{4}{5}$ of a bushel of wheat, at $\frac{7}{8}$ of a dollar per bushel: how much did he pay for it?

Analysis.—Since 1 bushel costs $\frac{7}{8}$ of a dollar, $\frac{1}{5}$ of a bushel must cost $\frac{1}{5}$ of $\frac{7}{8}$, which is $\frac{7}{40}$ of a dollar; for, multiplying the denominator, divides the value of the fraction. (Art. 188.) Now, if $\frac{1}{5}$ of a bushel costs $\frac{7}{40}$ of a dollar, $\frac{4}{5}$ of a bushel will cost 4 times as much · and 4 times $\frac{7}{40}$ are $\frac{28}{40}$, or $\frac{7}{10}$ dolls. (Art. 195.)

Ans. $\frac{7}{10}$ of a dollar.

Or, we may reason thus: since 1 bushel costs $\frac{7}{8}$ of a dollar, $\frac{4}{5}$ of a bushel must cost $\frac{4}{5}$ of $\frac{7}{8}$ of a dollar. Now $\frac{4}{5}$ of $\frac{7}{8}$ is a *compound* fraction, whose value is found by multiplying the numerators together for a new numerator, and the denominators for a new denominator. (Art. 198.)

Solution.—$\frac{7}{8}\times\frac{4}{5}=\frac{28}{40}$, or $\frac{7}{10}$ dollars, *Ans.* Hence,

219. To multiply a fraction by a fraction.

Multiply the numerators together for a new numerator, and the denominators together for a new denominator.

OBS. 1. It will be seen that the process of multiplying one fraction by another, is precisely the same as that of reducing compound fractions to simple ones. (Art. 198.)

2. The *reason* of this rule may be thus explained. Multiplying by a fraction is taking a *certain part* of the multiplicand as many times, as there are *like parts* of a *unit* in the multiplier. (Art. 210.) Now multiplying the denominator of the multiplicand by the denominator of the multiplier, gives the value of only *one* of the parts denoted by the given multiplier; (Art. 188;) we therefore multiply this new product by the numerator of the multiplier, to find the *number* of parts denoted by the given multiplier. (Art. 186.)

QUEST.—219. How is a fraction multiplied by a fraction? *Obs.* To what is the process of multiplying one fraction by another similar?

2. Multiply $\frac{2}{3}$ by $\frac{3}{4}$. *Ans.* $\frac{6}{12}=\frac{1}{2}$.
3. Multiply $\frac{5}{8}$ by $\frac{7}{10}$.
4 Multiply $\frac{11}{12}$ by $\frac{15}{28}$.
5. Multiply $\frac{16}{23}$ by $\frac{44}{60}$.
6. Multiply $\frac{25}{30}$ by $\frac{42}{70}$.
7. Multiply $\frac{37}{56}$ by $\frac{75}{88}$.
8. Multiply $\frac{68}{95}$ by $\frac{100}{250}$.
9. What is the product of $\frac{4}{5}$ into $\frac{5}{8}$ into $\frac{10}{11}$ into $\frac{1}{3}$ into $\frac{7}{2}$?
10. What cost $6\frac{2}{3}$ yards of cloth, at $4\frac{1}{2}$ dollars per yard?

Analysis.—$4\frac{1}{2}$ dollars$=\frac{9}{2}$, and $6\frac{2}{3}$ yards$=\frac{20}{3}$. (Art. 197.) Now $\frac{9}{2}\times\frac{20}{3}=\frac{180}{6}$, or 30. (Art. 196.) *Ans.* 30 dollars. Hence,

220. When the multiplier and multiplicand are both *mixed* numbers, they should be reduced to *improper fractions*, and then be multiplied according to the rule above.

OBS. Mixed numbers may also be multiplied together, *without reducing* them to improper fractions.

Take, for instance, the last example. We first multiply by 4, the whole number. Thus, 4 times $\frac{2}{3}$ are $\frac{8}{3}$, equal to 2 and $\frac{2}{3}$; set down the $\frac{2}{3}$, and carry the 2. Next, 4 times 6 are 24, and 2 to carry are 26. We then multiply by $\frac{1}{2}$, the fractional part. Thus, $\frac{1}{2}$ of 6 is 3; and $\frac{1}{2}$ of 2 thirds is $\frac{1}{3}$. The sum of the two partial products is 30 dollars, the same as before.

Operation

$6\frac{2}{3}$
$4\frac{1}{2}$
$26\frac{2}{3}$
$3\frac{1}{3}$
30 dolls.

11. Multiply $6\frac{3}{4}$ by $2\frac{14}{27}$.
12. Multiply $8\frac{5}{11}$ by $6\frac{2}{7}$.
13. Multiply $13\frac{2}{3}$ by $17\frac{7}{9}$.
14. Multiply $15\frac{3}{4}$ by $20\frac{8}{9}$.
15. Multiply $30\frac{5}{6}$ by $44\frac{4}{20}$.
16. Multiply $63\frac{21}{45}$ by $50\frac{2}{5}$.
17. Multiply $17\frac{8}{13}$ by $25\frac{11}{17}$.
18. Multiply $47\frac{30}{51}$ by $17\frac{13}{24}$.
19. Multiply $61\frac{7}{15}$ by $32\frac{21}{37}$.
20. Multiply $71\frac{24}{37}$ by $45\frac{7}{63}$.
21. Multiply $83\frac{9}{28}$ by $61\frac{35}{48}$.
22. Multiply $96\frac{45}{67}$ by $72\frac{34}{86}$.
23. Multiply $246\frac{1}{15}$ by $\frac{51}{75}$.
24. Multiply $91\frac{38}{63}$ by $\frac{2}{3}$ of $\frac{7}{8}$.
25. Multiply 1475 by $\frac{7}{9}$ of 21.
26. Multiply $34\frac{1}{3}$ by $\frac{1}{5}$ of 68.
27. Multiply 800 by $\frac{3}{8}$ of 1000.
28. Multiply $\frac{1}{2}$ of 75 by $\frac{2}{3}$ of 28.
29. Multiply $2\frac{1}{3}$ by $\frac{2}{3}$ of $\frac{3}{4}$ of 85.
30. Multiply $\frac{6}{7}$ of $2\frac{1}{2}$ by $\frac{3}{5}$ of 61.
31. Multiply $\frac{4}{5}$ of $10\frac{3}{4}$ by $\frac{2}{3}$ of $8\frac{1}{4}$.
32. Multiply $\frac{2}{3}$ of $16\frac{1}{8}$ by $\frac{6}{7}$ of $9\frac{1}{3}$.
33. Multiply $\frac{7}{8}$ of $\frac{9}{10}$ of 20 by $25\frac{1}{2}$.
34. Multiply $\frac{23}{47}$ of $65\frac{1}{3}$ by $46\frac{1}{9}$.
35. What cost $125\frac{1}{2}$ bbls. of flour, at $7\frac{3}{4}$ dollars per barrel?
36. What cost $250\frac{3}{7}$ acres of land, at $25\frac{1}{4}$ dollars per acre?
37. If a man travels $40\frac{3}{5}$ miles per day, how far will he travel in $135\frac{1}{2}$ days?

QUEST.—220. When the multiplier and multiplicand are mixed numbers how proceed?

CONTRACTIONS IN MULTIPLICATION OF FRACTIONS.

Ex. 1. Multiply $\frac{3}{5}$ by $\frac{5}{8}$ and $\frac{8}{11}$ and $\frac{1}{3}$ and $\frac{7}{2}$.

Operation.

$$\frac{\not 3}{\not 5}\times\frac{\not 5}{\not 8}\times\frac{\not 8}{11}\times\frac{1}{\not 3}\times\frac{7}{2}=\frac{7}{22}$$

Since the factors 3, 5 and 8 are common to the numerators and denominators, we may *cancel* them; (Art. 191;) and then multiply the remaining factors together, as in reduction of compound fractions to simple ones. (Art. 199.) Hence,

221. To multiply fractions by CANCELATION.

Cancel all the factors common both to the numerators and denominators; then multiply together the factors remaining in the numerators for a new numerator, and those remaining in the denominators for a new denominator, as in reduction of compound fractions. (Art. 199.)

OBS. 1. The *reason* of this process may be seen from the fact that the product of the numerators is divided by the *same numbers* as that of the denominators, and therefore the *value* of the answer is not *altered.* (Art. 191.)

2. Care must be taken that the factors canceled in the numerators are *exactly equal* to those canceled in the denominators.

2. Multiply $\frac{2}{3}$ by $\frac{3}{4}$ and $\frac{4}{5}$. *Ans.* $\frac{2}{5}$.
3. Multiply $\frac{3}{4}$ by $\frac{5}{8}$ into $\frac{4}{5}$.
4. Multiply $\frac{3}{7}$ by $\frac{1}{15}$ into $\frac{5}{6}$.
5. Multiply $\frac{3}{7}$ by $\frac{1}{3}$ into $\frac{7}{9}$.
6. Multiply $\frac{7}{8}$ by $\frac{3}{2}$ of $\frac{3}{4}$.
7. Multiply $\frac{2}{3}$ of $\frac{3}{4}$ by $\frac{12}{15}$.
8. Multiply $\frac{9}{11}$ by $\frac{33}{27}$ of $\frac{9}{13}$.
9. Multiply $\frac{11}{12}$ of 4 by $\frac{3}{44}$.
10. Multiply $3\frac{3}{4}$ by $\frac{13}{15}$ of 8.
11. Multiply $\frac{4}{5}$ by $\frac{7}{8}$ and $\frac{5}{6}$ and $\frac{8}{9}$ and $\frac{6}{7}$.
12. Multiply $\frac{4}{9}$ by $\frac{5}{7}$ and $\frac{3}{12}$ and $\frac{9}{25}$ and $\frac{35}{40}$.
13. Multiply $\frac{16}{25}$ by $\frac{4}{3}$ and $\frac{75}{64}$ and $\frac{5}{12}$ and $\frac{6}{10}$.
14. Multiply $\frac{9}{14}$ into $\frac{42}{63}$ into $\frac{5}{8}$ into $\frac{7}{3}$ into $\frac{8}{40}$ into $\frac{2}{7}$.
15. Multiply $\frac{3}{13}$ into $\frac{25}{75}$ into $\frac{60}{81}$ into $\frac{78}{55}$ into $\frac{5}{6}$ into $\frac{11}{12}$.
16. Multiply $\frac{44}{63}$ into $\frac{9}{11}$ into $\frac{7}{4}$ into $\frac{8}{96}$ into $\frac{24}{27}$ into $\frac{8}{9}$.
17. What must a man pay for $3\frac{1}{3}$ barrels of flour, when flour is worth 6 dollars a barrel?

QUEST.—221. How are fractions multiplied by cancelation? *Obs.* How does it appear that this process will give the true answer? What is necessary to be observed with regard to canceling factors?

Analysis.—$3\frac{1}{3}$ bbls. is $\frac{1}{3}$ of 10 bbls.; now since 1 bbl. costs 6 dollars, 10 bbls. will cost 10 times as much, or 60 dollars. But we wished to find the cost of only $3\frac{1}{3}$ barrels, which is $\frac{1}{3}$ of 10 bbls. Therefore if we take $\frac{1}{3}$ of the cost of 10 bbls., it will of course be the price of $3\frac{1}{3}$ bbls.

Operation.

dolls.	6	price of 1 bbl.
	10	
	3)60	" of 10 bbls.
dolls.	20	" of $3\frac{1}{3}$ bbls.

PROOF.—6 dolls.$\times 3\frac{1}{3}=20$ dolls., the same as before.

Note.—In like manner, when the multiplier is $33\frac{1}{3}$, $333\frac{1}{3}$, &c., if we multiply by 100, 1000, &c., $\frac{1}{3}$ of the product will be the answer. Hence,

222. To multiply a whole number by $3\frac{1}{3}$, $33\frac{1}{3}$, $333\frac{1}{3}$, &c.

Annex as many ciphers to the multiplicand as there are 3s *in the integral part of the multiplier; then take $\frac{1}{3}$ of the number thus produced, and the result will be the answer required.*

OBS. 1. The *reason* of this contraction is evident from the principle that annexing a cipher to a number multiplies it by 10, annexing two ciphers multiplies it by 100, &c. (Art. 98.) But $3\frac{1}{3}$ is $\frac{1}{3}$ of 10; $33\frac{1}{3}$ is $\frac{1}{3}$ of 100, &c.; therefore annexing as many ciphers to the multiplicand, as there are 3s in the integral part of the multiplier, gives a product 3 times too large; consequently $\frac{1}{3}$ of this product must be the *true answer.*

2. When the multiplicand is a *mixed* number, and the multiplier is $3\frac{1}{3}$, $33\frac{1}{3}$ &c., it is evident we may multiply by 10, 100, &c., as the case may be, and $\frac{1}{3}$ of the number thus produced will be the answer required.

18. Multiply 158 by $33\frac{1}{3}$. *Ans.* $5266\frac{2}{3}$.
19. Multiply 148 by $3\frac{1}{3}$.
20. Multiply 256 by $33\frac{1}{3}$.
21. Multiply 1728 by $33\frac{1}{3}$.
22. Multiply 297 by $333\frac{1}{3}$.
23. Multiply $561\frac{1}{2}$ by $3\frac{1}{3}$.
24. Multiply $426\frac{2}{3}$ by $33\frac{1}{3}$.

223. To multiply a whole number by $6\frac{2}{3}$, $66\frac{2}{3}$, $666\frac{2}{3}$, &c.

Annex as many ciphers to the multiplicand as there are 6s *in the integral part of the multiplier; then take $\frac{2}{3}$ of the number thus produced, and the result will be the answer required.*

OBS. The *reason* of this contraction is manifest from the fact that $6\frac{2}{3}$ is $\frac{2}{3}$ of 10; $66\frac{2}{3}$ is $\frac{2}{3}$ of 100, &c.

25. What will $6\frac{2}{3}$ tons of iron cost, at 75 dollars per ton?

QUEST.—222. How may a whole number be multiplied by $3\frac{1}{3}$, $33\frac{1}{3}$, &c.? 223. How may a whole number be multiplied by $6\frac{2}{3}$, $66\frac{2}{3}$, &c.

Analysis.—$6\frac{2}{3}$ tons is $\frac{2}{3}$ of 10 tons. Now, if 1 ton costs 75 dollars, 10 tons will cost 10 times as much, or 750 dollars; and $\frac{2}{3}$ of 750 dollars, ($6\frac{2}{3}=\frac{2}{3}$ of 10,) are 500 dollars, which is the answer required.

Operation.

```
dolls.  75, price of 1 ton.
        10
      3)750     "  of 10 tons.
        250
          2
dolls.  500,    "  of 6⅔ tons.
```

PROOF.—75 dolls.$\times 6\frac{2}{3}=500$ dolls., the same as above.

26. Multiply 320 by $6\frac{2}{3}$.
27. Multiply 277 by $66\frac{2}{3}$.
28. Multiply 837 by $6\frac{2}{3}$.
29. Multiply 645 by $666\frac{2}{3}$.
30. What will $12\frac{1}{2}$ acres of land cost, at 46 dollars per acre?

Analysis.—$12\frac{1}{2}$ acres is $\frac{1}{8}$ of 100 acres; now since 1 acre costs 46 dollars, 100 acres will cost 100 times as much, or 4600 dollars. But we wished to find the cost of only $12\frac{1}{2}$ acres, which is $\frac{1}{8}$ of 100 acres. Therefore $\frac{1}{8}$ of the cost of 100 acres, will obviously be the cost of $12\frac{1}{2}$ acres.

Operation.

```
dolls.  46, price of 1 A.
       100
     8)4600    "  100 A.
dolls.  575    "   12 A.
```

PROOF.—46 dolls.$\times 12\frac{1}{2}=575$ dolls., the same as before.

Note.—In like manner, if the multiplier is $37\frac{1}{2}$, $62\frac{1}{2}$, or $87\frac{1}{2}$, we may multiply by 100, and $\frac{3}{8}$, $\frac{5}{8}$, or $\frac{7}{8}$ of the product will be the answer. Hence,

224. To multiply a whole number by $12\frac{1}{2}$, $37\frac{1}{2}$, $62\frac{1}{2}$, or $87\frac{1}{2}$.

Annex two ciphers to the multiplicand, then take $\frac{1}{8}$, $\frac{3}{8}$, $\frac{5}{8}$, or $\frac{7}{8}$, of the number thus produced, as the case may be, and the result will be the answer required.

OBS. The *reason* of this contraction may be seen from the fact that $12\frac{1}{2}$ is $\frac{1}{8}$ $37\frac{1}{2}$ is $\frac{3}{8}$, $62\frac{1}{2}$ is $\frac{5}{8}$, and $87\frac{1}{2}$ is $\frac{7}{8}$ of 100.

31. Multiply 275 by $37\frac{1}{2}$. *Ans.* $10312\frac{1}{2}$.
32. Multiply 381 by $12\frac{1}{2}$.
33. Multiply 425 by $37\frac{1}{2}$.
34. Multiply 643 by $62\frac{1}{2}$.
35. Multiply 748 by $87\frac{1}{2}$.

225. To multiply a whole number by $1\frac{2}{3}$, $16\frac{2}{3}$, $166\frac{2}{3}$, &c.

Annex as many ciphers to the multiplicand as there are integral figures in the multiplier, then $\frac{1}{6}$ of the number thus produced will be the product required.

OBS. The *reason* of this contraction is evident from the fact that $1\frac{2}{3}$ is $\frac{1}{6}$ of 10; $16\frac{2}{3}$ is $\frac{1}{6}$ of 100; $166\frac{2}{3}$ is $\frac{1}{6}$ of 1000, &c.

36. What will $16\frac{2}{3}$ bales of Swiss muslin cost, at 735 dollars per bale?

Solution.—Annexing two ciphers to 735 dolls., it becomes 73500 dolls.; and $73500 \div 6 = 12250$ dolls. *Ans.*

37. Multiply 767 by $1\frac{2}{3}$.

38. Multiply 245 by $16\frac{2}{3}$.

39. Multiply 489 by $16\frac{2}{3}$.

40. Multiply 563 by $166\frac{2}{3}$.

Note.—Specific rules might be added for multiplying by $1\frac{1}{9}$, $11\frac{1}{9}$, $111\frac{1}{9}$, $8\frac{1}{3}$, $83\frac{1}{3}$, $833\frac{1}{3}$, $6\frac{1}{4}$, &c., but they will naturally be suggested to the inquisitive mind, from the contractions already given.

DIVISION OF FRACTIONS.

CASE I.

226. *Dividing a fraction by a whole number.*

Ex. 1. If 4 yards of calico cost $\frac{8}{9}$ of a dollar, what will 1 yard cost?

Analysis.—1 is 1 *fourth* of 4; therefore 1 yard will cost 1 fourth part as much as 4 yards. And 1 fourth of 8 ninths of a dollar, is 2 ninths. *Ans.* $\frac{2}{9}$ of a dollar.

Operation.
$\frac{8}{9} \div 4 = \frac{2}{9}$ *Ans.*

We divide the numerator of the fraction by 4, and the quotient 2, placed over the denominator, forms the answer required.

2. If 5 bushels of apples cost $\frac{11}{12}$ of a dollar, what will 1 bushel cost?

Operation.
$\frac{11}{12} \div 5 = \frac{11}{12 \times 5}$, or $\frac{11}{60}$ *Ans.*

Since we cannot divide the numerator by the divisor 5, without a remainder, we multiply the denominator by it, which, in effect, divides the fraction. (Art. 188.)

PROOF.—$\frac{11}{60}$ dolls. $\times 5 = \frac{11}{12}$ dolls., the same as above. Hence,

227. To divide a fraction by a whole number.

Divide the numerator by the whole number, when it can be done without a remainder; but when this cannot be done, multiply the denominator by the whole number.

3. What is the quotient of $\frac{15}{20}$ divided by 5?

First Method.

$\frac{15}{20} \div 5 = \frac{3}{20}$ *Ans.*

Second Method.

$\frac{15}{20} \div 5 = \frac{15}{20 \times 5} = \frac{15}{100}$, or $\frac{3}{20}$ *Ans.*

4. Divide $\frac{18}{23}$ by 9.
5. Divide $\frac{24}{31}$ by 7.
6. Divide $\frac{64}{70}$ by 16.
7. Divide $\frac{84}{75}$ by 12.
8. Divide $\frac{125}{30}$ by 25.
9. Divide $\frac{256}{100}$ by 29.

CASE II.

228. *Dividing a fraction by a fraction.*

10. At $\frac{1}{5}$ of a dollar a basket, how many baskets of peaches can you buy for $\frac{4}{5}$ of a dollar?

Analysis.—Since $\frac{1}{5}$ of a dollar will buy 1 basket, $\frac{4}{5}$ of a dollar will buy as many baskets as $\frac{1}{5}$ is contained times in $\frac{4}{5}$; and $\frac{1}{5}$ is contained in $\frac{4}{5}$, 4 times. *Ans.* 4 baskets.

11. At $\frac{2}{3}$ of a dollar per yard, how many yards of cloth can be bought for $\frac{7}{8}$ of a dollar?

OBS. 1. Reasoning as before, $\frac{7}{8}$ of a dollar will buy as many yards, as $\frac{2}{3}$ is contained times in $\frac{7}{8}$. But since the fractions have different denominators, it is plain we cannot divide one numerator by the other, as we did in the last example. This difficulty may be remedied by reducing the fractions to a common denominator. (Art. 200.)

First Operation.

$\frac{7}{8}$ and $\frac{2}{3}$ reduced to a common denominator, become $\frac{21}{24}$ and $\frac{16}{24}$. (Art. 200.) Now $\frac{21}{24} \div \frac{16}{24} = \frac{21}{16}$; and $\frac{21}{16} = 1\frac{5}{16}$. *Ans.* $1\frac{5}{16}$ yards.

OBS. 2. It will be perceived that no use is made of the *common denominator*, after it is obtained. If, therefore, we *invert the divisor*, and then multiply the two fractions together, we shall have the same result as before.

Second Operation.

$\frac{7}{8} \times \frac{3}{2}$ (divisor inverted) $= \frac{21}{16}$, or $1\frac{5}{16}$ yards, the same as above

QUEST.—227. How is a fraction divided by a whole number?

229. Hence, to divide a fraction by a fraction.

I. *If the given fractions have a common denominator, **divide the** numerator of the dividend by the numerator of the divisor.*

II. *When the fractions have not a common denominator, **invert** the divisor, and proceed as in multiplication of fractions.* (Art. 219.)

OBS. 1. When two fractions have a *common denominator*, it is plain *one numerator* can be divided by the *other*, as well as *one whole* number by *another;* for, the *parts* of the two fractions are of the *same denomination.*

2. When the fractions do not have a *common* denominator, the *reason* that inverting the divisor and proceeding as in multiplication, will produce the *true answer*, is because this process, in effect, reduces the two fractions to a *common denominator*, and then the numerator of the dividend is divided by the numerator of the divisor. Thus, reducing the two fractions to a common denominator, we multiply the numerator of the dividend by the denominator of the divisor, and the numerator of the divisor by the denominator of the dividend; (Art. 200;) and, then dividing the *former* product by the *latter*, we have the *same combination* of the *same numbers* as in the rule above, which will consequently produce the *same result.*

We do not multiply the two denominators together for a common denominator; for, in dividing, no use is made of a common denominator when found, therefore it is unnecessary to obtain it. (Art. 228. Obs. 2.)

The object of *inverting the divisor* is simply for *convenience* in multiplying.

3. *Compound fractions* occurring in the divisor or dividend, must be reduced to simple ones, and *mixed numbers* to improper fractions.

230. The principle of dividing a fraction by a fraction may also be illustrated in the following manner. Thus, in the last example,

Operation.

$\frac{7}{8}\div 2=\frac{7}{16}$

$\frac{7}{16}\times 3=\frac{21}{16}$

And $\frac{21}{16}=1\frac{5}{16}$ *Ans.*

Dividing the dividend $\frac{7}{8}$ by 2, the quotient is $\frac{7}{16}$. (Art. 188.) But it is required to divide it by 1 *third* of two; consequently the $\frac{7}{16}$ is 3 *times too small* for the *true* quotient; therefore multiplying $\frac{7}{16}$ by 3, will give the quotient required; and $\frac{7}{16}\times 3=\frac{21}{16}$, or $1\frac{5}{16}$.

Note.—By examination the learner will perceive that this process is precisely

---

QUEST.—229. How is one fraction divided by another when they have a common denominator? How, when they have not common denominators? *Obs.* When the fractions have a common denominator, how does it appear that dividing one numerator by the other will give the true answer? When the fractions have not a common denominator, how does it appear that inverting the divisor and proceeding as in multiplication will give the true answer? What is the object of inverting the divisor? How proceed when the divisor or dividend are compound fractions, or mixed numbers?

the same in effect as the preceding; for, in both cases the denominator of the dividend is multiplied by the numerator of the divisor, and the numerator of the dividend, by the denominator of the divisor.

12. Divide $\frac{3}{4}$ of $\frac{2}{3}$ by $2\frac{1}{3}$. *Ans.* $\frac{18}{84}$, or $\frac{3}{14}$.
13. Divide $8\frac{2}{3}$ by $3\frac{1}{2}$. *Ans.* $\frac{52}{21}$, or $2\frac{10}{21}$.
14. Divide $\frac{42}{54}$ by $\frac{24}{35}$.
15. Divide $\frac{63}{56}$ by $\frac{18}{45}$.
16. Divide $55\frac{1}{2}$ by $16\frac{3}{4}$.
17. Divide $46\frac{2}{3}$ by $68\frac{5}{6}$.

231. The process of dividing fractions may often be *contracted* by *canceling* equal factors in the divisor and dividend; (Art. 146;) or, after the divisor is inverted, by canceling factors which are common to the numerators and denominators. (Art. 191)

18. Divide $\frac{1}{3}$ of $\frac{4}{7}$ of $\frac{2}{11}$ by $\frac{4}{5}$ of $\frac{2}{3}$ of $\frac{1}{7}$.

Operation.

$$\begin{array}{r|l} \not{3} & 1 \\ \not{7} & \not{4} \\ 11 & \not{2} \\ \not{4} & 5 \\ \not{2} & \not{3} \\ 1 & \not{7} \\ \hline 11 & 5 = \frac{5}{11} \text{ Ans.} \end{array}$$

For convenience we arrange the numerators, (which answer to dividends,) on the right of a perpendicular line, and the denominators, (which answer to divisors,) on the left; then canceling the factors, 2, 3, 4, and 7, which are common to both sides, (Art. 151;) we multiply the remaining factors in the numerators together, and those remaining in the denominators, as in the rule above. Hence,

232. To divide fractions by CANCELATION.

Having inverted the divisor, cancel all the factors common both to the numerators and denominators, and the product of those remaining on the right of the line placed over the product of those remaining on the left, will be the answer required.

OBS. 1. Before arranging the terms of the divisor for cancelation, it is always necessary to invert them, or suppose them to be inverted.

2. The *reason* of this contraction is evident from the principle, that if the numerator and denominator of a fraction are both divided by the *same number*, the value of the fraction is not *altered.* (Arts. 148, 191.)

19. Divide $18\frac{3}{4}$ by $6\frac{2}{8}$. *Answer* 3.

QUEST.—232. How divide fractions by cancelation? How arrange the terms of the given fractions? *Obs.* What must be done to the divisor before arranging its terms? How does it appear that this contraction will give the true answer?

20. Divide $\frac{3}{4}$ of $\frac{8}{9}$ by $\frac{5}{3}$ of $\frac{9}{11}$.
21. Divide $\frac{5}{7}$ of $\frac{14}{20}$ by $6\frac{3}{4}$.
22. Divide $15\frac{3}{4}$ by $\frac{9}{10}$ of $\frac{3}{4}$.
23. Divide $\frac{6}{7}$ of $7\frac{3}{6}$ by $\frac{2}{7}$ of $\frac{5}{9}$.
24. Divide $\frac{2}{3}$ of $\frac{5}{7}$ of $\frac{3}{5}$ by $\frac{4}{7}$.
25. Divide $\frac{25}{42}$ of 7 by $\frac{50}{63}$ of 42.
26. Divide $\frac{45}{28}$ of $\frac{21}{15}$ of $\frac{9}{12}$ of $\frac{10}{13}$ by $\frac{3}{28}$ of $\frac{30}{36}$ of $\frac{3}{4}$ of 5.

CASE III.

233. *Dividing a whole number by a fraction.*

27. How many pounds of tea, at $\frac{3}{4}$ of a dollar a pound, can be bought for 15 dollars?

Analysis.—Since $\frac{3}{4}$ of a dollar will buy 1 pound, 15 dollars will buy as many pounds as $\frac{3}{4}$ is contained times in 15. Reducing the dividend 15, to the form of a fraction, it becomes $\frac{15}{1}$; (Art. 197. Obs. 1;) then inverting the divisor and proceeding as before, we have $\frac{15}{1}\times\frac{4}{3}=\frac{60}{3}$, or 20. *Ans.* 20 pounds.

Or, we may reason thus: $\frac{1}{4}$ is contained in 15, as many times as there are *fourths* in 15, viz: 60 times. But 3 fourths will be contained in 15, only a *third* as many times as 1 fourth, and $60\div3=20$, the same result as before. Hence,

234. To divide a whole number by a fraction.

Reduce the whole number to the form of a fraction, (Art. 197. Obs. 1,) *and then proceed according to the rule for dividing a fraction by a fraction.* (Art. 229.)

Or, multiply the whole number by the denominator, and divide the product by the numerator.

OBS. 1. When the divisor is a *mixed* number, it must be reduced to an improper fraction; then proceed as above.

Or, reducing the dividend to a fraction having the *same denominator*, (Art. 197. Obs. 2,) we may divide one numerator by the other. (Art. 229. I.)

2. If the divisor is a *unit* or 1, the quotient is *equal* to the dividend; if the divisor is *greater* than a unit, the quotient is *less* than the dividend; and if the divisor is *less* than a unit, the quotient is *greater* than the dividend.

28. How much cloth, at $3\frac{1}{2}$ dollars per yard, can you buy for 28 dollars?

QUEST.—234. How is a whole number divided by a fraction? *Obs.* How by a mixed number?

Operation.
$3\frac{1}{2})28$
2 2
7) 56 halves.
Ans. 8 yards.

Since the divisor is a mixed number, we reduce it to *halves;* we also reduce the dividend to the *same* denominator; (Art. 197. Obs. 2;) then divide one numerator by the other. (Art. 229. I.)

29. Divide 75 by $\frac{5}{9}$.
30. Divide 96 by $\frac{4}{7}$.
31. Divide 120 by $10\frac{3}{4}$.
32. Divide 145 by $12\frac{1}{6}$.
33. Divide 237 by $25\frac{3}{7}$.
34. Divide 425 by $31\frac{3}{8}$.

CONTRACTIONS IN DIVISION OF FRACTIONS.

235. When the divisor is $3\frac{1}{3}$, $33\frac{1}{3}$, $333\frac{1}{3}$, &c.

Multiply the dividend by 3, *divide the product by* 10, 100, *or* 1000, *as the case may be, and the result will be the true quotient.* (Art. 131.)

OBS. The *reason* of this contraction will be understood from the principle, that if the divisor and dividend are both multiplied by the *same* number, the quotient will not be *altered.* (Art. 146.) Thus $3\frac{1}{3}\times3=10$; $33\frac{1}{3}\times3=100$, $333\frac{1}{3}\times3=1000$, &c.

35. At $3\frac{1}{3}$ dollars per yard, how many yards of cloth can be bought for 561 dollars?

Operation.
dolls. 561
3
1|0)168|3
Ans. $168\frac{3}{10}$ yds.

We first multiply the dividend by 3, then divide the product by 10; for, multiplying the divisor $3\frac{1}{3}$ by 3, it becomes 10. (Art. 146.)

36. Divide 687 by $33\frac{1}{3}$. *Ans.* $20\frac{61}{100}$.
37. Divide 453 by $33\frac{1}{3}$,
38. Divide 2783 by $333\frac{1}{3}$.

236. When the divisor is $1\frac{2}{3}$, $16\frac{2}{3}$, $166\frac{2}{3}$, &c.

Multiply the dividend by 6, *and divide the product by* 10, 100, *or* 1000, *as the case may be.*

OBS. This contraction also depends upon the principle, that if the divisor and dividend are both multiplied by the same number, the quotient will not be altered. (Art. 146.) Thus, $1\frac{2}{3}\times6=10$; $16\frac{2}{3}\times6=100$; $166\frac{2}{3}\times6=1000$, &c.

39. What is the quotient of 725 divided by $16\frac{2}{3}$?

Solution.—$725\times6=4350$; and $4350\div100=43\frac{1}{2}$ *Ans.*

40. Divide 367 by $1\frac{2}{3}$.
41. Divide 507 by $16\frac{2}{3}$.
42. Divide 849 by $16\frac{2}{3}$.
43. Divide 1124 by $166\frac{2}{3}$.

237. When the divisor is $1\frac{1}{9}$, $11\frac{1}{9}$, $111\frac{1}{9}$, &c.

Multiply the dividend by 9, *and divide the product by* 10, 100, *or* 1000, *as the case may be.*

OBS. This contraction depends upon the same principle as the preceding. Thus, $1\frac{1}{9}\times9=10$; $11\frac{1}{9}\times9=100$; $111\frac{1}{9}\times9=1000$, &c.

44. Divide 587 by $11\frac{1}{9}$.

Solution.—$587\times9=5283$, and $5283\div100=52\frac{83}{100}$ *Ans.*

45. Divide 861 by $1\frac{1}{9}$. *Ans.* $774\frac{9}{10}$.
46. Divide 4263 by $11\frac{1}{9}$.
47. Divide 6037 by $111\frac{1}{9}$.

Note.—Other methods of contraction might be added, but they will naturally suggest themselves to the student, as he becomes familiar with the principles of fractions.

238. From the definition of *complex* fractions, and the manner of expressing them, it will be seen that they arise from *division* of fractions. (Art. 183.) Thus, the complex fraction $\frac{4\frac{1}{2}}{1\frac{1}{4}}$, is the same as $\frac{9}{2}\div\frac{5}{4}$; for, the numerator, $4\frac{1}{2}=\frac{9}{2}$, and the denominator $1\frac{1}{4}=\frac{5}{4}$; but the numerator of a fraction is a dividend, and the denominator a divisor. (Art. 184.) Now, $\frac{9}{2}\div\frac{5}{4}=\frac{36}{10}$, which is a simple fraction. Hence,

239. To reduce a complex fraction to a simple one.

Consider the denominator as a divisor, and proceed as in division of fractions. (Arts. 229, 232.)

OBS. The *reason* of this rule is evident from the fact that the *denominator* of a fraction denotes a *divisor*, and the *numerator*, a *dividend;* (Art. 184;) hence the process required, is simply *performing the division* which is expressed by the given fraction.

QUEST.—238. From what do complex fractions arise? 239. How reduce them to simple fractions?

48. Reduce $\frac{4\frac{2}{3}}{7\frac{1}{4}}$ to a simple fraction.

Solution.—$4\frac{2}{3}=\frac{14}{3}$, and $7\frac{1}{4}=\frac{29}{4}$. (Art. 197.)
Now $\frac{14}{3}\div\frac{29}{4}=\frac{14}{3}\times\frac{4}{29}$, or $\frac{56}{87}$ *Ans.*

Reduce the following complex fractions to simple ones:

49. Reduce $\frac{8}{3\frac{1}{4}}$.
50. Reduce $\frac{5\frac{1}{6}}{7}$.
51. Reduce $\frac{2\frac{2}{5}}{3\frac{3}{7}}$.
52. Reduce $\frac{6\frac{1}{3}}{7\frac{3}{5}}$.
53. Reduce $\frac{5\frac{7}{8}}{\frac{3}{16}}$.
54. Reduce $\frac{\frac{7}{12}}{15\frac{3}{4}}$.
55. Reduce $\frac{\frac{25}{45}}{\frac{35}{90}}$.
56. Reduce $\frac{\frac{42}{54}}{\frac{28}{36}}$.

240. To multiply complex fractions together.

First reduce the complex fractions to simple ones; (Art. 239;) *then arrange the terms, and cancel the common factors, as in multiplication of simple fractions.* (Art. 219.)

OBS. The terms of the complex fractions may be arranged for reducing them to simple ones, and for multiplication at the same time.

57. Multiply $\frac{3\frac{1}{2}}{2\frac{2}{5}}$ by $\frac{1\frac{5}{7}}{4\frac{1}{2}}$.

Operation.

~~2~~	~~7~~
~~12~~	5
~~7~~	~~12~~
9	~~2~~
9	5 $=\frac{5}{9}$. *Ans.*

The numerator $3\frac{1}{2}=\frac{7}{2}$. (Art. 197.) Place the 7 on the right hand and 2 on the left of the perpendicular line. The denominator $2\frac{2}{5}=\frac{12}{5}$, which must be inverted; (Art. 239;) i. e. place the 12 on the left and the 5 on the right of the line, $1\frac{5}{7}=\frac{12}{7}$, and $4\frac{1}{2}=\frac{9}{2}$, both of which must be arranged in the same manner as the terms of the multiplicand. Now, canceling the common factors, we divide the product of those remaining on the right of the line by the product of those on the left, and the answer is $\frac{5}{9}$. (Art. 219.)

QUEST.—240. How are complex fractions multiplied together? 241. How is one complex fraction divided by another?

58. Multiply $\frac{2\frac{1}{3}}{2\frac{1}{4}}$ by $\frac{4\frac{1}{2}}{2\frac{1}{2}}$.

59. Multiply $\frac{4\frac{1}{3}}{8\frac{2}{3}}$ by $\frac{4\frac{1}{5}}{17\frac{1}{2}}$.

60. Multiply $\frac{\frac{3}{7}}{\frac{3}{4}}$ by $\frac{\frac{1}{9}}{\frac{5}{7}}$ into $\frac{\frac{6}{11}}{\frac{2}{5}}$.

61. Multiply $\frac{2\frac{1}{2}}{\frac{5}{12}}$ by $\frac{2\frac{1}{3}}{1\frac{3}{4}}$ into $\frac{1\frac{1}{5}}{1\frac{4}{5}}$.

241. To divide one complex fraction by another.

Reduce the complex fractions to simple ones, then proceed as in division of simple fractions. (Arts. 229, 239.)

62. Divide $\frac{4\frac{1}{2}}{2\frac{1}{4}}$ by $\frac{\frac{1}{3}}{1\frac{3}{4}}$.

Solution.—$\frac{4\frac{1}{2}}{2\frac{1}{4}}=\frac{9}{2}\times\frac{4}{9}=\frac{36}{18}$, and $\frac{\frac{1}{3}}{1\frac{3}{4}}=\frac{1}{3}\times\frac{4}{7}=\frac{4}{21}$. (Art. 239.)

Now, $\frac{36}{18}\div\frac{4}{21}=\frac{36}{18}\times\frac{21}{4}=\frac{756}{72}=10\frac{1}{2}$. *Ans.*

Or, since the given dividend$=\frac{9\times4}{2\times9}$ and the divisor$=\frac{1\times4}{3\times7}$;

then $\frac{9\times4}{2\times9}\times\frac{3\times7}{1\times4}=$the answer. (Art. 231.)

But, (Art. 232,) $\frac{9\times4}{2\times9}\times\frac{3\times7}{1\times4}=\frac{9\times4\times3\times7}{2\times9\times1\times4}=\frac{21}{2}$, or $10\frac{1}{2}$ *Ans.*

63. Divide $\frac{\frac{1}{2}}{\frac{3}{4}}$ by $\frac{\frac{1}{3}}{\frac{1}{4}}$.

64. Divide $\frac{1\frac{3}{4}}{4\frac{1}{2}}$ by $\frac{2\frac{1}{3}}{2\frac{1}{4}}$.

APPLICATION OF FRACTIONS.

242. Ex. 1. A merchant bought $15\frac{7}{8}$ yards of domestic flannel of one customer, $19\frac{1}{4}$ of another, $12\frac{1}{2}$ of another, and $41\frac{3}{16}$ of another: how many yards did he buy of all?

2. A grocer sold $16\frac{1}{4}$ lbs. of sugar to one customer, $112\frac{1}{2}$ to another, and $33\frac{1}{3}$ to another: how many pounds did he sell?

3. A clerk spent $26\frac{3}{8}$ dollars for a coat, $9\frac{3}{4}$ dollars for pants, $6\frac{3}{4}$ dollars for a vest, $5\frac{1}{2}$ dollars for a hat, and $6\frac{1}{4}$ dollars for a pair of boots: how much did his suit cost him?

4. A man having bought a bill of goods amounting to $85\frac{3}{14}$ dollars, handed the clerk a bank note of 100 dollars: how much change ought he to receive back?

5. A lady went a shopping with $135\frac{1}{4}$ dollars in her purse; she paid $17\frac{1}{16}$ dollars for silk, $3\frac{5}{8}$ dollars for trimmings, $37\frac{1}{2}$ dollars for a shawl, and $14\frac{3}{8}$ dollars for a muff: how much money had she left?

6. A man having $1563\frac{5}{16}$ dollars, spent $365\frac{3}{8}$ dollars, and lost $562\frac{1}{2}$ dollars: how much had he left?

7. What will 563 sheep cost, at $2\frac{3}{4}$ dollars per head?

8. What cost 748 barrels of flour, at $7\frac{3}{8}$ dollars per barrel?

9. What cost $378\frac{3}{4}$ yards of cloth, at 4 dollars per yard?

10. What cost $1121\frac{5}{16}$ lbs. of tea, at 5 shillings per pound?

11. What cost 430 gallons of oil, at $1\frac{1}{8}$ dollar per gallon?

12. What cost $\frac{4}{6}\frac{2}{3}$ of an acre of land, at 150 dollars per acre?

13. A man worth 25000 dollars, lost $\frac{2}{4}\frac{2}{5}$ of it by fire: what was the amount of his loss?

14. A garrison had 856485 pounds of flour; after being blockaded 60 days, it was found that $\frac{1}{2}\frac{2}{3}\frac{4}{5}$ of it were consumed: how many pounds of flour were left?

15. At $17\frac{1}{2}$ dollars per ton, what cost $103\frac{1}{4}$ tons of hay?

16. How many bushels of corn will $115\frac{3}{5}$ acres produce, at $31\frac{1}{4}$ bushels per acre?

17. What cost $675\frac{1}{3}$ tons of iron, at $45\frac{6}{7}$ dollars per ton?

18 If a ship sails $140\frac{3}{15}$ miles per day, how far will she sail in $49\frac{1}{2}$ days?

19. If a Railroad car should run $41\frac{1}{2}$ miles per hour, how far would it go in 12 days, running $10\frac{1}{2}$ hours per day?

20. A young man having a patrimony of 12234 dollars, spent $\frac{1}{4}$ of it in dissipation: how much had he left?

21. At $\frac{3}{4}$ of a dollar per yard, how many yards of satinet can be bought for 124 dollars?

22. How many pounds of tea, at $\frac{3}{5}$ of a dollar a pound can you buy for 131 dollars?

23. How many gallons of molasses, at $\frac{3}{8}$ of a dollar per gallon can you buy for 235 dollars?

24. At 8 pence a pound, how many pounds of sugar can you buy for $163\frac{1}{2}$ pence?

25. At $5\frac{1}{4}$ pence a yard, how many yards of lace can be bought for 279 pence?

26. A dairy-man has $229\frac{1}{2}$ pounds of butter which he wishes to pack in boxes containing $8\frac{1}{2}$ pounds each: how many boxes will it require?

27. A farmer wishes to put 384 bushels of apples into barrels, each containing $2\frac{1}{2}$ bushels: how many barrels will it require?

28. If $4\frac{3}{4}$ yards of cloth make a suit of clothes, how many suits will $141\frac{1}{2}$ yards make?

29. One rod contains $5\frac{1}{2}$ yards: how many rods are there in 210 yards?

30. A merchant paid $204\frac{5}{8}$ dollars for 57 yards of cloth: how much was that per yard?

31. A grocer sold 50 barrels of flour for $311\frac{1}{9}$ dollars: what did he get per barrel?

32. A merchant wishes to lay out $657\frac{1}{2}$ dollars for wheat, which is worth $1\frac{1}{8}$ of a dollar a bushel: how much can he buy?

33. At $18\frac{3}{4}$ cents a dozen, how many dozen of eggs can you buy for $87\frac{1}{2}$ cents?

34. A grocer sold $15\frac{1}{2}$ pounds of coffee for $93\frac{3}{4}$ cents: how much was that a pound?

35. A shopkeeper sold $16\frac{1}{2}$ yards of satin for $163\frac{7}{12}$ shillings: how much was that per yard?

36. Bought 19 sacks of wool for $250\frac{3}{8}$ dollars: what was that per sack?

37. Paid $575\frac{3}{4}$ dollars for $96\frac{7}{8}$ yards of cloth: what was the cost per yard?

38. Paid $1565\frac{1}{6}$ dollars for iron, valued at $37\frac{1}{4}$ dollars per ton: how many tons were bought?

39. Paid $1315\frac{5}{8}$ dollars for the transportation of 1286 barrels of pork: what was that per barrel?

40. Bought $375\frac{1}{2}$ pounds of indigo for $652\frac{3}{4}$ dollars: what was the cost per pound?

41. Paid $1679\frac{1}{2}$ dollars for 475 kegs of lard: how much was that per keg?

42. If an army consumes $563\frac{7}{8}$ pounds of meat per day, how long will 150000 pounds supply it?

43. The cost of making $25\frac{1}{3}$ miles of Railroad was $856235\frac{1}{2}$ dollars: what was the cost per mile?

SECTION VIII.

COMPOUND NUMBERS.

ART. **243.** Numbers which express things of the *same* kind or denomination, are called *Simple Numbers.* Thus, 3 oranges, 7 books, 12 chairs, &c., are simple numbers.

Numbers which express things of different kinds or denominations, as the divisions of money, weight, and measure, are called COMPOUND NUMBERS. Thus, 15 shillings 6 pence; 10 bushels 3 pecks, &c., are compound numbers.

OBS. The origin of Compound Numbers is ascribed to the wants and necessities of the earlier ages of the world. Their divisions and subdivisions are generally irregular, and seem to have been suggested by the caprice, or the limited business transactions of the rude ages of antiquity. It is much to be regretted, both on account of simplicity and their adaptation to scientific purposes, that their different denominations were not graduated according to the *law of increase* in the *decimal* notation.

Note.—Compound Numbers, by some authors, are called Denominate Numbers

FEDERAL MONEY.

244. *Federal Money* is the currency of the *United States.* The denominations are, *Eagles, Dollars, Dimes, Cents,* and *Mills.*

10 mills (*m.*)	make 1 cent,	marked	*ct.*
10 cents	" 1 dime,	"	*d.*
10 dimes	" 1 dollar,	"	*doll.* or $.
10 dollars	" 1 eagle,	"	*E.*

OBS. 1. Federal money was established by Congress, Aug. 8, 1786. It is based upon the principles of the *decimal notation.* The *law* of *increase* or *radix*, is the same as that of simple numbers, and it is confessedly one of the most *simple* and *comprehensive* systems of currency in the civilized world. Previous to its adoption, English or sterling money was the principal currency of the country.

QUEST.—243. What are simple numbers? What are compound numbers? 244. What is Federal money? Recite the table. *Obs.* When and by whom was it established?

2. The names of the coins or denominations less than a dollar, are significant of their value. The term *dime*, is derived from the French *disme*, which signifies *ten;* the terms *cent* and *mill*, are from the Latin *centum* and *mille*, the former of which signifies a *hundred*, and the latter a *thousand*. Thus 10 dimes, 100 cents, or 1000 mills, make 1 dollar.

3. The sign ($), which is prefixed to Federal money, is called the *Dollar mark*. It is said to be a contraction of "U. S.," the initials of *United States*, which were originally prefixed to sums of money expressed in the Federal currency. At length the two letters were moulded or merged into a single character by dropping the curve of the *U*, and writing the *S* over it. Thus, the sum of seventy-five dollars, which was originally written "U. S. 75 dollars," is now written $75.

245. The national coins of the United States are of three kinds; viz: gold, silver, and copper.

1. The gold coins are the *eagle*, the *double eagle*,* *half eagle*, *quarter eagle*, and *gold dollar*.*

The eagle contains 258 grains of *standard* gold; the half eagle and quarter eagle like proportions.†

2. The silver coins are the *dollar*, *half dollar*, *quarter dollar*, the *dime*, and *half dime*.

The dollar contains 412½ grains of *standard* silver; the others like proportions.†

3. The copper coins are the *cent*, and *half cent*.

The cent contains 168 grains of *pure* copper; the half cent, a like proportion.† *Mills* are not *coined*.

OBS. The fineness of gold used for coin, jewelry, and other purposes, also the gold of commerce, is estimated by the number of parts of gold which it contains. Pure gold is commonly supposed to be divided into 24 equal parts, called *carats*. Hence, if it contains 10 parts of *alloy*, or some *baser* metal, it is said to be 14 carats fine; if 5 parts of alloy, 19 carats fine; and when absolutely pure, it is 24 carats fine.

246. The *present standard* for both *gold* and *silver coin* of the United States, by Act of Congress, 1837, is 900 parts of pure

QUEST.—245. Of how many kinds are the coins of the United States? What are they? What are the gold coins? The silver coins? The copper? *Obs.* How is the fineness of gold estimated? Into how many carats is pure gold supposed to be divided? When it contains 10 parts of alloy, how fine is it said to be? 5 parts of alloy? 246. What is the present standard for the gold and silver coin of the United States? What is the alloy of gold coin? What of silver coin?

* Added by Act of Congress, 1849.

† According to Act of Congress, 1837.

metal by weight to 100 parts of alloy. The alloy of gold coin is composed of silver and copper, the silver not to exceed the copper in weight. The alloy of silver coin is pure copper.

Note.—The *original* standard for the gold coin of the United States by Act of Congress, 1792, was 22 parts of pure gold to 2 parts of alloy; the alloy consisting of 1 part silver and 1 part copper.

The *original* standard for the silver coin was 1489 parts of pure silver to 179 parts of alloy; the alloy being of pure copper.

The *eagle* by the same act contained 270 grains of *standard* gold. The *dollar* contained 416 grains of *standard* silver. The *cent* contained 11 pennyweights, or 264 grains of *pure* copper.

STERLING MONEY.

247. *English* or *Sterling Money* is the national currency of *Great Britain.*

4 farthings (*qr.* or *far.*)	make	1 penny, marked	*d.*
12 pence	"	1 shilling, "	*s.*
20 shillings	"	1 pound, or sovereign,	£.
21 shillings	"	1 guinea.	

OBS. 1. It is customary, at the present day, to express farthings in fractions of a penny. Thus, 1 qr. is written ¼ d.; 2 qrs. ½ d.; 3 qrs. ¾ d.

2. The Pound Sterling is represented by a gold coin, called a *Sovereign*. According to *Act of Congress*, 1842, its value is 4 *dollars* and 84 *cents*. Hence, the value of a shilling is $24\frac{1}{5}$ cents; that of a penny 2 cents, very nearly.

3. The letters £. *s. d.* and *q.* are the initials of the Latin words, *libra, solidus, denarius,* and *quadrans,* which respectively signifiy a *pound, shilling, penny,* and *farthing* or *quarter.* The mark /, which is often placed between shillings and pence, is a corruption of the long *ſ*.

Note.—1. Sterling money is supposed by some to have received its name from the *Easterlings*, who it is said first coined it; others think it is so called to distinguish it from *stocks*, &c., whose value is *nominal.*

2. The *pound* is so called, because in ancient times the silver for it weighed a pound Troy. A pound Troy of silver is now worth 66 shillings, or £3, 6s.

The *Guinea* is so called, because the gold of which it was originally made, was brought from Guinea, on the coast of Africa.

248. The following denominations are frequently met with, viz: the Groat=4*d.*; the Crown=5*s.*; the Noble=6*s.* 8*d.*;

QUEST.—247. What is Sterling Money? Repeat the Table? *Obs.* How are farthings usually expressed? How is a pound sterling represented? What is its value in dollars and cents?

the Angel=10*s.*; the Mark=13*s.* 4*d.*; the Pistole=16*s.* 10*d.*; the Moidore=27*s.*

OBS. The *present* standard gold coin of Great Britain, consists of 22 parts *pure gold*, and 2 parts of *copper*.*

The weight of a Sovereign or £, is 5 pwts., $3\frac{171}{623}$ grains.

The *standard silver* coin consists of 37 parts of *pure* silver, and 3 parts of copper. The weight of a shilling is 3 pwts. $15\frac{3}{11}$ grs.

In *copper* coin, 24 pence weigh 1 pound avoirdupois.

TROY WEIGHT.

249. *Troy Weight* is used in weighing gold, silver, jewels, liquors, &c., and is generally adopted in philosophical experiments.

24 grains (*gr.*)	make 1 pennyweight,	marked	*pwt.*
20 pennyweights	" 1 ounce	"	*oz.*
12 ounces	" 1 pound,	"	*lb.*

OBS. 1. The abbreviation *oz.*, is derived from the Spanish *onza*, which signifies an *ounce*.

2. The *standard* of Weights and Measures is different in different countries, and in different States of the Union. In 1834, the Government of the United States adopted a uniform standard, for the use of the several Custom-houses and other purposes.

250. The *standard Unit* of *Weight* adopted by the Government, is the *Troy Pound* of the United States Mint. It is equal to 22.794422 cubic inches of distilled water, at its maximum density,† the barometer standing at 30 inches, and is identical with the Imperial *Troy* pound of Great Britain, established by Act of Parliament, in 1826.‡

OBS. The weights and measures in present use, were derived from very *imperfect* and *variable* standards. A *grain* of wheat, taken from the middle of the ear or head, and being thoroughly dried, was the original element of all weights used in England; it was thence called a *grain*. At first, a weight

QUEST.—249. In what is Troy Weight used? Repeat the Table? *Obs.* Do all the States have the same standard of weights and measures? 250. What is the standard unit of weight adopted by the Government of the United States? *Note.* When was Troy Weight introduced into Europe? From what was its name derived?

* Hind's Arithmetic; also, Hutton's Mathematics.

† The maximum density of water, according to Mr. Hassler, is at the temperature of 39·83 deg. Fahrenheit.

‡ The *Troy pound of the U. S. Mint*, is an exact copy, by Captain Kater, of the British Imperial Troy pound. Report of the Secretary of the Treasury, March 3, 1831

equal to 32 grains, was called a *pennyweight*, from its being the weight of the silver *penny* then in circulation. At a later period the *pennyweight* was divided into 24 equal parts instead of 32, which are still called grains, being the smallest weight now in common use.

Note.—Troy Weight was formerly used in weighing articles of every kind. It was introduced into Europe from Cairo in Egypt, about the time of the Crusades, in the 12th century. Some suppose its name was derived from *Troyes*, a city in France, which first adopted it; others think it was derived from *Troy-novant*, the former name of London.*

AVOIRDUPOIS WEIGHT.

251. *Avoirdupois Weight* is used in weighing groceries and all coarse articles; as, sugar, tea, coffee, butter, cheese, flour, hay, &c., and all metals except gold and silver.

16 drams (*dr.*)	make	1 ounce,	marked	*oz.*
16 ounces	"	1 pound,	"	*lb.*
25 pounds	"	1 quarter,	"	*qr.*
4 quarters, or 100 lbs.	"	1 hundred weight,	"	*cwt.*
20 hundred weight	"	1 ton,	"	*T.*

Note.—In weighing wool in England, 7 pounds make 1 clove; 2 cloves, 1 stone; 2 stone, 1 tod; 6½ tods, 1 wey; 2 weys, 1 sack; 12 sacks, 1 last; 240 pounds, 1 pack.

OBS. 1. Formerly it was the custom to allow 112 pounds for a hundred weight, and 28 pounds for a quarter; but this practice has become nearly or quite obsolete. In buying and selling all articles of commerce estimated by weight, the laws of most of the States as well as general usage, call 100 pounds a hundred weight, and 25 pounds a quarter.

2. *Gross* weight is the weight of goods with the boxes, casks, or bags which contain them.

Net weight is the weight of the goods only.

252. The *Avoirdupois Pound* of the *United States*, is equal to 27.7554 cubic inches of distilled water, at the maximum density, and at 30 inches barometer.† It is determined from the Troy Pound, by the legal proportions of 5760 grains, which con-

QUEST.—251. In what Avoirdupois Weight used? Repeat the Table? *Obs.* How many pounds were formerly allowed for a hundred weight? For a quarter? What is gross weight Net weight? 252. How is the Avoirdupois pound of the United States determined?

* Hind's Arithmetic, Art. 224. Also, North American Review, Vol. XLV.

† Reports of Secretary of Treasury, March 3, 1832: June 30, 1832. Also, Congressional Documents of 1833.

stitute the Troy pound, to 7000 grains Troy, which constitute the Avoirdupois pound. That is,

5760 grains	Troy	make 1 pound Troy.
7000 grains	"	" 1 pound Avoirdupois.
$437\frac{1}{2}$ grains	"	" 1 ounce "
$27\frac{11}{32}$ grains	"	" 1 dram "

OBS. 1. The *British Imperial* Pound Avoirdupois is equal to 27·7274 cubic inches of distilled water, at the temperature of 62° Fahrenheit, when the barometer stands at 30°. It is determined from the Imperial Troy pound, which contains 5760 grains, while the former contains 7000 grains.

2. Since the Troy pound of the United States is identical with the Troy pound of England, the Avoirdupois pound of the former must be equal to that of the latter; for both bear the same ratio to the Troy pound. But the English avoirdupois pound is said to contain 27.7274 cu. in. of distilled water, while that of the United States, according to Mr. Hassler, contains 27.701554 cu. in. This slight difference may be accounted for by the fact that the former was measured at the temperature of 62°, while the latter was measured at its maximum density, which is 39.83 degrees.

3. The *standard of weight* adopted by the State of New York, in 1827, is the *avoirdupois pound*, whose magnitude is such that a cubic foot of distilled water, at the maximum density, in a vacuum, will weigh $62\frac{1}{2}$ pounds, or 1000 ounces.

Note.—The term *avoirdupois*, is thought by some to be derived from the French *avoir du poids*, a phrase signifying to have weight. Others think it is from *avoirs*, the ancient name of *goods* or *chattels*, and *poids* signifying *weight* in the Norman dialect.*

APOTHECARIES' WEIGHT.

253. *Apothecaries' Weight* is used by apothecaries and physicians in *mixing* medicines.

20 grains (*gr.*)	make 1 scruple,	marked *sc.*, or ℈.
3 scruples	" 1 dram,	" *dr.*, or ʒ.
8 drams	" 1 ounce,	" *oz.*, or ℥.
12 ounces	" 1 pound,	" ℔.

OBS. 1. The *pound* and *ounce* in this weight are the same, as the *Troy* pound and ounce; the other denominations are different.

2. Drugs and medicines are bought and sold by *avoirdupois* weight.

QUEST.—253. In what is Apothecaries' Weight used? Recite the Table? *Obs.* To what are the apothecaries' ounce and pound equal? How are drugs and medicines bought and sold?

* President John Quincy Adams on Weights and Measures; also, Hind's Arithmetic.

LONG MEASURE.

254. *Long Measure* is used in measuring distances where *length* only is considered, without regard to breadth or depth. It is frequently called *linear* or *lineal* measure.

12 inches (*in.*)	make	1 foot,	marked	*ft.*
3 feet	"	1 yard,	"	*yd.*
5½ yards, or 16½ feet	"	1 rod, perch, or pole,	"	*r.* or *p.*
40 rods	"	1 furlong,	"	*fur.*
8 furlongs, or 320 rods	"	1 mile,	"	*m.*
3 miles	"	1 league,	"	*l.*
60 geographical miles, or 69½ statute miles	"	1 degree,	"	*deg.* or °.

360 degrees make a great circle, or the circumference of the earth.

Note.—4 inches make 1 hand; 9 inches, 1 span; 18 inches, 1 cubit; 6 feet, 1 fathom.

In measuring roads and land, surveyors use a chain which is 4 rods long, and which is divided into 100 links. Hence, 25 links make 1 rod, and $7\frac{92}{100}$ inches make 1 link. This chain is commonly called *Gunter's Chain*, from the name of its inventor.

OBS. 1. The inch is commonly divided either into *eighths* or *tenths;* sometimes, however, it is divided into *twelfths*, which are called *lines.* Formerly the inch was divided into 3 *barleycorns;* but the barleycorn is not employed as a measure at the present day. The term *barleycorn*, is derived from a *grain* of barley, which was the original element of Linear Measure.

2. The terms *rod*, *pole*, and *perch*, from the French *perche* signifying a rod, are each expressive of the instrument, which was originally used as a measure of this length.

255. The *standard Unit* of *Length* adopted by the *United States*, is the *Yard* of 3 feet, or 36 inches, and is identical with the British Imperial Yard. It is made of brass, at the temperature of 62° Fahrenheit, from the scale of eighty-two inches prepared by Troughton, a celebrated English artist, for the survey of the Coast of the United States.

OBS. 1. The *Imperial standard yard of Great Britain* is determined from the *pendulum* which vibrates seconds in a vacuum, at the level of the sea, in

QUEST.—254. In what is Long Measure used? What is Long Measure sometimes called? Recite the Table? *Obs.* How are inches usually divided? What is the origin of the measure called barleycorn? Is this measure now used? 255. What is the standard unit of Length adopted by the United States?

Greenwich or London. This pendulum is divided into 391393 equal parts, and 360000 of these parts are declared, by *act of Parliament*, to be the standard yard, at the temperature of 62°; consequently, since the yard is divided into 36 inches, it follows that the length of a pendulum *vibrating seconds*, under these circumstances, is 39.1393 inches.

The *English yard* is said to have been originally determined by the length of the arm of Henry I. King of England.

2. The *standard* of linear measure adopted by the *State* of *New York*, is the *pendulum* which vibrates seconds, in a vacuum, at Columbia College, in the city of New York, which is in the latitude of 40° 42′, 43″. The yard is declared to be $\frac{1000000}{1086141}$ of this pendulum; hence, the length of the pendulum is 39.101688 inches, at the temperature of 32°. Should the standard yard ever be lost, it could be recovered by resorting to the preceding experiment.

CLOTH MEASURE.

256. *Cloth Measure* is used in measuring cloth, lace, and all kinds of goods, which are bought and sold by the yard.

2¼ inches (*in.*)	make	1 nail,	marked	*na.*
4 nails, or 9 in.	"	1 quarter of a yard,	"	*qr.*
4 quarters	"	1 yard,	"	*yd.*
3 quarters, or ¾ of a yard	"	1 Flemish ell,	"	*Fl. e.*
5 quarters, or 1¼ yard	"	1 English ell,	"	*E. e*
6 quarters, or 1½ yard	"	1 French ell,	"	*F. e.*

OBS. *Cloth* measure is a species of *long* measure. Cloths, laces, &c., are bought and sold by the *linear* yard, without regard to their width.

SQUARE MEASURE.

257. *Square Measure* is used in measuring surfaces, or things whose *length* and *breadth* are considered without regard to height or depth; as, land, flooring, plastering, &c.

144 square inches (*sq. in.*)	make	1 square foot,	marked	*sq. ft.*
9 square feet	"	1 square yard,	"	*sq. yd.*
30¼ square yards, or 272¼ square feet	"	1 sq. rod, perch, or pole,		*sq. r.*
40 square rods	"	1 rood,	"	*R.*
4 roods, or 160 square rods	"	1 acre,	"	*A.*
640 acres	"	1 square mile,	"	*M.*

QUEST.—256. In what is cloth measure used? Repeat the Table. *Obs.* Of what is cloth measure a species? What is the kind of yard by which cloths, laces, &c., are bought and sold? 257. In what is Square Measure used? Recite the Table.

Note.—16 square rods make 1 square chain; 10 square chains, or 100,000 square links, make an acre. Flooring, roofing, plastering, &c., are frequently estimated by the "square," which contains 100 square feet.

A *hide* of land, which is spoken of by ancient writers, is 100 acres.

OBS. 1. A *square* is a figure which has *four equal* sides, and all its angles *right angles*, as seen in the diagram. Hence,

9 sq. ft.=1 *sq. yd.*

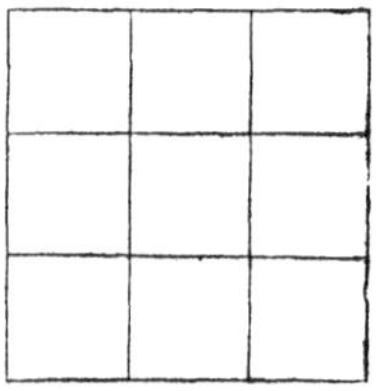

A *Square Inch* is a square, whose sides are each a *linear* inch in length.

A *Square Foot* is a square, whose sides are each a *linear* foot in length.

A *Square Yard* is a square, whose sides are each a *linear* yard, or three *linear* feet in length, and contains 9 *square feet*, as represented in the adjacent figure.

2. Square measure is so called, because its *measuring unit* is a *square*. The *standard* of *square* measure is derived from the standard *linear* measure. Hence,

A *unit* of square measure is a *square* whose sides are respectively equal, in length, to the *linear* unit of the same name.

CUBIC MEASURE.

258. *Cubic Measure* is used in measuring solid bodies, or things which have *length, breadth,* and *thickness;* such as timber stone, boxes of goods, the capacity of rooms, ships, &c.

1728 cubic inches (*cu. in.*)	make	1 cubic foot,	marked	*cu. ft.*
27 cubic feet	"	1 cubic yard,	"	*cu. yd.*
40 feet of round, or 50 ft. of hewn timber	"	1 ton, or load,	"	*T.*
42 cubic feet	"	1 ton of shipping,	"	*T.*
16 cubic feet	"	1 foot of wood, or a cord foot,	"	*c. ft.*
8 cord feet, or 128 cubic feet	"	1 cord,	"	*C.*

A pile of wood 8 feet long, 4 feet wide, and 4 feet high, contains 1 cord For, $8 \times 4 \times 4 = 128$.

QUEST.—*Obs.* What is a square? What is a square inch? A square foot? A square yard? 258. In what is Cubic Measure used? Recite the Table.

OBS. 1. A *Cube* is a solid body bounded by *six equal sides*. It is often called a *hexaedron*. Hence,

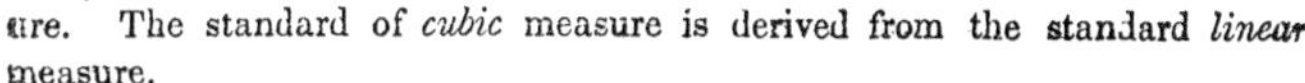

A *Cubic Inch* is a cube, each of whose sides is a *square* inch, as represented by the adjoining figure.

A *Cubic Foot* is a cube, each of whose sides is a *square* foot.

2. Cubic Measure is so called, because its *measuring unit* is a *cube*. It is often called *solid* measure. The standard of *cubic* measure is derived from the standard *linear* measure.

A *unit* of cubic measure, therefore, is a *cube* whose sides are respectively equal in length to the *linear* unit of the same name.

3. The *cubic ton*, sometimes called a *load*, is chiefly used for estimating the cartage and transportation of timber. By a *ton* of *round* timber is meant, such a quantity of timber in its rough or natural state, as when hewn, will make 40 cubic feet, and is supposed to be equal in weight to 50 feet of hewn timber.

The cubic *ton* or load, is by no means an *accurate* or *uniform* standard of estimating weight; for, different kinds of timber, are of very different degrees of density. But it is perhaps sufficiently accurate for the purposes to which it is applied.

Note.—For an easy method of forming *models* of the *Cube* and other regular *Solids*, see Thomson's Legendre's Geometry, p. 220.

WINE MEASURE.

259. *Wine Measure* is used in measuring wine, alcohol, molasses, oil, and all other liquids except beer, ale, and milk.

4 gills (*gi.*)	make	1 pint,	marked	*pt.*
2 pints	"	1 quart,	"	*qt.*
4 quarts	"	1 gallon,	"	*gal.*
31½ gallons	"	1 barrel,	"	*bar.* or *bbl.*
42 gallons	"	1 tierce	"	*tier.*
63 gallons, or 2 barrels	"	1 hogshead,	"	*hhd.*
2 hogsheads	"	1 pipe or butt,	"	*pi.*
2 pipes	"	1 tun,	"	*tun.*

OBS. 1. In England, 10 gallons make 1 anker; 18 gallons, 1 runlet; 2 tierces, or 84 gallons, 1 puncheon.

2. Liquids are generally bought and sold by the *gallon* or its *subdivisions*, as

---

QUEST.—*Obs.* What is a cube? What is a cubic inch? A cubic foot? What is meant by a ton of round timber? 259. In what is Wine Measure used? Recite the Table.

the quart, pint, &c. Cider and a few cheap articles are bought and sold by the *barrel.* The capacities of cisterns, vats, &c., are sometimes estimated in *hogsheads,* and the quotations or prices-current of oils in foreign markets, are usually made in *tuns.* But the *tierce,* and the *pipe* or *butt* are never used, as such, in business transactions; their contents are given in gallons, quarts, &c.

260. The *standard Unit of Liquid Measure* adopted by the *United States,* is the *Wine Gallon* of 231 cubic inches, which is equal to 58372.1754 grains of distilled water, at the maximum density, weighed in air at 30 inches barometer, or 8.339 lbs. avoirdupois, very nearly.*

OBS. The *British imperial standard measure* of capacity, both for *liquids* and *dry goods,* is the *imperial gallon,* which is equal to 10 pounds avoirdupois of distilled water, at 62° thermometer and 30 inches barometer, and contains 277.274 cubic inches. It is equal to 1.2 gal. wine measure U. S.

BEER MEASURE.

261. *Beer Measure* is used in measuring beer, ale, and milk.

2 pints (*pts.*)	make	1 quart,	marked	*qt.*
4 quarts	"	1 gallon,	"	*gal.*
36 gallons	"	1 barrel,	"	*bar.* or *bbl.*
1½ barrels, or 54 gallons	"	1 hogshead,	"	*hhd.*

OBS. 1. In England, 9 gallons make 1 firkin; 2 firkins, 1 kilderkin; 2 kilderkins, 1 barrel.

2. The *beer* gallon contains 282 cubic inches, and is equal to 10.179932? pounds avoirdupois of distilled water, at the maximum density. In many places milk is measured by wine measure.

DRY MEASURE.

262. *Dry Measure* is used in measuring grain, fruit, &c.

2 pints (*pt.*)	make	1 quart,	marked	*qt.*
8 quarts	"	1 peck,	"	*pk.*
4 pecks, or 32 qts.	"	1 bushel,	"	*bu.*
8 bushels	"	1 quarter,	"	*qr.*
32 bushels, or 4 qrs.	"	1 chaldron,	"	*ch.*

QUEST.—260. What is the standard unit of Liquid Measure of the United States? How many cubic inches in a wine gallon? 261. In what is Beer Measure used? Recite the Table. 262. In what is Dry Measure used? Repeat the Table.

* Reports of the Secretary of the Treasury, March, 1831, and June, 1832. Also, Hassler on Weights and Measures.

OBS. In England flour is often sold by weight. A sack is equal to 280 lbs., and contains about five imperial bushels.

The following denominations, are sometimes used, viz: 2 quarts make 1 pottle; 2 bushels, 1 strike; 2 strikes or 4 bu., 1 coom; 2 cooms or 8 bu., 1 quarter; 5 quarters, 1 wey or load; 2 loads, 1 last.

In London 36 bushels of coal make a chaldron, but in New Castle 79½ bushels are said to be allowed for a chaldron. But coal in England and in this country, is now usually bought and sold by weight.

Note.—Wine, Beer, and *Dry* Measures are often called *capacity* measures, and are evidently a species of *cubic* measure.

263. The *standard Unit of Dry Measure* adopted by the *United States,* is the *Winchester Bushel,* which is equal to 77.627413 pounds avoirdupois of distilled water, at the maximum density, weighed in air at 30 inches barometer, and contains 2150.4 cubic inches, nearly.

The Winchester bushel is so called, because the standard measure was formerly kept at *Winchester,* England. By statute, it is an upright cylinder, 18½ inches in diameter, and 8 inches deep.

OBS. 1. The *imperial bushel* of Great Britain is equal to 80 lbs. avoirdupois of distilled water, at 62° Fahrenheit, and 30 inches barometer, and contains 2218.192 cubic inches; consequently, it is equal to 1.032 bushel U. S., nearly. It is an upright cylinder, whose internal diameter is 18.789 inches, and its depth 8 inches.

The use of *heaped measure* was abolished by Act of Parliament, in 1835.

2. The *standard bushel* of the State of New York, is equal to 80 pounds avoirdupois of distilled water, at the maximum density, at the mean pressure of the atmosphere, and contains 2218.192 cubic inches.*

It is customary, at the present day, to determine *capacity measures* by the weight of distilled water which they contain. This is evidently more accurate than the former method of measurement by cubic inches.

3. In buying and selling grain, when no special agreement as to measurement or weight, is made by the parties, a bushel, in the State of New York, by Act of 1836, consists of 60 lbs. of wheat, 56 lbs. rye or Indian corn, 48 lbs. of barley, and 32 lbs. of oats.

There are similar statutes in most of the other States of the Union. This is the most impartial method by which the value of grain can be estimated.

QUEST.—263. What is the standard unit of Dry Measure adopted by the Government?

* By the same Act it was declared, that the standard *liquid gallon* should be 8 lbs., and the standard *dry gallon* 10 lbs. avoirdupois of distilled water, at its maximum density But this part of the statute was subsequently repealed, and the previous standard gallon in the office of the Secretary of State, was continued in use

TIME.

264. *Time* is naturally divided into *days* and *years;* the former are caused by the revolution of the Earth on its axis, the latter by its revolution round the sun.

60 seconds (*sec.*)	make	1 minute,	marked	*min.*
60 minutes	"	1 hour,	"	*hr.*
24 hours	"	1 day,	"	*d.*
7 days	"	1 week,	"	*wk.*
4 weeks	"	1 lunar month,	"	*mo.*
12 calendar months, or 365 days and 6 hrs., (nearly,)		1 civil year,	"	*yr.*

The following are the names of the 12 calendar months into which the civil or legal year is divided, with the number of days in each.

January,	written	(Jan.)	the	*first*	month,	has	31	days
February,	"	(Feb.)	"	*second*	"	"	28	"
March,	"	(Mar.)	"	*third*	"	"	31	"
April,	"	(Apr.)	"	*fourth*	"	"	30	"
May,	"	(May)	"	*fifth*	"	"	31	"
June,	"	(June)	"	*sixth*	"	"	30	"
July,	"	(July)	"	*seventh*	"	"	31	"
August,	"	(Aug.)	"	*eighth*	"	"	31	"
September,	"	(Sept.)	"	*ninth*	"	"	30	"
October,	"	(Oct.)	"	*tenth*	"	"	31	"
November,	"	(Nov.)	"	*eleventh*	"	"	30	"
December,	"	(Dec.)	"	*twelfth*	"	"	31	"

The number of days in each month may be easily remembered from the following lines:

" Thirty days hath September,
April, June, and November;
February twenty-eight alone,
All the rest have thirty-one;
Except in Leap year, then is the time,
When February has twenty-nine."

OBS. 1. A *Solar* year is the exact time in which the earth revolves round the sun, and contains 365 days, 5 hours, 48 minutes, and 48 seconds.

2. Since the civil year contains 365 days and 6 hours, (nearly,) it is plain that in four years a whole day will be gained, and therefore every fourth year must have 366 days. This day was originally added to the year, by *repeating* the *sixth* of the *Calends of March* in the Roman calendar, which corresponds with

QUEST.—264. How is time naturally divided? Recite the Table. *Obs.* What is a solar year? How is leap year occasioned? To which month is the odd day added?

the 24th of *February* in ours. It was called the *intercalary* day, from the Latin *intercalo*, to *insert*.

The year in which this day is added, is called *Bissextile*, from the Latin *bis*, *twice*, and *sextilis*, the *sixth*. It is also called "*Leap Year*," because it leaps over a day more than a common year.

3. The *civil* or *legal* year is often called the *Julian* year, from Julius Cæsar emperor of Rome, who adapted the *calendar* or *register* of the *civil* year to the *supposed* length of the *solar* year, by adding 1 day to every *fourth* year.

265. In process of time, as mathematical and astronomical science advanced, it was found that the length of a solar year was only 365 d. 5 hrs. 48 min. 48 sec., or 11 min. 12 sec. less than 365¼ days, which in 400 years amounted to about 3 days; consequently, the Julian calendar was behind the solar time. This error at the time of Pope Gregory XIII., amounted to 10 days, which he corrected in 1582 by suppressing 10 days in the month of October, the day after the 4th being called the 15th. Hence this calendar is sometimes called the *Gregorian calendar*.

OBS. 1. This correction was not adopted in England till 1752, when the error amounted to 11 days. By *Act of Parliament*, 11 days, after the 2d of September, were therefore omitted; and the *civil* year by the same Act, was made to commence on the 1st of January, instead of the 25th of March, as it had done previously.

2. Dates reckoned by the *old method* or Julian calendar, are called *Old Style;* and those reckoned by the *new method*, are called *New Style*.

To change any date from *Old* to *New Style*, we must add 11 days to it; and if the given date in *Old Style*, is between the 1st of January and the 25th of March, we must add 1 to the year in *New Style*.

Russia still reckons dates according to *Old Style*. The *difference* now amounts to 12 days.

266. To ascertain whether a year is LEAP YEAR.

Divide the given year by 4, and if there is no remainder, it is Leap year. The remainder, if any, shows how many years have elapsed since a Leap year occurred. Thus, dividing the year 1847 by 4, the remainder is 3; hence it is 3 years since the last leap year, and the ensuing year will be leap year.

OBS. 1. To this rule there is an exception. For, we have seen, that a *solar* year is 11 min. and 12 sec. less than a Julian year, which is 365¼ days. This error, in 400 years, amounts to about 3 days; consequently, if 1 day is added

---

QUEST.—266. How do you ascertain whether a year is leap year?

every *fourth* year; that is, if we have 100 leap years in 400 years, according to the Julian calendar, the reckoning would fall 3 days behind the *solar* time. Thus, reckoning from the commencement of the Christian era, when it was January 1st, 401 by the Julian time, it was January 4th by the *solar* time.

2. To remedy this error only 1 *centennial* year in *four* is regarded a *leap year;* or, which is the same in effect, whenever the *centennial* year, or the *number* expressing the *century*, is not divisible by 4, that year is not a leap year, while the other centennial years are. Thus, 17, 18, 19, denoting 1700, 1800, and 1900, are *not divisible* by 4, consequently they are *not* leap years, though according to the rule above they would be; on the other hand 16 and 20, denoting 1600 and 2000, *are divisible* by 4, and are therefore leap years. There is still a slight error, but it is so small that in 5000 years it scarcely amounts to a day

CIRCULAR MEASURE, OR MOTION.

267. *Circular Measure* is applied to the divisions of the circle, and is used in reckoning *latitude* and *longitude*, and the *motion* of the heavenly bodies.

60 seconds (″)	make	1 minute,	marked	′
60 minutes	"	1 degree,	"	°
30 degrees	"	1 sign,	"	s.
12 signs, or 360°	"	1 circle,	"	c.

This measure is often called *Angular Measure*, and is chiefly used by astronomers, navigators, and surveyors.

OBS. 1. The circumference of every circle is divided or supposed to be divided, into 360 equal parts, called *degrees*, as in the subjoined figure.

2. Since a degree is $\frac{1}{360}$ part of the circumference of a circle, it is obvious that its length must depend on the size of the circle

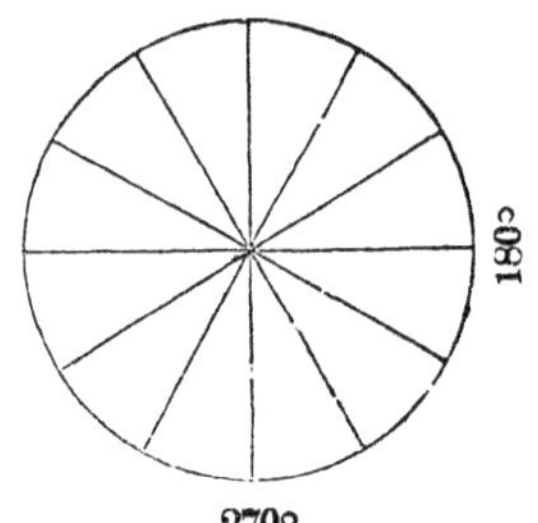

Note.—The division of the circumference of the circle into 360 equal parts took its origin from the length of the year, which, (in round numbers) was supposed to contain 360 days, or 12 months of 30 days each. The 12 *signs*

QUEST.—267. In what is Circular Measure used? Repeat the Table. *Obs.* How is the circumference of every circle divided? On what does the length of a degree depend?

correspond to the 12 months. The term *minutes*, is from the Latin *minutum*, which signifies a small part. The term *seconds*, is an abbreviated expression for *second* minutes, or minutes of the *second order*.

268. Since the earth turns on its axis from west to east *once* in 24 hours, it evidently revolves 15° per hour; or 1° in 4 minutes, and 1′ in 4 seconds of time. Hence,

When the difference of longitude between two places is 1′, the difference in the time, or the hour of the day at these two places, is 4 seconds; if the difference of longitude is 1°, the difference of time is 4 minutes; if 2°, the difference of time is 8 minutes, &c.

Thus, when it is noon at London, in Philadelphia, which is about 75° west from London, it is only 7 o'clock, A. M. For, if the earth revolves 1° in 4 minutes, to revolve 75°, it will require 75 times as long, and $4 \times 75 = 300$ min., or 5 hours.

OBS. 1. Since the earth revolves from west to east, it is manifest, that the time is *earlier* as we go eastward, and *later* as we go westward.

2. This principle affords navigators and others a convenient and useful method of ascertaining the *difference* of *time* between two places, when the difference of their *longitude* is known; also, for ascertaining the *difference* of *longitude* between two places, when the difference in their *time* is known.

MISCELLANEOUS TABLE.

269. The following denominations not included in the preceding Tables, are frequently used.

12 units	make	1 dozen, (*doz.*)
12 dozen, or 144	"	1 gross.
12 gross, or 1728	"	1 great gross.
20 units	"	1 score.
56 pounds	"	1 firkin of butter.
100 pounds	"	1 quintal of fish.
30 gallons	"	1 bar. of fish in Mass.
200 lbs. of shad or salmon	"	1 bar. in N. Y. and Conn.
196 pounds	"	1 bar. of flour.
200 pounds	"	1 bar. of pork.
14 pounds of iron or lead	"	1 stone.
21½ stone	"	1 pig.
8 pigs	"	1 fother.

Note.—Formerly it was customary to allow 112 lbs. for a quintal.

QUEST.—268. When the difference of longitude between two places is 1′ what is the difference of time? When 1°, what is the difference of time?

PAPER AND BOOKS.

270. The terms *folio, quarto, octavo,* &c., applied to books, denote the *number* of leaves into which a sheet of paper is folded.

24 sheets of paper	make	1 quire.
20 quires	"	1 ream.
2 reams	"	1 bundle.
5 bundles	"	1 bale.

A sheet folded in two leaves	forms	a *folio.*
A sheet " " four leaves	"	a *quarto*, or 4to.
A sheet " " eight leaves	"	an *octavo*, or 8vo.
A sheet " " twelve leaves	"	a *duodecimo*, or 12mo.
A sheet " " eighteen leaves	"	an 18mo.
A sheet " " thirty-six leaves	"	a 36mo.

DIMENSIONS OF DIFFERENT KINDS OF ENGLISH PAPER.

Names.	*Writing.*	*Drawing.*	*Printing.*
Pott,	15½ by 12½ in.	15½ by 12½ in.	
Small Post,	16½ by 13½ in.		
Fool's Cap,	16¾ by 13½ in.	16¾ by 13½ in.	
Crown,		20 by 15 in.	20 by 15 in
Demy,	20 by 15½ in.	22 by 17 in.	
Medium,	22½ by 17½ in.		23 by 18 in
Royal,	24 by 19¼ in.	24 by 19¼ in.	26 by 20 in.
Super Royal,	27½ by 19¼ in.	27½ by 19¼ in.	
Elephant,		28 by 23 in.	
Double Crown,			30 by 20 in.
Imperial,	30¼ by 22 in.	30¼ by 22 in.	
Atlas,		34 by 26¼ in.	
Columbier,		34½ by 23½ in.	
Double Demy,			38½ by 26 in
Double Elephant,		40 by 26¾ in.	
Antiquarian,		52 by 31 in.	
Double Atlas,		55 by 31½ in.	
Emperor,		68 by 48 in.	

Note.—*American* paper is usually rather larger than English paper of the same name.

QUEST.—270. What do the terms, folio, quarto, &c., denote, when applied to books? What is a folio? A quarto? An octavo? A duodecimo? An 18mo.? A 36mo.?

FRENCH MONEY, WEIGHTS, AND MEASURES.

271. The new system of Money, Weights, and Measures of France, adopted in 1795, was formed according to the *decimal* Notation.

FRENCH MONEY.

272. The *Franc* is the *unit* money of the new system of French currency. It is a silver coin, consisting of $\frac{9}{10}$ pure silver, and $\frac{1}{10}$ of alloy.

10 centimes	make	1 decime.
10 decimes	"	1 franc.

Note.—The *value* of a *franc* by Act of Congress in 1843, is $.186. The *value* of the *livre tournois*, the former unit of money, is $.185.

FRENCH LINEAR MEASURE.

273. The *standard unit* of the French *Linear Measure*, is the *Metre*. Its length, according to the *mean* of the several comparisons of Troughton, Nicollet and Hassler, is equal to 39.3809171 English, or United States inches.

10 metres	make	1 decametre	=	32.817431	U. S.	feet.
10 decametres	"	1 hectometre	=	328.17431	"	"
10 hectometres	"	1 kilometre	=	3281.7431	"	"
10 kilometres	"	1 myriametre	=	32817.431	"	"

Note.—1. The *standard* by which the new *French measures* of length are determined, is the *quadrant* of a meridian of the earth, or the *terrestrial arc* from the equator to the pole, in the meridian of Paris. The *ten-millionth* part of this arc is called a *metre*, which is equal to 39.381 U. S. in., nearly.

2. The *metre* is divided into 10 decimetres; the *decimetre* into 10 centimetres; the *centimetre* into 10 millimetres.

3. The denominations of the old system of linear measure were the toise, foot, inch, line, and point. 12 points=1 line; 12 lines=1 inch; 12 in.=1 foot; 6 ft.=1 toise. The old French foot was equal to 1.066 U. S. feet.

4. By a decree of 1812, the Toise, Aune, Foot, &c., are allowed to be used, having the following ratios to the *metre*, viz: the toise=2 metres; the foot=$\frac{1}{3}$ metre; the inch=$\frac{1}{12}$ metre; the aune or ell=$1\frac{1}{5}$ metre; the bushel=$\frac{1}{8}$ hectolitre.

FRENCH SQUARE MEASURE.

274. The *unit* of French *Superficial Measure,* is the *Are,* whose sides are each a decametre in length; consequently, it contains 100 square metres, or 119.6648496 U. S. sq. yds.

10 ares	make	1 decare	= 1196.648496	U. S. sq. yds.	
10 decares	"	1 hectare	= 11966.48496	"	"
10 hectares	"	1 kilare	= 119664.8496	"	"
10 kilares	"	1 myriare	= 1196648.496	"	"

Note.—The *are* is divided into 10 deciares; the *deciare* into 10 centiares; the *centiare* into 10 milliares.

FRENCH CUBIC MEASURE.

275. The *unit* of French *Cubic Measure,* is the *Stere,* which is a *cubic metre,* and is equal to 61074.1564445 cu. in. U. S.

10 decisteres	make	1 stere	= 35.34384	cu. ft. U. S.	
10 steres	"	1 decastere	= 353.4384	"	"

FRENCH LIQUID AND DRY MEASURE.

276. The *unit* of French *Liquid* and *Dry Measures,* is called the *Litre,* which is a *cubic decimetre,* and is equal to 61.0741564445 cu. in. U. S., or 1.05756 qts. wine measure.

10 litres	make	1 decalitre	= 2.6439	gals.	wine	meas.
10 decalitres	"	1 hectolitre	= 26.439	"	"	"
10 hectolitres	"	1 kilolitre	= 264.39	"	"	"

Note.—The *litre* is divided into 10 decilitres; the *decilitre* into 10 centilitres; the *centilitre* into 10 millilitres.

FRENCH WEIGHTS.

277. The *unit* of French *Weights,* is the weight of a *cubic centimetre* of distilled water, at the maximum density, and is called the *Gramme.* It is equal to 15.433159 grains Troy.

10 grammes	make	1 decagramme	= 154.33159	grs.	Troy.
10 decagrammes	"	1 hectogramme	= 1543.3159	"	"
10 hectogrammes	"	1 kilogramme	= 15433.159	"	"
10 kilogrammes,	"	1 myriagramme	= 154331.59	"	"

Note.—1. The *gramme* is divided into 10 decigrammes; the *decigramme* into 10 centigrammes; the *centigramme* into 10 milligrammes.

2. The denomination chiefly used in making out invoices of goods sold by weight, and in business transactions, is the *kilogramme*, which is equal to 1000 grammes, or 2.21 lbs. avoirdupois, very nearly.

3. In the *old system* of French weight, the livre-poids=2 marcs; the marc =8 onces; the once=8 gros; the gros=72 grains. The livre is equal to one-half the kilogramme.

FRENCH CIRCULAR MEASURE.

278. The *circle* is divided into 400 equal parts, called *grades*, and the *quadrant* into 100 grades. The grade is again divided into 100 equal parts, and each of these parts is subdivided into 100 other equal parts, according to the *centesimal* scale. Hence,

The seconde	=	.00009	English	deg.
The minute	=	.009	"	"
The grade	=	.9	"	"

Note.—The names of the denominations *larger* than the *unit* in the French Compound Numbers, are formed by prefixing to the name of the unit, the Greek words, *deca*, *hecto*, *kilo*, and *myria;* those *less* than the *unit*, are formed by prefixing to the name of the unit, the Latin words, *deci*, *centi*, and *milli*.

279. *Foreign Weights and Measures compared with those of the United States.**

Amsterdam.—100 lbs. (1 centner)=108.923 lbs.†; 1 last=85.25 bu.; 1 ahm=41 gals.; 1 foot Amsterdam=11$\frac{1}{7}$ in.; 1 foot Antwerp=11$\frac{1}{4}$ in.; 1 ell Amsterdam=2.26 ft.; 1 ell Brabant=2.3ft.; 1 ell Hague=2.28 ft.

Batavia.—1 picul=136 lbs.; 1 kann=.39 gal.; 1 ell=2.25 ft.

Bengal.—1 haut=1.5 ft.; 1 guz=3 ft.; 1 coss or mile=1.24 miles; 1 bazar maud=82.14 lbs.; 1 factory maud=74.66 lbs.

Bencoolen.—1 bahar=560 lbs.; 1 bamboo=1 gal.; 1 coyang=8 gals.

Bombay.—1 maud=28 lbs.; 1 covid=1.5 ft.; 1 candy=25 bu.

Bremen.—1 pound=1.1 lb.; 1 centner=116 lbs.; 1 last=80.7 bu.; 1 ft.=11$\frac{3}{8}$ in.

Canton.—1 tael=1$\frac{1}{3}$ oz.; 1 catty=1$\frac{1}{3}$ lbs.; 1 picul=133$\frac{1}{3}$ lbs.; 1 covid=14$\frac{5}{8}$ in.

Denmark.—100 lbs. (1 centner)=110.25 lbs.; 1 bbl. (toende)=3.95 bu.; 1 viertel=2.04 gals.; 1 foot Copenhagen, or Rhineland=12$\frac{1}{3}$ in.

Florence and Leghorn.—100 lbs. (1 cantaro)=74.86 lbs.; 1 moggio=16.59 bu.; 1 barile=12.04 gals.; 1 palmo=9$\frac{3}{4}$ in.

* M'Culloch's Commercial Dictionary; also Kelly's Universal Cambist.

† The pounds in this and the following comparisons are avoirdupois.

Genoa.—100 lbs. (1 peso grosso)=76⅝ lbs.; 1 peso sottile=69.89 lbs.; 1 mina =3.43 bu.; 1 mezzarola=39.22 gals.; 1 palmo=9⅕ in.

*Hamburg.**—1 foot=11.3 in.; 1 ell=22.6 in. nearly; 1 ell Brabant=27.6 in.; 1 mile=4.68 miles; 1 fass=1½ bu.; 1 last=89.64 bu.; 1 ahm=38¼ gals.

Japan.—1 catti=1.3 lbs; 1 picul=130 lbs.; 1 ichan=3½ ft.; 1 inc or tetamy =6¼ ft.; 1 balec=16¼ gals.

Madras.—1 covid=1½ ft.; 1 catty=1⅓ lbs.; 1 picul=133⅓ lbs.; 1 maud= 25 lbs.; 1 candy=500 lbs.; 1 garee=140 bu.

Malta.—1 foot=10⅙ in.; 100 lbs. (1 cantaro)=174.5 lbs.; 1 salma=8.22 bu.

Manilla.—1 arroba=26 lbs.; 1 picul=143 lbs.; 1 palmo=10.38 in.

Naples.—1 cantaro grosso=196.5 lbs.; 1 cantaro piccolo=106 lbs.; 1 palmo= 10⅜ in.; 1 tomolo=1.45 bu.; 1 carro=52.24 bu.; 1 carro of wine=264 gals.

Netherlands.—1 ell=3.28 ft.; 1 mudde=2.84 bu.; 1 kan litre=2.11 pints; 1 vat hectolitre=26.42 gals.; 1 pond kilogramme=2.21 lbs.

Portugal.—100 lbs.=101.19 lbs.; 1 arroba=22.26 lbs.; 1 quintal=89.05 lbs.; 1 almude=4.37 gals.; 1 alquiere=4¾ bu.; 1 moyo=23.03 bu.; 1 last=70 bu.; 1 pe or foot=12⅚ in.; 1 mile=1¼ mile.

Prussia.—100 lbs=103.11 lbs.; 1 quintal (110 lbs.)=113.42 lbs.; 1 eimar= 18.14 gal.; 1 scheffel=1.56 bu.; 1 foot=1.03 ft.; 1 ell=2.19 ft.; 1 mile= 4.68 miles.

Rome.—100 libras=74.77 lbs.; 1 rubbio=9.36 bu.; 1 barile=15.31 gals.; 1 foot =11¾ in.; 1 canna=6½ ft.; 1 mile=7⅖ fur.

Russia.—100 lbs.=90.26 lbs.; 1 berquit=361.04 lbs.; 40 lbs. (1 pood)= 36 lbs.; 1 vedro=3¼ gals.; 1 chetwert=5.95 bu.; 1 foot Petersburg=1.18 ft.†; 1 foot Moscow=1.1 ft.; 1 arsheen=2⅓ ft.; 1 mile (verst)=5.3 fur.

Sicily.—100 lbs. (libras)=70 lbs.; 1 cantaro grosso=192.5 lbs.; 1 cantaro sottile=175 lbs.; 1 salma generale=7.85 bu.; 1 salma grossa=9.77 bu.; 1 salma of wine=23.06 gals.; 1 palmo=9½ in.; 1 canna=6⅓ ft.

Spain.—1 arroba=25.36 lbs.; 1 quintal=101.44 lbs.; 1 arroba of wine= 4¼ gals.; 1 moyo=68 gals.; 1 fanega =1.6 bu.; 1 foot=11.128 in.; 1 vara= 2.78 ft.; 1 league (leagua)=4.3 m., nearly.

Sweden.—100 lbs. (victualie)=73.76 lbs.; 1 foot=11.69 in.; 1 ell=1.95 ft.; 1 mile=6.64 m.; 1 kann=7.42 bu.; 1 last=75 bu.; 1 kann of wine= 69.09 gals.

Smyrna.—100 lbs. (1 quintal)=129.48 lbs.; 1 oke=2.83 lbs.; 1 quillot= 1.46 bu.; 1 quillot of wine=13.5 gals.; 1 pic=2¼ ft.

Trieste.—100 lbs.=123.6 lbs.; 1 stajo=2⅓ bu.; 1 orna, or eimer=14.94 gals.; 1 ell (for silk)=2.1 ft.; 1 ell (for woollen)=2.2 ft.; 1 foot Austrian=1.037 ft.; 1 mile Austrian=4.6 m.

Venice.—100 lbs. (1 peso grosso)=105.18 lbs.; 1 peso sottile=64.42 lbs.; 1 stajo =2.27 bu.; 1 moggio=9.08 bu.; 1 anifora=137 gals.; 1 foot=1.14 ft.; 1 braccio (for silk)=24.8 in.; 1 braccio (for woollen)=26.6 in.

---

* New system of weights and measures adopted in 1843.

† In measuring timber English feet and inches are chiefly used throughout Russia.

REDUCTION.

280. The process of changing *compound numbers* frcm one denomination into another, without altering their *value,* is called REDUCTION.

Ex. 1. Reduce £5, 2s. 7d. and 3 far. to farthings.

Analysis.—Since in £1 there are 20s., in £5 there are 5 times as many, which is 100s., and 2, (the given shillings,) make 102s. Again, since there are 12d. in 1s., in 102s. there are 102 times as many, which is equal to 1224d., and 7 (the given pence) make 1231d. Finally, since in 1d. there are 4 far., in 1231d. there are 1231 times as many, or 4924 far., and 3, (the given far.,) make 4927 far. *Ans.* 4927 farthings.

Operation.

Operation	
£ *s.* *d.* *far.*	We first reduce the given pounds to shillings, by multiplying them by 20, because 20s. make £1. (Art. 247.) We next reduce the shillings to pence, by multiplying them by 12, because 12d. make 1s. Finally, we reduce the pence to farthings by multiplying them by 4, because 4 far. make 1d.
5 2 7 3.	
20s. in £1.	
102 shillings.	
12d. in 1s.	
1231 pence.	
4 far in 1d.	
4927 far. *Ans.*	

Note.—1. In this example it is required to reduce higher denominations to lower; as pounds to shillings, shillings to pence, &c. This is done by *successive multiplications.*

2. In 4927 farthings, how many pounds, shillings, and pence?

Analysis.—Since 4 far. make 1d., in 4927 farthings, there are as many pence as 4 is contained times in 4927, which is 1231d., and 3 far. over. Again, since 12d. make 1s., in 1231d. there are as many shillings as 12 is contained times in 1231, which is 102s., and 7d. over. Finally, since 20s. make £1, in 102s. there

QUEST.—280. What is Reduction? How are pounds reduced to shillings? Why multiply by 20? How are shillings reduced to pence? Why? How pence to farthings? Why?

are as many pounds as 20 is contained times in 102, which is £5, and 2s. over. *Ans.* £5, 2s. 7d. 3 far.

Operation.

```
 4)4927 far.
12)1231d. 3 far. over.
 20)102s. 7d. over.
    £5, 2s. over.
Ans. £5, 2s. 7d. 3. far.
```

We first reduce the given farthings to pence, the next higher denomination, by dividing them by 4, because 4 far. make 1d. (Art. 247.) Next we reduce the pence to shillings by dividing them by 12, because 12d. make 1s. Finally, we reduce the shillings to pounds by dividing them by 20, because 20s. make £1. The last quotient and the several remainders constitute the answer.

Note.—2. The last example is exactly the reverse of the first; that is, lower denominations are reduced to higher, which is done by *successive divisions.*

281. From the preceding illustrations we derive the following

GENERAL RULE FOR REDUCTION.

I. To reduce compound numbers to lower denominations.

Multiply the highest denomination given, by that number which it takes of the next lower denomination to make ONE *of this higher; to the product, add the number expressed in this lower denomination in the given example. Proceed in this manner with each successive denomination, till you come to the one required.*

II. To reduce compound numbers to higher denominations.

Divide the given denomination by that number which it takes of this denomination to make ONE *of the next higher. Proceed in this manner with each successive denomination, till you come to the one required. The last quotient, with the several remainders, will be the answer sought.*

282. PROOF.—*Reverse the operation; that is, reduce back the answer to the original denominations, and if the result corresponds with the numbers given, the work is right.*

QUEST.—How are farthings reduced to pence? Why divide by 4? How reduce pence to shillings? Why? How reduce shillings to pounds? Why? 281. How are compound numbers reduced to lower denominations? How to higher denominations? 282. How is Reduction proved?

OBS. 1. Each remainder is of the *same denomination* as the dividend from which it arose. (Art. 113. Obs. 1.)

2. Reducing compound numbers to *lower* denominations may, with propriety, be called *Reduction by Multiplication;* reducing them to *higher* denominations, *Reduction by Division*. The former is often called *Reduction Descending;* the latter, *Reduction Ascending*. They mutually prove each other.

EXAMPLES FOR PRACTICE.

1. In 136 rods and 2 yards, how many feet?

Operation.	*Proof.*
rods. yds.	3)2250 ft.
2)136 2	5½)750 yds.
5½ yds. 1 r.	2
682	11)1500
68	136 r. 4 rem.=2 yards.
750 yds.	Now 136 r. 2 yds. is the
3 ft. 1 yd.	given number.
2250 ft. *Ans.*	

2. In £71, 13s. 6½d., how many farthings?
3. In £90, 7s. 8d., how many farthings?
4. In £295, 18s. 3¾d., how many farthings?
5. In 95 guineas, 17s. 9¾d., how many farthings?
6. How many pounds, shillings, &c., in 24651 farthings?
7. How many pounds, shillings, &c., in 415739 farthings?
8. How many guineas, &c., in 67256 pence?
9. In £36, 4s., how many six-pences?
10. In £75, 12s. 6d., how many three-pences?
11. Reduce 29 lbs. 7 oz. 3 pwts. to grains.
12. Reduce 37 lbs. 6 oz. to pennyweights.
13. Reduce 175 lbs. 4 oz. 5 pwts. 7 grs. to grains.
14. Reduce 12256 grs. to pounds, &c.
15. Reduce 42672 pwts. to pounds, &c.
16. In 15 cwt. 3 qrs. 21 lbs., how many pounds?
17. In 17 tons 12 cwt. 2 qrs., how many ounces?

QUEST.—*Obs.* Of what denomination is each remainder? What may reducing compound numbers to lower denominations be called? To higher denominations? Which of the fundamental rules is employed by the former? Which by the latter?

18. In 52 tons 3 cwt., how many pounds?
19. In 140 tons, how many drams?
20. In 16256 ounces, how many hundred weight, &c.?
21. In 267235 pounds, how many tons, &c.?
22. In 563728 drams, how many tons, pounds, &c.?
23. Reduce 95 pounds (apothecaries' weight) to drams.
24. Reduce 130 pounds to scruples.
25. Reduce 6237 drams (apothecaries' weight) to pounds.
26. Reduce 25463 scruples to ounces, &c.
27. How many feet in 27 miles?
28. How many inches in 45 leagues?
29. How many yards in 3000 miles?
30. In 290375 feet, how many miles?
31. In 1875343 inches, how many leagues?
32. In 15 m. 5 fur. 31 r., how many rods?
33. In 1081080 inches, how many miles, &c.?
34. How many feet in the circumference of the earth?
35. How many nails in 160 yards?
36. How many quarters in 1000 English ells?
37. In 102345 nails, how many yards, &c.?
38. In 223267 nails, how many French ells?
39. In 634 yards, 3 qrs., how many nails?
40. In 28 hhds. 15 gals. wine measure, how many quarts?
41. In 5 pipes, 1 hhd., how many gallons?
42. In 3 tuns, 1 hhd. 10 gals., how many gills?
43. In 12256 pints, how many barrels, wine measure?
44. In 475262 gills, how many pipes, &c.?
45. In 50 hhds. 1 bbl. 10 gals., how many gills, wine measure?
46. In 45 bbls., how many pints, beer measure?
47. How many barrels of beer in 25264 pints?
48. How many hogsheads of beer in 136256 quarts?
49. How many pints in 45 hhds. 10 gals. of beer?
50. In 15 bushels, 1 peck, how many quarts?
51. In 763 bushels, 3 pecks, how many quarts?
52. In 56 quarters, 5 bushels, how many pints?
53. In 45672 quarts, how many bushels, &c.?
54. In 260200 pints, how many quarts?

55. Reduce 25 days, 6 hours to minutes.

56. Reduce 365 days, 6 hours to seconds.

57. Reduce 847125 minutes to weeks.

58. Reduce 5623480 seconds to days.

59. How many seconds in a solar year?

60. How many seconds in 30 years, allowing 365 days 6 hours to a year?

61. How many years of Sabbaths are there in 70 years?

62. In 110 degrees, 20 minutes, how many seconds?

63. In 11 signs, 45 degrees, how many seconds?

64. In 7654314 seconds, how many degrees?

65. In 1000000000 minutes, how many signs?

66. Reduce 1728 sq. rods, 23 yds. 5 feet to feet.

67. Reduce 100 acres, 37 rods to square feet.

68. Reduce 832590 sq. rods to sq. inches.

69. Reduce 25363896 sq. feet to acres, &c.

70. In 150 cubic feet, how many inches?

71. In 97 yds. 15 ft., how many cubic inches?

72. In 49 cords, 23 feet, how many cubic inches?

73. In 84673 cubic inches, how many feet?

74. In 39216 cubic feet, how many cords?

75. In 65 tons of round timber, how many cubic inches?

76. In 4562100 cubic inches, how many tons of hewn timber?

APPLICATIONS OF REDUCTION.

283. To reduce Troy to Avoirdupois weight.

First reduce the given pounds, ounces, &c., to grains; then divide by the number of grains in a dram, and the quotient will be the answer in drams. (Art. 252.)

OBS. If the answer is required to be in *pounds* and a *fraction* of a pound, divide the grains by 7000.

Ex. 1. In 175 pounds Troy, how many pounds avoirdupois?

Solution.—$175\times12\times20\times24=1008000$ grs., and 1008000 grs.$\div27\frac{11}{32}=36864$ drams, or 144 lbs. avoirdupois. *Ans.*

QUEST.—283. How is Troy weight reduced to avoirdupois?

2. In 700 lbs. Troy of silver, how many pounds avoirdupois?

3. In 840 lbs. 6 oz. 10 pwts., how many pounds, &c., avoirdupois?

4. An apothecary bought 1000 lbs. of opium by Troy weight, and sold it by avoirdupois: how many pounds did he lose?

5. A merchant bought 1500 pounds of lead Troy weight, and sold it by avoirdupois: how many pounds did he lose?

284. To reduce Avoirdupois to Troy weight.

First reduce the given pounds, ounces, &c., to drams, then multiply by the number of grains in a dram, and the product will be the answer in grains. (Art. 252.)

OBS. 1. When the given example contains *pounds* only, we may multiply them by 7000, and the product will be grains.

2. If the answer is required to be in *pounds* and a *fraction* of a pound, divide the grains by 5760.

6. In 32 lbs. avoirdupois, how many pounds Troy?

Solution.—$32 \times 16 \times 16 \times 27\frac{11}{32} = 224000$ grs., and 224000 grs. $=38$ lbs. 10 oz. 13 pwts. 8 grs. *Ans.*

7. In 48 lbs. avoirdupois, how many pounds Troy?

8. A merchant bought 100 lbs. 10 oz. of tea avoir. and sold it by Troy weight: how many pounds did he gain?

9. A druggist bought 1260 lbs. of alum avoirdupois, and retailed it by Troy weight: how many more pounds did he sell than he bought?

285. The *area* of a floor, a piece of land, or any surface which has *four sides* and *four right-angles*, is found by *multiplying its length and breadth together*.

Note.-1. The *area* of a figure is the *superficial contents* or *space* contained within the line or lines, by which the figure is bounded. It is reckoned in *square* inches, feet, yards, rods, &c.

2. A figure which has four sides and four right-angles, like the following diagram, is called a *Rectangle* or *Parallelogram*.

QUEST.—284. How is avoirdupois weight reduced to Troy? 285. How do you find the area or superficial contents of a surface having four sides and four right-angles? *Note.* What is meant by the term area? How is it reckoned? What is a figure which has four sides and four right-angles called?

10. How many square yards of carpeting will it take to cover a room, 4 yards long and 3 yards wide?

Suggestion.—Let the given room be represented by the subjoined figure, the length of which is divided into 4 equal parts, and the breadth into 3 equal parts, which we will call *linear yards.* Now it is plain that the room will contain as many square yards as there are squares in the given figure. But the number of squares in the figure is equal to the number of equal parts (linear yards) which its length contains, repeated as many times as there are equal parts (linear yards) in its breadth; that is, it is equal to 4×3, or 12. *Ans.* 12 yds.

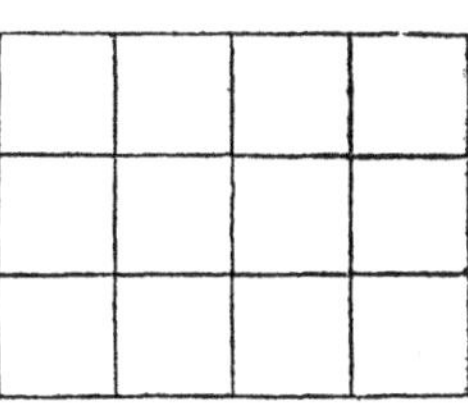

11. How many sq. feet in a floor, 20 feet long, 18 feet wide?

12. How many acres in a field, 50 rods long, 45 rods wide?

13. How many square yards in a ceiling, 35 feet long and 28 feet wide?

14. How many acres in a farm, 420 rods long and 170 rods wide?

15. What is the area of a square field, whose sides are 80 rods in length?

16. How many yards of carpeting a yard wide, will it take to cover a floor 18 feet square.

17. How many yards of plastering are required to cover four sides of a room, 18 ft. long, 15 feet wide, and 9 ft. high?

18. How many square yards of shingling will cover both sides of a roof, whose rafters are 20 feet, and whose ridge pole is 25 feet long?

286. The *cubical contents,* or *solidity* of boxes of goods, piles of wood, &c., are found by *multiplying* the *length, breadth,* and *thickness together.*

19. How many cubic feet in a box 5 feet long, 4 feet wide and 3 feet deep?

Solution.—5×4=20, and 20×3=60. *Ans.* 60 cu. ft.

QUEST.—286. How are the cubical contents of a box of goods, a pile of wood, &c., found?

20. How many cubic feet in a block of granite, 65 in. long, 42 in. wide, and 36 in. thick?

21. How many cubic feet in a load of wood, 8 ft. long, $4\frac{1}{2}$ ft. high, and $3\frac{1}{2}$ ft. wide?

22. How many cords of wood in a pile, 46 ft. long, 16 ft. high, and 15 feet wide?

23. How many cubic feet in a vat, 12 ft. long, $8\frac{1}{2}$ ft. wide, and $7\frac{1}{3}$ ft. deep?

24. How many cubic feet in a bin, 12 ft. long, 9 ft. deep, and 7 ft. wide?

25. How many cubic yards in a cellar, 18 ft. long, 12 ft. wide, and 9 ft. deep?

26. How many cubic feet in a stick of timber, 2 ft. square, and 40 ft. long?

27. How many cubic feet in a cistern 15 ft. long, 12 ft. wide, and 10 ft. deep?

287. To reduce Cubic to Dry, or Liquid Measure.

First reduce the given yards, feet, &c., to cubic inches; then divide by the number of cubic inches in a gallon, or bushel, as the case may be, and the quotient will be the answer required. (Arts. 260, 263.)

28. In 10752 cubic feet, how many bushels?

Solution. — $10752 \times 1728 = 18579456$ cubic inches; and $18579456 \div 2150\frac{4}{10} = 8640$ bushels.

29. In 21504 cubic feet, how many bushels?

30. In 462 cubic feet, how many wine gallons?

31. In 1155 cubic feet and 33 inches, how many wine gallons?

32. In 846 cubic feet, how many beer gallons?

33. In 1128 cubic feet and 141 in., how many beer gallons?

34. How many bushels will a bin contain, which is 5 ft. long, 5 ft. wide, and 4 ft. deep?

35. How many bushels will a bin contain, which is 8 ft. long, $4\frac{1}{2}$ ft. wide, and $3\frac{1}{2}$ ft. deep?

QUEST.—287. How reduce cubic to dry, or liquid measure?

36. How many bushels will a bin contain, which is 14 ft. long, 10 ft. 8 in. wide, and 6 ft. 8 in. deep?

37. How many wine gallons in a cistern, which is 6 ft. long, 5 ft. wide, and 4 feet deep?

38. How many barrels of water (wine meas.) will a cistern hold, which is 20 ft. long, 15 ft. wide, and 10 ft. deep?

39. The distributing reservoir of the Croton Water Works in the City of New York, is 436 ft. square and 40 feet high: how many hogsheads of water will it hold?

288. To reduce Dry, or Liquid, to Cubic Measure.

First find the number of bushels, if dry measure, or gallons, if liquid measure, in the given example; then multiply by the number of cubic inches in a gallon, or bushel, as the case may be, and the product will be the answer required. (Art. 263.)

40. How many cubic feet in a bin, which contains 100 bushels?

Solution. — $100 \times 2150\frac{4}{10} = 215040$, and $215040 \div 1728 = 124\frac{768}{1728}$, or $124\frac{4}{9}$ cubic feet. *Ans.*

41. How many cubic feet in a lime kiln, which holds 500 bushels?

42. How many cubic feet in the hold of a ship, which contains 1000 bushels of grain?

43. How many cubic feet in 1 hogshead, wine measure?

44. How many cubic feet in a cistern, which holds 50 barrels of water?

45. How many cubic feet in a vat, which contains 100 hogsheads wine measure?

289. To reduce Liquid to Dry Measure, or Dry to Liquid Measure.

First find the cubic inches in the given example; then divide hem by the number of cubic inches in a gallon, or bushel, as the case may be, and the quotient will be the answer required.

QUEST.—288. How reduce dry, or liquid measure to cubic? 289. How reduce liquid dry measure? How dry to liquid measure?

46. In 40 gallons wine measure, how many bushels?

Solution.—$40\times231=9240$ cu. in., and 9240 cu. in.$\div2150\frac{4}{10}=4\frac{19}{54}$ bushels. *Ans.*

47. In 6 hogsheads, 16 gallons, how many bushels?

48. In 5 bushels, how many gallons wine measure?

49. In 3200 quarts dry measure, how many hogsheads wine measure?

290. To reduce Wine to Beer Measure, or Beer to Wine Measure.

First find the number of cubic inches in the given example; then divide them by the number of cubic inches which it takes to make a gallon in the required measure.

50. In 94 wine gallons, how many beer gallons?

Solution.—$94\times231=21714$ cu. in., and 21714 cu. in.$\div282=77$ gallons. *Ans.*

51. In 1 hhd. wine measure, how many beer gallons?

52. A tavern-keeper bought 4 hhds. of cider wine measure, and retailed it by beer measure: how many gallons did he lose?

53. In 20 beer gallons, how many wine gallons?

54. A grocer bought 7238 gallons of milk beer measure, and retailed it by wine measure: how many gallons did he gain?

55. A druggist bought 10000 gallons of alcohol beer measure, and sold it by wine measure: how many gallons did he gain?

56. A grocer bought 65 hhds. 29 gals. and 2 quarts of milk by beer measure, and sold it to his customers by wine measure: how many quarts more did he sell than he bought?

57. A liquor dealer bought 120 pipes of wine which his clerk retailed by beer measure: how many gallons more did he buy than he sold?

291. Since the earth revolves on its axis 1° in 4 minutes, or 1′ in 4 seconds of time, (Art. 268,) it is evident that *longitude* may be reduced to *time*. That is, multiplying degrees of longitude by 4 reduces them to minutes of time, multiplying minutes of longitude by 4 reduces them to seconds of time, &c.

QUEST.—290. How reduce wine to beer measure? How beer to wine measure?

By reversing this process it is evident that *time* may be reduced to *longitude*. Thus, dividing seconds of time by 4, will reduce them to minutes of longitude; dividing minutes of time by 4, will reduce them to degrees, &c. Hence,

292. To find the difference of *time* between two places from the difference of their *longitude*.

Reduce the difference of longitude to minutes; multiply them by 4, *and the product will be the difference of time in seconds, which may be reduced to hours and minutes.*

OBS. When the difference of longitude consists of *degrees only*, we may multiply them by 4, and the product will be the answer in *minutes*.

58. The difference of longitude between New York and Cincinnati is 10° 26′: what is the difference in their time?

Solution.—10° and 26′=626′; (Art. 281;) now 626′×4= 2504 seconds of time; and 2504 sec. ÷60=41 min. 44 sec. *Ans.*

59. The difference of longitude between Albany and Boston is 2° 9′: what is the difference in their time?

60. The difference of longitude between Albany and Detroit is 9° 45′: what is the difference in their time?

61. The difference of longitude between New Haven and New Orleans is 17° 10′: what is the difference in their time?

62. The difference of longitude between Charleston, S. C. and Mobile is 8° 27′: what is the difference in their time?

63. The difference of longitude between New York and Canton is 187° 3′: what is the difference in their time?

293. To find the difference of *longitude* between two places from the difference in their *time*.

Reduce the given difference of time to seconds; divide them by 4, *and the quotient will be the difference of longitude in minutes, which may be reduced to degrees.* (Art. 281.)

OBS. When there are no seconds in the difference of time, we may divide the minutes by 4, and the quotient will be the answer in degrees.

QUEST.—292. How find the difference of time between two places from their difference of longitude? 293. How find the difference of longitude from the difference of time?

64. A ship sailed from Boston to Liverpool; on the fourth day the master took an observation of the sun at noon, and found by his chronometer that it was 1 hr. 5 min. and 40 sec. earlier than the Boston time: how many degrees east of Boston was the ship?

Solution.—1 hr. 5 m. 40 sec.=3940 sec., (Art. 281,) and 3940 sec.÷4=985′. The ship had therefore sailed 985′ east, which is equal to 16° 25′. *Ans.*

65. The difference of time between Albany and Buffalo is 19 minutes: what is the difference of their longitude?

66. The difference of time between Richmond and New Orleans is 51 min. 4 sec.: what is the difference of their longitude?

67. The difference of time between Boston and Cincinnati is 53 min. 32 sec.: what is the difference of their longitude?

COMPOUND NUMBERS REDUCED TO FRACTIONS.

294. That one *concrete* number may properly be said to be a *part* of another, the two numbers must necessarily express objects of the *same kind*, or objects which can be *reduced* to the same kind or denomination. Thus, 1 *penny* is $\frac{1}{240}$ of a *pound*, but 1 penny cannot properly be said to be a part of a *foot*, or of a *year;* for, feet and years cannot be reduced to pence. So, 1 orange is $\frac{1}{5}$ of 5 oranges; but 1 orange cannot be said to be $\frac{1}{5}$ of 5 apples, or 5 pumpkins; for apples and pumpkins cannot be reduced to oranges.

Ex. 1. Reduce 2s. 7d. to the fraction of a pound.

Analysis.—The object in this example is to find what part of 1 pound, 2s. 7d. is equal to. To ascertain this, we must reduce both the given numbers to the same denomination, viz: pence. Now 2s. 7d.=31d., and £1=240d. (Art. 281. I.) The question, therefore, resolves itself into this: what part of 240 is 31? The answer is $\frac{31}{240}$; consequently 2s. 7d. (31d.) is $\frac{31}{240}$ of a pound. Hence,

QUEST.—294. When can one concrete number be said to be a part of another?

295. To reduce a compound number to a common **fraction**, of a higher denomination.

First reduce the given compound number to the lowest denomination mentioned for the numerator; then reduce a UNIT *of the denomination of the required fraction to the same denomination as the numerator, and the result will be the denominator.* (Art. 281.)

OBS. 1. The given number, and that of which it is said to be a *part*, must, in all cases, be reduced to the same denomination. (Art. 294.)

2. When the given number contains but one denomination, it of course requires no reduction.

If the given number contains a fraction, the denominator of the fraction is the lowest denomination mentioned. Thus, in $6\frac{3}{4}$s., the lowest denomination is *fourths* of a shilling; in $\frac{3}{5}$far., the lowest denomination is *fifths* of a farthing.

2. Reduce $\frac{6}{7}$ of a penny to the fraction of a pound.

Solution.—Since sevenths of a penny is the *lowest* and *only* denomination given, we simply reduce £1 to sevenths of a penny for the denominator. Now £1=240d., and 240d.×7=1680. *Ans.* £$\frac{6}{1680}$, or £$\frac{1}{280}$. Hence,

296. To reduce a fraction of a *lower* denomination to an equivalent fraction of a *higher* denomination.

Reduce a unit of the denomination of the required fraction to the same denomination as the given fraction, and the result will be the denominator.

Or, divide the given fraction by the same numbers as in reducing whole compound numbers to higher denominations. (Art. 281. II.) Thus in the last example, $\frac{6}{7}$d.÷12=$\frac{6}{84}$s., (Art. 227,) and $\frac{6}{84}$s.÷20=£$\frac{6}{1680}$,=£$\frac{1}{280}$. *Ans.*

OBS. When factors common to the numerator and denominator occur, the operation may be *shortened* by *canceling* those factors. (Art. 221.)

3. Reduce $\frac{4}{7}$ of a penny to the fraction of a pound.

Solution.—By the last article, $\frac{4}{7\times12\times20}$ = the answer.

By *Cancelation* $\frac{4}{7\times12\times20}=\frac{\not{4}}{7\times12\times\not{2}\not{0},5}=£\frac{1}{420}$. *Ans.*

QUEST.—295. How is a compound number reduced to a common fraction? 296. How is a fraction of a lower denomination reduced to the fraction of a higher?

4. Reduce $4\frac{2}{3}$s. to the fraction of a pound. *Ans.* £$\frac{14}{60}$, or £$\frac{7}{30}$.
5. Reduce 4s. 7d. to the fraction of a pound.
6. Reduce 9d. $2\frac{1}{2}$ far. to the fraction of a pound.
7. What part of £1 is $\frac{7}{8}$ of 1 penny?
8. What part of 1 lb. Troy is 7 ounces?
9. What part of 1 lb. Troy is 16 pwts. 3 grs?
10. What part of 1 lb. avoirdupois is 8 oz. and 12 drams?
11. What part of 1 ton is 14 cwt. and 15 lbs?
12. What part of 1 yd. is 2 ft. and 4 inches?
13. What part of 1 mile is $82\frac{1}{3}$ rods?
14. What part of 1 acre is $45\frac{1}{2}$ rods?
15. What part of 1 square rod is 63 square feet?
16. Reduce $\frac{4}{5}$ of 1 qt. to the fraction of a gallon.
17. Reduce 7 gallons to the fraction of a hogshead.
18. Reduce $\frac{7}{8}$ of 1 hour to the fraction of a day.
19. Reduce $\frac{5}{9}$ of 1 minute to the fraction of an hour.
20. Reduce $\frac{3}{4}$ of 1 second to the fraction of a week.
21. What part of £3, 5s. 6d. 1far. is £2, 1s. 3d.?

Solution.—Reducing both numbers to farthings, £3, 5s. 6d. 1far. =3145 far., and £2, 1s. 3d.=1980 far. (Art. 295. Obs.1.) Now 1980 is $\frac{1980}{3145}$ of 3145, which is equal to $\frac{396}{629}$. *Ans.*

22. What part of £2 is 7s. 6d.?
23. What part of £7, 3s. is £3?
24. What part of 2 bushels is 3 pecks?
25. What part of 10 bushels is 10 quarts?
26. What part of 16 rods is 40 feet?
27. What part of 3 weeks is 2 days and 7 hours?
28. What part of 2 hhds. 10 gals. is 45 gals.?
29. What part of 2 tons, 3 cwt. is 15 cwt. 65 lbs.?
30. What part of 1 ton is 7 lbs. 10 ounces?
31. What part of 90° is 1° 15′ 30″?
32. What part of 360° is 45° 15′ 10″?
33. What part of 3 lbs. Troy is 1 lb. 3 oz.?
34. What part of 25 lbs. Troy is 10 lbs. 7 oz. 10 pwts.?
35. What part of 1 acre is 40 rods?
36. What part of 5 acres is $1\frac{1}{2}$ acres?

FRACTIONAL COMPOUND NUMBERS

REDUCED TO WHOLE NUMBERS OF LOWER DENOMINATIONS.

Ex. 1. Reduce $\frac{5}{8}$ of £1 to shillings and pence.

Analysis.—$\frac{5}{8}$ of 1s.$=\frac{1}{8}$ of 5s. or $\frac{5}{8}$s., consequently $\frac{5}{8}$ of 20s. (£1) is 20 times as much, and $\frac{5}{8}$s.$\times 20=\frac{100}{8}$s. or 12s. and $\frac{4}{8}$ of a shilling.

Reasoning as before, $\frac{4}{8}$ of 1d.$=\frac{1}{8}$ of 4d., or $\frac{4}{8}$d., and $\frac{4}{8}$ of 12d. (1s.) is 12 times as much; but $\frac{4}{8}$d.$\times 12=\frac{48}{8}$d., or 6d. Therefore £$\frac{5}{8}$=12s. 6d. *Ans.* Hence,

297. To reduce fractional compound numbers to whole numbers of lower denominations.

First reduce the given numerator to the next lower denomination; then divide the product by the denominator, and the quotient will be an integer of the next lower denomination. (Art. 281. I.)

Proceed in like manner with the remainder, and the several quotients will be the whole numbers required.

OBS. This operation is the same in principle as reducing *higher* denominations of whole numbers to *lower*. (Art. 281. I.) Whenever the fraction becomes *improper*, it is reduced to a whole or mixed number. (Art. 196.)

2. Reduce $\frac{4}{5}$ of £1 to shillings. *Ans.* 16s.
3. Reduce $\frac{7}{8}$ of £1 to shillings and pence.
4. Reduce $\frac{3}{5}$ of 1s. to pence and farthings.
5. Reduce $\frac{3}{7}$ of 1 lb. Troy to ounces, &c.
6. Reduce $\frac{5}{8}$ of 1 ounce Troy to pennyweights.
7. Reduce $\frac{2}{3}$ of 1 lb. avoirdupois to ounces, &c.
8. Reduce $\frac{4}{7}$ of 1 cwt. to pounds, &c.
9. Reduce $\frac{5}{8}$ of 1 ton to pounds, &c.
10. Reduce $\frac{4}{5}$ of 1 yard to feet and inches.
11. Reduce $\frac{3}{8}$ of 1 rod to feet and inches.
12. Reduce $\frac{5}{9}$ of 1 mile to rods, feet, &c.
13. Reduce $\frac{2}{3}$ of 1 gallon wine measure to quarts, &c.
14. Reduce $\frac{7}{8}$ of 1 hogshead wine measure to gallons, &c.
15. Reduce $\frac{5}{6}$ of 1 peck to quarts, &c. *Ans.* 6 qts $1\frac{1}{3}$ pts.
16. Reduce $\frac{4}{5}$ of 1 bushel to quarts, &c.
17. Reduce $\frac{7}{9}$ of 1 hour to minutes and seconds.

QUEST.—297. How are fractional compound numbers reduced to whole ones?

18. Reduce $\frac{2}{10}$ of 1 day to hours, &c.
19. Reduce $\frac{3}{8}$ of 1 minute to seconds.
20. Reduce $\frac{2}{7}$ of 1 degree to minutes, &c.
21. Reduce £$\frac{2}{720}$ to the fraction of a penny.

Solution.—We reduce the numerator to pence, the denomination required, and divide it by the denominator, as in the last article. Thus, $2\times20\times12=480$; and $480\div720=\frac{480}{720}$. Therefore £$\frac{2}{720}=\frac{480}{720}$d.$=\frac{48}{72}=\frac{4}{6}$ or $\frac{2}{3}$d. *Ans.* Hence,

298. To reduce a fraction of a *higher* denomination to an equivalent fraction of a *lower* denomination.

Reduce the given numerator to the denomination of the required fraction, and place the result over the given denominator.

OBS. 1. This process is the same in principle as to reduce a *whole* compound number to a lower denomination. (Art. 281. I.)

2. When factors common to the numerator and denominator occur, the operation may be shortened by canceling those factors. (Art. 221.)

Thus, in the last example, $\frac{2\times20\times12}{720}=$the answer.

By *Cancelation*, $\frac{2\times20\times12}{720}=\frac{2\times\not{2}\not{0}\times\not{1}\not{2}}{\not{7}\not{2}\not{0},3}=\frac{2}{3}$d. *Ans.*

22. Reduce $\frac{1}{417}$ of £1 to the fraction of a penny.
23. Reduce $\frac{4}{217}$ of 1 lb. avoirdupois to the fraction of an ounce.
24. Reduce $\frac{6}{2784}$ of 1 mile to the fraction of a rod.
25. Reduce $\frac{29}{87}$ of a day to the fraction of an hour.
26. Reduce $\frac{4}{15}$ of 1 week to the fraction of a minute.
27. Reduce $\frac{45}{69}$ of 1 yard to the fraction of a nail.
28. Reduce $\frac{63}{114}$ of 1 bushel to the fraction of a quart.
29. Reduce $\frac{25}{137}$ of 1 hhd. wine measure to the fraction of a quart.
30. Reduce $\frac{75}{185}$ of 1 lb. Troy to the fraction of an ounce.
31. Reduce $\frac{275}{1000}$ of 1 pound Troy to the fraction of a pwt.
32. Reduce $\frac{15}{327}$ of an acre to the fraction of a rod.
33. Reduce $\frac{4}{120}$ of a square yard to the fraction of a foot.
34. Reduce $\frac{7}{360}$ of a degree to the fraction of a second.

QUEST.—298. How is a fraction of a higher denomination reduced to the fraction of lower denomination?

ADDITION OF COMPOUND NUMBERS.

299. *The process of adding numbers of different denominations, is called* Compound Addition.

1. What is the sum of £6, 11s. 5d. 1 far.; £4, 9s. 6d. 2 far.; £3, 12s. 8d. 3 far.; and £8, 6s. 9d. 1 far.?

Operation.

£	s.	d.	far.	
6 ″	11 ″	5 ″	1	
4 ″	9 ″	6 ″	2	
3 ″	12 ″	8 ″	3	
8 ″	6 ″	9 ″	1	
23 ″	0 ″	5 ″	3	*Ans.*

Having placed the farthings under farthings, the pence under pence, &c., we add the column of farthings together, as in simple addition, and find the sum is 7, which is equal to 1d. and 3 far. over. Set the 3 far. under the column of farthings, and carry the 1d. to the column of pence. The sum of the pence is 29, which is equal to 2s. and 5d. over. Place the 5d. under the column of pence, and carry the 2s. to the column of shillings. The sum of the shillings is 40, which is equal to £2, and nothing over. Write a cipher under the column of shillings, and carry the £2 to the column of pounds The sum of the pounds is 23. *Ans.* £23, 0s. 5d. 3 far.

300. Hence, we derive the following general

RULE FOR ADDING COMPOUND NUMBERS.

I. *Write the numbers so that the same denominations shall stand under each other.*

II. *Beginning with the lowest denomination, find the sum of each column separately, and divide it by that number which it requires of the column added, to make* one *of the next higher denomination. Set the remainder under the column added, and carry the quotient to the next column.*

III. *Proceed in this manner with all the other denominations except the highest, whose entire sum is set down.*

Proof.—*The proof is the same as in Simple Addition.* (Art. 55.)

Obs. 1. *Fractional* compound numbers should be reduced to whole numbers of lower denominations, then added as above. (Art. 166.)

Quest.—299. What is Compound Addition? 300. How do you write compound numbers for addition? Which denomination do you add first? When the sum of any column is found, what is to be done with it? What is done with the last column?

2. Compound Addition is the same in principle as *Simple* Addition. In the latter, it is true, we uniformly carry the *tens*, and in the former we carry for *different* numbers; yet in each we always carry for that number which it takes of the order or denomination we are adding to make *one* in the next higher order or denomination.

2.				3.			4.			
£	*s.*	*d.*		£	*s.*	*d.*	£	*s.*	*d.*	*far.*
16	8	9		25	17	11	68	17	10	3
8	5	6		30	12	10	10	9	6	0
25	6	8		13	15	7	43	10	11	2
50	0	11	*Ans.*	35	16	9	65	14	8	1

5. A farmer sold to one customer 3 tons, 5 cwt. 17 lbs. 13 oz. f hay; to another, 4 tons, 7 cwt. 35 lbs. 12 oz.; to another, 1 ton, 15 cwt. 63 lbs. 7 oz.: how much hay did he sell to all?

6. What is the sum of 15 tons, 6 cwt. 45 lbs. 5 oz.; 3 tons, 17 cwt. 80 lbs. 6 oz.; 26 tons, 31 lbs. 7 oz.?

7. What is the sum of 21 lbs. 7 oz. 12 pwts. 10 grs.; 28 lbs. 5 oz. 8 pwts. 7 grs.; 7 lbs. 6 pwts. 15 grs.; 41 lbs. 6 oz. 20 grs.; 9 lbs. 7 grs.?

8. What is the sum of 16 lbs. 3 oz. 6 pwts. 19 grs.; 100 lbs. 8 oz. 16 pwts.; 97 lbs. 5 oz. 10 grs.; 115 lbs. 9 oz.?

9. Add together 19 rods, 12 ft. 8 in.; 64 rods, 13 ft. 3 in; 28 rods, 10 ft. 5 in.; 60 rods, 9 ft. 11 in.

10. Add together 5 leagues, 2 m. 4 fur. 7 rods, 4 yds.; 18 leagues, 2 m. 3 fur. 21 rods, 3 yds.; 85 leagues, 6 fur. 10 rods, 4 yds. 1 ft.

11. Add together 19 yds. 3 qrs. 3 na.; 21 yds. 2 qrs. 1 a.; 42 yds. 1 qr. 2 na.; 30 yds. 3 qrs. 2 na.

12. Add together 65 yds. 3 qrs. 1 na.; 91 yds. 2 qrs. 2 na; 100 yds. 3 qrs. 1 na.; 95 yds. 1 qr. 1 na.; 15 yds. 3 na.; 28 yds. 2 qrs.

13 Add together 17 A. 25 r. 29 sq. ft.; 49 A. 15 r. 4 sq. ft.; 62 A. 29 r. 31 sq. ft.; 10 A. 45 r. 16 sq. ft.

14. Add together 100 A. 3 R. 12 r.; 115 A. 2 R. 20 r; 160 A. 1 R. 15 r.; 91 A. 2 R. 26 r.

QUEST.—*Obs.* Does Compound Addition differ from Simple Addition?

15. One room in a house contains 15 sq. yds. 5 ft. 7 in. of plastering; another 10 yds. 7 ft. 30 in.; another 9 yds. 6 ft. 25 in.; another 7 yds. 5 ft. 63 in.: how much plastering is there in all of them?

16. A merchant bought one cask of oil containing 73 gals. 3 qts.; another 60 gals. 2 qts.; another 40 gals. 1 qt.; another 65 gals. 2 qts.: how much oil did he buy?

17. What is the sum of 20 hhds. 41 gals. 3 qts. 1 pt. 3 gi.; 31 hhds. 20 gals. 1 qt. 1 pt. 3 gi.; 48 hhds. 19 gals. 2 qts. 1 pt. 2 gi.; 81 hhds. 40 gals. 1 gi.?

18. What is the sum of 10 wks. 5 d. 12 hrs. 40 min.; 21 wks. 3 d. 9 hrs. 15 min.; 40 wks. 4 d. 17 hrs. 30 min.; 42 wks. 1 d.?

19. What is the sum of $40\frac{1}{2}$ bu. $1\frac{1}{2}$ pks. 4 qts.; 63 bu. $2\frac{1}{4}$ pks. 5 qts.; 80 bu. $7\frac{1}{4}$ pks. 1 qt.; 45 bu. 2 pks. 3 qts.; 90 bu. 1 pk.?

20. What is the sum of 7 qrs. 6 bu. 1 pk. 3 qts.; 27 qrs. 6 bu. 6 qts.; 34 qrs. 1 bu. 6 qts.; 65 qrs. 6 bu. 3 qts.?

SUBTRACTION OF COMPOUND NUMBERS.

301. *The process of finding the difference between numbers of different denominations, is called* COMPOUND SUBTRACTION.

1. From £35, 17s. 6d. 3 far., subtract £16, 9s. 8d. 2 far.

Operation.

£	s.	d.	far.	
35 "	17 "	6 "	3	
16 "	9 "	8 "	2	
19 "	7 "	10 "	1	*Ans.*

Having placed the less number under the greater, with farthings under farthings, pence under pence, &c., we subtract 2 far. from 3 far., and set the remainder 1 far. under the column of farthings. But 8d. cannot be taken from 6d.; we therefore borrow 1 from the next higher denomination, which is shillings; and 1s. or 12d. added to the 6d. make 18d. Now 8d. from 18d. leaves 10d. Since we borrowed, we must carry 1 to the next denomination in the lower number, as in simple subtraction. (Art. 72.) 1 added to 9 makes 10; and 10 from 17, leaves 7. Finally, 16 from 35, leaves 19.

Ans. £19, 7s. 10d. 1 far.

QUEST.—301. What is Compound Subtraction?

302. Hence, we derive the following general

RULE FOR SUBTRACTING COMPOUND NUMBERS.

I. *Write the less number under the greater, so that the same denominations may stand under each other.*

II. *Beginning with the lowest denomination, subtract the number in each denomination of the lower line from the number above it, and set the remainder below.*

III. *When a number in any denomination of the lower line is larger than the number above it, borrow one of the next higher denomination and add it to the number in the upper line. Subtract as before, and carry* 1 *to the next denomination in the lower line, as in subtraction of simple numbers.* (Art. 72.)

PROOF.—*The proof is the same as in Simple Subtraction.*

OBS. 1. *Fractional* compound numbers should be reduced to whole numbers of lower denominations, then subtracted as above. (Art. 166.)

2. Compound Subtraction is the same in principle as Simple Subtraction. In both cases, when the number in the lower line is larger than that above it, we borrow as many units as it takes of the order or denomination we are subtracting to make *one* of the next higher order or denomination, and in both, we carry 1 to the next figure in the lower number.

2. From £48, 17s. 6d. 2 far., take £39, 14s. 9d. 3 far.
3. From £160½, 6⅓s. 3¾d., take £100⅕, 8s.
4. From £1000, take £500, 6s. 7d. 2 far.
5. From 16 cwt. 3 qrs. 15 lbs., take 8 cwt. 2 qrs. 8 lbs. 6 oz.
6. From 85 tons 16 cwt. 39 lbs., take 61 tons 14 cwt. 68 lbs.
7. Subtract 69 m. 41 r. 12 ft. from 89 m. 10 r. 14 ft.
8. Subtract 17 l. 2 m. 3 fur. 4 r. 4 ft. from 19 l. 1 m. 2 fur. 15 r
9. Subtract 49 bu. 3 pks. 6 qts. from 85 bu. 2 pks. 4 qts.
10. Subtract 95 qrs. 4 bu. 3 pks. from 115 qrs. 3 bu. 1 pk.
11. Subtract 29 yds. 2 qrs. 3 na. from 85 yds. 1 qr. 2 na.
12. Subtract 55 yds. 2 qrs. 1 na. from 100 yds.
13. Subtract 75 gals. 3 qts. 1 pt. from 82 gals. 2 qts.

QUEST.—302. How do you write compound numbers for subtraction? Where begin to subtract? When the number in the lower line is larger than that above it, what is to be done? *Obs.* Does Compound Subtraction differ from Simple Subtraction?

15. A man having 140 A. 17 r. of land, sold 54 A. 58 r.: how much had he left?

16. Two men having bought 465 A. 48 r. of land, one of them wished to take 230 A.: how much would the other have?

17. A farmer having 144 cords, 55 ft. of wood, sold 87 c. 93 ft.: how much had he left?

18. In a certain village there are two public cisterns; one contains 446 cu. ft. 69 in., the other 785 cu. ft. 95 in.: what is the difference in their capacity?

19. The latitude of the Cape of Good Hope is 30° 55′ 15″, and that of Cape Horn, 55° 58′ 30″: what is their difference?

20. The latitude of the Straits of Gibraltar is 36° 6′ 30″, and that of the North Cape, 71° 10′: required their difference.

21. The longitude of New York is 74° 1′, and that of Cincinnati 84° 27′: required their difference.

22. From 160 yrs. 11 mo. 2 wks. 5 ds. 16 hrs. 30 min. 40 sec., take 106 yrs. 8 mo. 3 wks. 6 ds. 13 hrs. 45 min. 34 sec.

23. What is the time from Feb. 22d, 1845, to May 21st, 1847?

Operation.

yr.	*mo.*	*d.*
1847 ″	5 ″	21
1845 ″	2 ″	22
Ans. 2 ″	2 ″	29

May is the 5th month, and Feb. the 2d. Since 22 days cannot be taken from 21 d., we borrow 1 mo. or 30 d.; then say 22 from 51 leaves 29. 1 to carry to 2 makes 3, and 3 from 5 leaves 2. 5 from 7 leaves 2. Hence,

303. To find the time between two dates.

Write the earlier date under the later, placing the years on the left, the number of the month next, and the day of the month on the right, and subtract as before. (Art. 302.)

OBS. 1. The number of the month is easily determined by reckoning from January, the 1st month, February the 2d, &c. (Art. 264.)

2. In finding the time between two dates, and in casting interest, 30 days are considered a month, and 12 months a year.

3. Instead of setting down the *ordinal number* of the month, as in the solution above, some prefer to write the number of *whole* months that have

---

QUEST.—303. How do you find the time between two dates? *Obs.* In finding time between two dates, and in casting interest, how many days are considered a month? How many months a year?

elapsed in the given year. E. g., if the date is Feb. 22d, 1845, they would write 1 in the place of months; because, it is said, 2 whole months have not elapsed in the year 1845. But it may be doubted whether this method would not lead to frequent mistakes.

Besides, it may be urged with equal reason, that 1 ought to be deducted from the *day* of the month, and 1 from the *year;* for neither 22 *whole days*, nor 1845 *whole years* had elapsed at the time of the date, but the 22d day and the 1845th year were then passing. In this way, the subject, which in itself is simple, becomes intricate and perplexing.

24. General Washington was born Feb. 22d, 1732, and died Dec. 14th, 1799: how old was he?

25. The Independence of the United States was declared, July 4th, 1776: how long is it since?

26. A note was given Aug. 25th, 1840, and paid Feb. 6th, 1842: how long did it run?

27. The United States Exploring Expedition sailed from Norfolk on the 18th of Aug., 1838, and returned to New York on the 10th of June, 1842: how long was the voyage?

COMPOUND MULTIPLICATION.

304. *The process of multiplying numbers of different denominations, is called* COMPOUND MULTIPLICATION.

Ex. 1. What will 6 cows cost, at £5, 2s. $7\frac{3}{4}$d. apiece?

Operation.

£	*s.*	*d.*	*far.*	
5 ″	2 ″	7 ″	3	
			6	
30 ″	15 ″	10 ″	2	*Ans.*

Analysis.—Since 1 cow costs £5, 2s. $7\frac{3}{4}$d., 6 cows will cost 6 times as much. Beginning with the lowest denomination, 6 times 3 far. are 18 far., equal to 4d. and 2 far. over. Set the 2 far. under the denomination multiplied and carry the 4d. to the next product. 6 times 7d. are 42d. and 4d. make 46d., equal to 3s. and 10d. Set the 10d. under the pence, and carry the 3s. to the next product. 6 times 2s. are 12s. and 3s. make 15s. As the product 15s. does not make *one* in the next denomination, we set it under the column multiplied. Finally, 6 times £5 are £30. The answer is £30, 15s. $10\frac{1}{2}$d.

QUEST.—304. What is Compound Multiplication?

305. Hence, we deduce the following general

RULE FOR MULTIPLYING COMPOUND NUMBERS.

Multiply each denomination separately, beginning with the lowest, and divide each product by that number which it takes of the denomination multiplied, to make ONE *of the next higher; set down the remainder, and carry the quotient to the next product, as in addition of compound numbers.* (Art. 300.)

OBS. 1. When the multiplier is a composite number, it is advisable to multiply first by one factor and that product by the other. (Art. 97.)

2. Compound Multiplication is the same in principle as Simple Multiplication. In each we carry for *that number* which it takes of the order or denomination we are multiplying, to make *one* of the next higher order or denomination.

2. What will 28 horses cost, at £21, 3s. $7\frac{1}{4}$d. apiece?

Operation.

£	s.	d.	far.	
21 "	3 "	7 "	1	
			7	
148 "	5 "	2 "	3	
			4	
593 "	0 "	11 "	0	*Ans.*

We multiply by the factors of 28, which are 7 and 4, and set down each result as above.

3. What cost 7 acres of land, at £35, 6s. 7d. per acre?

4. What cost 18 barrels of flour, at £1, 6s. $8\frac{1}{2}$d. per barrel?

5. A man bought 15 loads of hay, each weighing 1 T. $270\frac{1}{3}$lbs.: what was the weight of the whole?

6. Multiply 16 tons, 3 cwt. $10\frac{1}{5}$lbs. by 25.

7. Multiply 12 lbs. 3 oz. 16 pwts. by 56.

8. If 1 dollar weighs 17 pwts. $4\frac{1}{2}$ grs., how much will 96 dollars weigh?

9. Multiply 48 hhds. 15 gals. 2 qts. 1 pt. by 63.

10. Multiply 56 pipes, 1 hhd. 23 gals. by 100.

QUEST.—305. Where do you begin to multiply a compound number? What is done with each product? *Obs.* When the multiplier is a composite number, how proceed? Does it differ from Simple Multiplication?

11. Bought 72 pieces of cloth, each containing $32\frac{3}{4}$ yards: how much did they all contain?

12. If 1 cloak requires 10 yds. 3 qrs., how much will 500 cloaks require?

13. Multiply 175 miles, 7 fur. 18 rods by 84.

14. Multiply 40 leagues, 2 m. 5 fur. 15 r. by 50.

15. Multiply 149 bu. 12 qts. by 60.

16. Multiply 26 qrs. 7 bu. 3 pks. 5 qts. by 110.

17. Multiply 150 acres, 65 rods by 52.

18. Multiply 310 acres, 3 roods, 3 rods by 81.

19. Multiply 265 cu. ft. 10 in. by 93.

20. Multiply 148 cords, $29\frac{8}{13}$ ft. by 650.

21. Multiply 365 d. 5 hrs. 48 min. 48 sec. by 35.

22. Multiply 70 yrs. 6 mo. 3 wks. 5 d. by 17.

23. Multiply 75° 40′ 21″ by 210.

24. If a ship sails 3° 24′ 10″ per day, how far will she sail in 60 days?

25. If 1 acre produce 45 bu. 26 qts., how much will 100 acres produce?

26. If 1 barrel of flour requires 4 bu. 3 pks. 5 qts. of wheat, how much will 500 barrels require?

27. What cost a chest of tea containing 17 lbs., at 6s. $10\frac{1}{2}$d. per pound?

28. What is the duty on 1000 gals. of brandy, at 13s. 7d. per gallon?

29. What is the duty on 10560 lbs. of sugar, at 6d. 3 far. per pound?

30. What is the duty on 1500 yards of broadcloth, at 6s. $9\frac{1}{4}$d. per yd.?

31. If 1 load of wood measures 117 ft. 110 in., how much will 40 loads of the same size measure?

32. If 1 quarter of beef weighs 216 lbs. 7 oz., how much will 4 quarters weigh?

33. If 1 bushel of salt weighs 72 lbs. 10 oz., how much will 350 bushels weigh?

34. If 1 cask of oil contains 86 gals. 2 qts. 1 pt., how much will 100 casks of the same size contain?

COMPOUND DIVISION.

306. *The process of dividing numbers of different denominations, is called* COMPOUND DIVISION.

Ex. 1. Divide £25, 3s. 4d. 2 far. by 6.

Operation.

£	s.	d.	far.	
6)25	3	4	2	
4	3	10	3	*Ans.*

Beginning with the pounds, we find 6 is contained in £25, 4 times and 1 over. Set the 4 under the pounds, and reduce the remainder £1 to shillings, which added to the 3s. make 23s. 6 in 23s. 3 times and 5s. over. Set the 3 under the shillings, and reduce the remainder 5s. to pence, which added to the 4d. make 64d. 6 in 64d., 10 times and 4d. over. Set the 10 under the pence, reduce the 4d. to farthings, and divide as before *Ans.* £4, 3s. 10d. 3 far.

307. Hence, we deduce the following general

RULE FOR DIVIDING COMPOUND NUMBERS.

Begin with the highest denomination, and divide each separately. Reduce the remainder, if any, to the next lower denomination, to which add the number of that denomination contained in the given example, and divide the sum as before. Proceed in this manner through all the denominations.

OBS. 1. Each partial quotient will be of the same denomination, as that part of the dividend from which it arose.

2. When the divisor exceeds 12, and is a composite number, it is advisable to divide first by one factor and that quotient by the other. (Art. 129.) If the divisor exceeds 12, but is not a composite number, long division may be employed. (Art. 120. II.)

3. Compound Division is the same in principle, as Simple Division. *Prefixing* the remainder to the next figure of the dividend in simple division, is the same as *reducing* it to the next lower order or denomination, and *adding* the next figure to it.

QUEST.—306. What is Compound Division? 307. Where do you begin to divide a compound number? What is done with the remainder? *Obs.* Of what denomination is each partial quotient? When the divisor is a composite number, how proceed? Does it differ from Simple Division?

2. A man wished to divide 75 cwt. 2 qrs. 10 lbs. of beef equally among 35 families: how much could he give to each?

Operation.

cwt.	*qrs.*	*lbs.*
7)75 ″	2 ″	10
5)10 ″	3 ″	5
2 ″	0 ″	16 *Ans.*

We divide by the factors of 35, which are 7 and 5, and set down each result as above.

3. Divide 312 lbs. 9 oz. 18 pwts. by 43.

Ans. 7 lbs. 3 oz. 6 pwts.

4. Divide 410 lbs. 4 oz. 5 pwts. 6 grs. by 8.

5. Divide 786 bu. 18 qts. by 25.

6. A farmer raised 1000 bu. 3 pks. 6 qts. of wheat on 40 acres: how much was that per acre?

7. A man bought 10 horses for £200, 15s.: how much did he give apiece?

8. Divide £87, 10s. $7\frac{1}{2}$d. by 18.

9. A merchant tailor put 216 yds. 3 qrs. of cloth into 20 cloaks: how much cloth did each cloak contain?

10. Divide 500 yds. 3 qrs. 2 na. by 54.

11. A man traveled 1000 miles in 12 days: at what rate did he travel per day?

12. Divide 1500 m. 2 fur. 30 r. 12 ft. by 7.

13. Divide 120 gals. 3 qts. 1 pt. by 72.

14. Divide 400 hhds. 10 gals. 2 qts. 1 pt. by 9.

15. Divide 365 d. 10 hr. 40 min. by 15.

16. Divide 111 yrs. 20 d. 13 hrs. 25 min. 10 sec. by 11.

17. Divide 45° 17′ 10″ by 25.

18. Divide 65 signs 12° 47′ by 41.

19. Divide 164 cords, 30 ft. by 17.

20. Divide 410 cords, 10 ft. 21 in. by 61.

21. If a chest of tea weighing 96 pounds cost £33, what will 1 pound cost?

22. If the duty on a pipe of wine is £50, 6s. 6d., what is the duty per gallon?

23. If a person spends £200 a year, what are his expenses per day?

SECTION IX.

DECIMAL FRACTIONS.

308. *Fractions which decrease in a tenfold ratio, or which express simply tenths, hundredths, thousandths, &c., are called* DECIMAL FRACTIONS.

They arise from dividing a *unit* into *ten* equal parts, then dividing each of these parts into *ten other* equal parts, and so on. Thus, if a *unit* is divided into 10 equal parts, 1 of those parts is called a *tenth.* (Art. 178.) If a *tenth* is divided into 10 equal parts, 1 of those parts will be a *hundredth;* for, $\frac{1}{10}\div10=\frac{1}{100}$. If a *hundredth* is divided into 10 equal parts, 1 of the parts will be a *thousandth;* for, $\frac{1}{100}\div10=\frac{1}{1000}$, &c. (Art. 227.)

OBS. Fractions of this class are called *decimals*, because they regularly *decrease* in a *tenfold ratio.* (Art. 37. Obs. 2.)

Decimal fractions are said to have been invented by Lord Napier, in 1602.

309. Each order of whole numbers, we have seen, *increases* in value from units towards the left in a tenfold ratio; and, conversely, each order must *decrease* from left to right in the same ratio, till we come to units' place again. (Art. 36.)

310. By extending this scale of notation below units towards the right hand, it is manifest that the *first* place on the right of units, will be *ten times* less in value than *units'* place; that the *second* will be ten times less than the *first;* the *third* ten times less than the *second,* &c.

Thus we have a series of *orders* below units, which *decrease* in a *tenfold ratio,* and exactly correspond in value with *tenths, hundredths, thousandths,* &c. (Art. 308.)

QUEST.—308 What are Decimal Fractions? From what do they arise? *Obs.* Why called decimals? 309. In what manner do whole numbers increase and decrease? 310. By extending this scale below units, what would be the value of the first place on the right of units? The second? The third? With what do these orders correspond in value?

311. *Decimal Fractions* are commonly expressed by writing the numerator with a point (.) before it.

The point placed before decimals is called the *Decimal Point*, or *Separatrix*. Its object is to distinguish the fractional parts from whole numbers.

If the numerator does not contain so many figures as there are ciphers in the denominator, the deficiency must be supplied by prefixing ciphers to it. For example, $\frac{1}{10}$ is written thus .1 ; $\frac{2}{10}$ thus .2 ; $\frac{3}{10}$ thus .3 ; &c. $\frac{1}{100}$ is written thus .01, putting the 1 in hundredths' place ; $\frac{5}{100}$ thus .05 ; &c. That is, tenths are written in the *first* place on the right of units ; hundredths in the *second* place ; thousandths in the *third* place, &c.

312. *The denominator of a decimal fraction is always* 1 *with as many ciphers annexed to it, as there are figures in the given numerator.* (Art. 308.)

313. The names of the different *orders of decimals*, or places below units, may be easily learned from the following

DECIMAL TABLE.

Hundred thousands.	Ten thousands.	Thousands.	Hundreds.	Tens.	*Units.*	(Decimal Point.)	Tenths.	Hundredths.	*Thousandths.*	Ten thousandths.	Hundred thousandths.	*Millionths.*	Ten millionths.	Hundred millionths.	*Billionths.*	Ten billionths.	Hundred billionths.	*Trillionths, &c.*
7	5	6	4	2	3	.	2	6	7	1	4	5	9	8	6	2	7	4

314. It will be seen from this table that the *value* of each figure in *decimals*, as well as in whole numbers, depends upon the *place* it occupies, reckoning from units. Thus, if a figure stands in the *first* place on the right of units, it expresses *tenths ;* if in

QUEST.—311. How are decimal fractions expressed ? What is the point placed before decimals called ? 312. What is the denominator of a decimal fraction ? 313. Repeat the Decimal Table, beginning units, tenths, &c. 314. Upon what does the value of a decimal depend ?

the *second*, *hundredths*, &c.; each successive place or order towards the right, decreasing in value in a tenfold ratio. Hence,

315. *Each removal of a decimal figure one place from units towards the right, diminishes its value ten times.*

Prefixing a cipher, therefore, to a decimal diminishes its value *ten times*; for, it removes the decimal one place farther from units' place. Thus, $.4=\frac{4}{10}$; but $.04=\frac{4}{100}$; and $.004=\frac{4}{1000}$, &c.; for the denominator to a decimal fraction is 1 with as many ciphers annexed to it, as there are figures in the numerator. (Art. 312.)

Annexing ciphers to decimals does not alter their value; for, each significant figure continues to occupy the same place from units as before. Thus, $.5=\frac{5}{10}$; so $.50=\frac{50}{100}$, or $\frac{5}{10}$, by dividing the numerator and denominator by 10; (Art. 191,) and $.500=\frac{500}{1000}$, or $\frac{5}{10}$, &c.

OBS. 1. It should be remembered that the *units'* place is always the *right hand* place of a whole number. The effect of annexing and prefixing ciphers to decimals, it will be perceived, is the *reverse* of annexing and prefixing them to whole numbers. (Art. 98.)

2. A whole number and a decimal, written together, is called a *mixed number*. (Art. 183.)

316. To read decimal fractions.

Beginning at the left hand, read the figures as if they were whole numbers, and to the last one add the name of its order. Thus,

.7	is	read	7 tenths.
.36	"	"	36 hundredths.
.475	"	"	475 thousandths.
.6342	"	"	6342 ten thousandths.
.57834	"	"	57834 hundred thousandths.
.284648	"	"	284648 millionths.
.8913329	"	"	8913629 ten millionths.

OBS. In reading decimals as well as whole numbers, the *units'* place should always be made the *starting* point. It is advisable for the learner to apply to

QUEST.—315. What is the effect of removing a decimal one place towards the right? What then is the effect of prefixing ciphers to decimals? What, of annexing them? *Obs.* Which is the units' place? What is a whole number and a decimal written together, called? 316. How are decimals read? *Obs.* In reading decimals, what should be made the starting point?

every figure the name of its order, or the place which it occupies, before attempting to read them. Beginning at the units' place, he should proceed towards the right, thus—*units*, *tenths*, *hundredths*, *thousandths*, &c., pointing to each figure as he pronounces the name of its order. In this way he will be able to read decimals with as much ease as he can whole numbers.

Read the following numbers:

(1.)	(2.)	(3.)	(4.)
.32	.46274	42.068	2.463126
.246	.03687	17.401	6.004534
3624	.00368	23.07	1.100492
.82344	.00046	81.4389	9.000028
.13236	.00009	90.0104	8.001249

(5.)	(6.)	(7.)	(8.)
12.683	6.00754	4.306702	9.2000076
20.064	3.0468	0.007006	8.0403842
35.0072	2.306843	1.13004	0.0000008
67.4008	1.710386	9.203167	4.3008004

Note.—Sometimes we pronounce the word *decimal* when we come to the separatrix, and then read the figures as if they were whole numbers; or, simply repeat them one after another. Thus, 125.427 is read, one hundred twenty-five, *decimal* four hundred twenty-seven; or, one hundred twenty-five, *decimal* four, two, seven.

Write the fractional part of the following numbers in decimals:

(9.)	(10.)	(11.)	(12.)
$25\frac{7}{10}$	$4\frac{7}{100}$	$43\frac{2143}{100000}$	$3\frac{312647}{10000000}$
$30\frac{62}{100}$	$6\frac{32}{1000}$	$13\frac{6}{100000}$	$8\frac{4}{10000000}$
$72\frac{80}{100}$	$7\frac{60}{1000}$	$41\frac{281}{100000}$	$9\frac{78234567}{100000000}$

13. Write 9 tenths; 25 hundredths; 45 thousandths.
14. Write 6 hundredths; 7 thousandths; 132 ten thousandths.
15. Write 462 thousandths; 2891 ten thousandths.
16. Write 25 hundred thousandths; 25 millionths.
17. Write 1637246 ten millionths; 65 hundred millionths.
18. Write 71 thousandths; 7 millionths.
19. Write 23 hundredths; 19 ten thousandths.
20. Write 261 hundred thousandths; 65 hundredths; 121 millionths; 751 trillionths.

QUEST.—*Note.* What other method of reading decimals is mentioned?

317. *Decimal Fractions*, it will be perceived, differ from *Common* Fractions both in their *origin* and in the *manner of expressing them*.

Common Fractions arise from dividing a *unit* into *any* number of *equal* parts; consequently, the *denominator* may be *any number whatever*. (Art. 182.) Decimals arise from dividing a *unit* into *ten equal* parts, then subdividing each of those parts into *ten other equal* parts, and so on; consequently, the denominator is always 10, 100, 1000, &c. (Arts. 308, 312.)

Again, *Common Fractions* are expressed by writing the *numerator* over the *denominator; Decimals* are expressed by writing the *numerator only*, with a point before it, while the denominator is understood. (Arts. 182, 311.)

318. Decimals are *added, subtracted, multiplied*, and *divided*, in the same manner as whole numbers.

OBS. The only thing with which the learner is likely to find any difficulty, is *pointing off* the answer. To this part of the operation he should give particular attention.

ADDITION OF DECIMAL FRACTIONS.

319. Ex. 1. What is the sum of 28.35; 345.329; 568.5; and 6.485?

Operation.

```
 28.35
345.329
568.5
  6.485
-------
948.664  Ans.
```

Write the *units* under *units, tenths* under *tenths, hundredths* under *hundredths*, &c.; then, beginning at the right hand or lowest order, proceed thus: 5 thousandths and 9 thousandths are 14 thousandths. Write the 4 under the column added, and carrying the 1 to the next column, proceed through all the orders in the same manner as in simple addition. (Art. 54.) Finally, place the decimal point in the amount directly under that in the numbers added.

QUEST.—317. How do decimals differ from common fractions? From what do common fractions arise? From what do decimals arise? How are common fractions expressed? How are decimals?

320. Hence, we deduce the following general

RULE FOR ADDITION OF DECIMALS.

Write the numbers so that the same orders may stand under each other, placing units under units, tenths under tenths, hundredths under hundredths, &c. Begin at the right hand or lowest order, and proceed in all respects as in adding whole numbers. (Art. 54.)

From the right hand of the amount, point off as many figures for decimals as are equal to the greatest number of decimal places in either of the given numbers.

PROOF.—*Addition of Decimals is proved in the same manner as Simple Addition.* (Art. 55.)

Note.—The decimal point in the *answer* will always fall directly under the decimal points in the given numbers.

EXAMPLES.

2. What is the sum of 25.7; 8.389; 23.056? *Ans.* 57.145.

3. What is the sum of 36.258; 2.0675; 382.45; and 7.3984?

4. What is the sum of 32.764; 5.78; 16.0037; and 49.3046?

5. What is the sum of 1.03041; 6.578034; 2.4178; and 4.72103?

6. Add together 4.25; 6.293; 4.612; 38.07; 2.056; 3.248; and 1.62.

7. Add together 35.7603; 47.0076; 129.03; 100.007; and 20.32.

8. Add together 467.3004; 28.78249; 1.29468; and 3.78241.

9. Add together 21.6434; 800.7; 29.461; 1.7506; and 3.45.

10. Add together 45.001; 163.4234; 20.3045; 634.2104; and 234.90213.

11. Add together 293.0072; 89.00301; 29.84567; 924.00369; and 72.39602.

12. Add together 1.721341; 8.620047; 51.720345; 2.684; and 62.304607.

13. Add together 1.293062; 3.00042; 9.7003146; 3.600426, 7.0040031; and 8.7200489.

QUEST.—320. How are decimals added? How point off the answer? How is addition of decimals proved?

14. Add together 394.61; 81.928; 3624.8103; 640.203; 8291.302; 721.004; and 3920.304.

15. Add together 25 hundredths, 8 tenths, 65 thousandths, 16 hundredths, 142 thousandths, and 39 hundredths.

16. Add together 9 tenths, 92 hundredths, 162 thousandths. 489 thousandths, and 92 millionths.

17. Add together 45 thousandths, 1752 millionths, 624 ten millionths, and 24368 millionths.

18. Add together 29 hundredths, 7 millionths, 62 thousandths, and 12567 ten millionths.

19. Add together 95 thousandths, 61 millionths, 6 tenths, 11 hundredths, and 265 hundred thousandths.

20. Add together 1 tenth, 2 hundredths, 16 thousandths, 7 millionths, 26 thousandths, 95 ten millionths, and 7 ten thousandths.

21. Add together 96 hundred thousandths, 92 millionths, 25 hundredths, 45 thousandths, and 7 tenths.

22. Add together 85 thousandths, 17 hundredths, 36 ten thousandths, 58 millionths, 363 hundred thousandths, 185 millionths, and 673 ten thousandths.

SUBTRACTION OF DECIMAL FRACTIONS.

321. Ex. 1. From 425.684 subtract 216.96.

Operation.

```
 425.684
 216.96
 -------
 208.724. Ans.
```

Having written the less number under the greater, so that units may stand under units, tenths under tenths, &c., we proceed exactly as in subtraction of whole numbers. (Art. 72.) Thus 0 thousandths from 4 thousandths leaves 4 thousandths. Write the 4 in the thousandths' place. Since the figure 9 in the lower line is larger than the one above it, we borrow 10. Now 9 from 16 leaves 7; set the 7 under the column and carry 1 to the next figure. (Art. 72.) Proceed in the same manner with the other figures in the lower number. Finally, place the decimal point in the remainder directly under that in the given number.

322. Hence, we deduce the following general

RULE FOR SUBTRACTION OF DECIMALS.

Write the less number under the greater, with units under units, tenths under tenths, hundredths under hundredths, &c. Subtract as in whole numbers, and point off the answer as in addition of decimals. (Art. 320.)

PROOF.—*Subtraction of Decimals is proved in the same manner as Simple Subtraction.* (Art. 73.)

Note.—When there are blank places on the right hand of the upper number, they may be supplied by ciphers without altering the value of the decimal (Art. 315.)

EXAMPLES.

2. From 456.0546 take 364.3123. *Ans.* 91.7423.
3. From 1460.39 take 32.756218.
4. From 21.67 take .682349.
5. From 81.6823401 take 9.163.
6. From 100.536 take 19.36723.
7. From .076345 take .009623478.
8. From 1 take .99.
9. From 10 take .000001.
10. From 65.00001 take .9682347.
11. From 24681 take .87623.
12. What is the difference between 25 and .25?
13. What is the difference between 3.29 and .999?
14. What is the difference between 10 and .0000001?
15. What is the difference between 9 and .999999?
16. What is the difference between 4636 and .4654?
17. What is the difference between 25.6050 and 567.392?
18. What is the difference between 76.2784 and 29.84234?
19. What is the difference between .0000001 and .0001?
20. What is the difference between .0000004 and .00004?
21. What is the difference between 32 and .00032?

QUEST.—322. How are decimals subtracted? How point off the answer? How is subtraction of decimals proved?

22. What is the difference between .00045 and 45?
23. What is the difference between .00000099 and 99?
24. From 1 thousandth take 1 millionth.
25. From 7 hundred take 7 hundredths.
26. From 29 thousand take 92 thousandths.
27. From 256 millions take 256 thousandths.
28. From 46 hundredths take 46 thousandths.
29. From 95 thousandths take 909 ten thousandths.
30. From 1 billionth take 1 trillionth.
31. From 2874 millionths take 211 billionths.
32. From 6231 hundred thousandths take 154 millionths.
33. From 7213 ten thousandths take 431 hundred thousandths.
34. From 8436 hundred millionths take 426 ten billionths.

MULTIPLICATION OF DECIMALS.

323. Ex. 1. If a man can reap .96 of an acre in a day, how much can he reap in .5 of a day?

Analysis.—Since he can reap 96 hundredths of an acre in a whole day, in 5 tenths of a day he can reap 5 tenths as much. But multiplying by a fraction we have seen, is taking a part of the multiplicand as many times as there are like parts of a unit in the multiplier. (Art. 210.) Hence, multiplying by .5, which is equal to $\frac{5}{10}$ or $\frac{1}{2}$, is taking *half* of the multiplicand *once.* Now .96, or $\frac{96}{100} \div 2 = \frac{48}{100}$. (Art. 227.) But $\frac{48}{100} = .48$. (Art. 311.)

Operation.

```
 .96
  .5
----
.480 Ans.
```

We multiply as in whole numbers, and pointing off as many decimals in the product as there are decimal figures in both factors, we have 480. But ciphers placed on the right of decimals do not affect their value; the 0 may therefore be omitted, and we have .48 for the answer.

	(2.)	(3.)	(4.)
Multiply	25.38	360.085	6843.02
By	.42	.0043	6.5
	10.6596 *Ans.*	1.5483655 *Ans.*	44479.630 *Ans.*

324. From the preceding illustrations we deduce the following general

RULE FOR MULTIPLICATION OF DECIMALS.

Multiply as in whole numbers, and point off as many figures from the right of the product for decimals, as there are decimal places both in the multiplier and multiplicand.

If the product does not contain so many figures as there are decimals in both factors, supply the deficiency by prefixing ciphers.

PROOF.—*Multiplication of Decimals is proved in the same manner as Simple Multiplication.*

OBS. The *reason* for pointing off as many decimal places in the product as there are decimals in both factors, may be illustrated thus:

Suppose it is required to multiply .25 by .5. Supplying the denominators $.25=\frac{25}{100}$, and $.5=\frac{5}{10}$. (Art. 312.) Now $\frac{25}{100}\times\frac{5}{10}=\frac{125}{1000}$. (Art. 219.) But $\frac{125}{1000}=.125$; (Art. 311;) that is, the product of $.25\times.5$, contains just as many decimals as the factors themselves. In like manner it may be shown that the product of any two or more decimal numbers, must contain as many decimal figures as there are places of decimals in the given factors.

EXAMPLES.

Ex. 1. In 1 rod there are 16.5 feet: how many feet are there in 41.3 rods?

2. In 1 degree there are 69.5 statute miles: how many miles are there in 360 degrees?

3. In 1 barrel there are 31.5 gallons: how many gallons in 65.25 barrels?

4. In 1 inch there are 2.25 nails: how many nails are there in 60.5 inches?

5. In 1 square rod there are 30.25 square yards: how many square yards are there in 26.05 rods?

6. In one square rod there are 272.25 square feet: how many square feet are there in 160 rods?

QUEST.—324. How are decimals multiplied together? How do you point off the product? When the product does not contain so many figures as there are decimals in both factors, what is to be done? How is multiplication of decimals proved?

7. How many square rods are there in a field 60.5 rods long and 40.75 rods wide?

Multiply the following decimals:

8. 1.0013×.25.
9. 44.046×.43.
10. 3.6051×4.1.
11. 0.1003×6.12.
12. 8.0004×.004.
13. 35.601×1.032.
14. 213.02×4.318.
15. 0.0006×.00012.
16. 0.3005×.0035.
17. 10.2106×38.26.
18. 164.023×1.678.
19. 9.40061×15.812.
20. 7.31042×10.021.
21. 40.4369×1.2904.
22. 100.0008×.000306.
23. 75.35060×62.3906.
24. 31.50301×17.0352.
25. 0.000713×2.30561.
26. 42.10062×3.821013.
27. 1.0142034×.0620034.
28. 25067823×.0000001
29. 64.301257×1.000402.
30. 394.20023×.00000003.
31. 2564.21035×4.300506.
32. 840003.1709×112.10371.
33. 0.834567834×.00000008.

CONTRACTIONS IN MULTIPLICATION OF DECIMALS.

CASE I.

325. When the multiplier is 10, 100, 1000, &c., the multiplication may be performed by simply *removing the decimal point* as many places towards the *right*, as there are ciphers in the multiplier. (Arts. 99, 324.)

1. Multiply 85.4321 by 100. *Ans.* 8543.21.
2. Multiply 42930.213401 by 10.
3. Multiply 1067.2350123 by 100.
4. Multiply 608.34017 by 1000.
5. Multiply 30.467214067 by 10000.
6. Multiply 446.3214032 by 100000.
7. Multiply 21.3456782106 by 100000.
8. Multiply 5 tenths by 1000.
9. Multiply 75 hundredths by 100000.
10. Multiply 65 ten thousandths by 1000.
11. Multiply 48 hundred thousandths by 100000.

Quest.—325. How proceed when the multiplier is 10, 100, &c.?

12. Multiply 248 thousandths by 10000.
13. Multiply 381 ten thousandths by 10000.
14. Multiply 6504 ten millionths by 100000.
15. Multiply 834 thousandths by 1000000.
16. Multiply 1 millionth by 10000000.

CASE II.

326. When the number of decimal places in the multiplier and multiplicand is large, the number of decimals in the product must also be large. But decimals below the fifth or sixth order, express so small parts of a unit, that when obtained, they are commonly rejected. It is therefore desirable to avoid the unnecessary labor of obtaining those which are not to be used.

17. It is required to multiply 1.3569 by .36742, and retain five places of decimals.

First Operation.

1.3569		
.36742		
.40707		
8141	4	
949	83	
54	276	
2	7138	
.49855	2198	*Ans.*

It is evident from the nature of decimal notation, that if the partial product of each figure in the multiplier is advanced one place to the right instead of the left, the operation will correspond with the descending scale, and at the same time will give the true product. (Art. 86. Obs.) But since only five decimals are required, those on the right of the perpendicular are useless. Our present object is to show how the answer can be obtained without them.

Contraction.

1.356	9	
.3674	2	
.4070	7	
814	1	
95	0	
5	4	
	3	
.4985	5	*Ans.*

Beginning at the right hand, we will first multiply the multiplicand by the tenths' figure of the multiplier, and place the first figure of the partial product under the figure multiplied. In obtaining the second partial product, (i. e. multiplying by 6,) it is plain we may omit the *right hand* figure of the multiplicand, for, if multiplied, its product will fall to the right of the perpendicular line, and therefore will not be used. But if we multiply 9 into 6, the product will be 54; consequently there would be 5 to carry to the next product; we therefore carry 5 to 36, which makes 41. Again, in the third partial product, (i. e. in multiplying by 7,) we may omit the *two* right hand figures of the multiplicand; for, their product will fall to the right of the perpendicular line. But by recurring to the rejected figures, it will be seen that the product of 7 into 6 is 42, and 6 to carry make 48; we therefore add 5 to the product of 7 into 5, because 48 is nearer 50 than 40; consequently it is nearer the truth to carry 5 than to carry 4. In the fourth partial product we may omit the *three* right hand figures, and in the fifth or last, the *four* right hand figures.

18. Multiply .2356 by .3765, and retain 4 decimals in the product.

Operation.

```
 .2356
 .3765
 -----
  0707
   165
    14
     1
 -----
 .0887 Ans.
```

Multiplying as before, the first figure of the partial product must be set in the fifth order, or one place to the right of the figure multiplied; for, there are 4 decimals in the multiplicand and the one by which we multiply makes 5. (Art. 324.) But since we wish to retain only 4 decimals in the product, we may omit this figure, carrying 2 to the next product. Proceed in the same manner with the other figures in the multiplier. Finally, the sum of the partial products which are retained, is the answer required. Hence,

327. To multiply decimals and retain only a given number of decimal figures in the product.

Count off in the multiplicand as many decimal places less one, as are required in the product. Then beginning at the right hand figure counted off, multiply the multiplicand by the tenths or first decimal figure of the multiplier, and set the first figure of the partial product one place to the right of the figure multiplied, increasing it by the nearest number of tens that would arise from the

QUEST.—327. How multiply decimals, and retain a given number of figures in the product?

rejected figure if multiplied. Next multiply by the second decimal figure, omitting the next right hand figure of the multiplicand and carrying as before. Proceed in the same manner with all the figures of the multiplier whose product will come under the decimal places counted off, omitting an additional figure on the right of the multiplicand, as you multiply by each successive figure, and set the first figure of each partial product under that of the preceding. Finally, from the sum of the partial products, cut off the required number of decimals, and the result will be the answer.

OBS. 1. In order to determine where to place the *decimal point* in the product, we have only to observe that the product of the right hand figure of the multiplicand into the *tenths* of the multiplier is of the order denoted by the *sum* of the orders of the two figures multiplied; (Art. 324;) and when the multiplier is tenths it is of the order *next lower* than the figure multiplied. For this reason the first partial product is set one place to the right of the figure multiplied. But since we count off one decimal less than is required in the product, the right hand figure in the sum of the partial products must consequently be the right hand decimal place in the answer.

2. If the multiplier contains *units, tens, hundreds*, &c., in multiplying by the units, we must begin *one* figure to the right of those counted off, and set the first figure of the partial product under the figure multiplied. In multiplying by the tens, we must begin *two* figures to the right of those counted off, and set the first figure of the partial product under that of the units; in multiplying by the hundreds, we must begin *three* figures to the right, and set the first figure of the partial product under that of the preceding, &c. This will bring the same orders under each other.

19. Multiply .72543414 by .24826421 retaining 5 decimal places in the product.

Operation.

```
.7254'3414
.2482 6421
----------
 1450 9
  290 2
   58 0
    1 4
      4
----------
.1800 9 Ans.
```

Having counted off 4 decimals in the multiplicand, increase the product of 2 into 4 by 1, because the product of the 3 rejected into 2 is nearer 10 than 0. Set the 9 one place to the right of the figure multiplied.

The 4 in the last partial product, is the number which would be carried to this order, if the 7 were multiplied by 6.

20. Multiply 67.1498601 by 92.4023553 retaining four decimals in the product.

Operation.

```
  67.149'8601
  92.402 3553
 ------------
 6043.487 4
  134 299 7
   26 859 9
      134 3
       20 1
        3 4
          3
 ------------
 6204.805 1  Ans.
```

In this operation we multiply first by the tens figure of the multiplier, beginning two places to the right of those counted off in the multiplicand. It is immaterial as to the result whether we multiply by the tenths first, or by the units, tens, or hundreds, provided we set the first figure of the partial product in its proper place. (Art. 327. Obs. 2.)

21. Multiply .863541 by .10983 retaining 5 decimal places.
22. Multiply 1.123674 by 1.123674 retaining 6 decimal places.
23. Multiply .26736 by .28758 retaining 4 decimal places.
24. Multiply .1347866 by .288793 retaining 7 decimal places.
25. Multiply .681472 by .01286 retaining 5 decimal places.
26. Multiply .053407 by .047126 retaining 6 decimal places.
27. Multiply .3857461 by .0046401 retaining 6 decimal places.

DIVISION OF DECIMAL FRACTIONS.

328. Ex. 1. How many bushels of oats, at .2 of a dollar a bushel, can you buy for .84 of a dollar?

Analysis.—Since 2 tenths of a dollar will buy 1 bushel, 84 hundredths of a dollar will buy as many bushels, as 2 tenths is contained times in 84 hundredths. Now $.84=\frac{84}{100}$; and $.2=\frac{2}{10}$, or $\frac{20}{100}$. (Art. 191.) And $\frac{84}{100}\div\frac{20}{100}=4\frac{4}{20}$, or $4\frac{2}{10}$. But, (Art. 311,) $4\frac{2}{10}=4.2$, which is the answer required.

Operation.

```
.2).84
   4.2 Ans.
```

We divide as in whole numbers, and point off one decimal figure in the quotient.

OBS. The *reason* for pointing off one decimal figure in the quotient may be thus explained.

We have seen in the multiplication of decimals, that the product has as many decimal figures, as the multiplier and multiplicand. (Art. 324.) Now since the dividend is equal to the product of the divisor and quotient, (Art. 112,) it follows that the *dividend* must have as many decimals as the *divisor* and *quotient together;* consequently, as the dividend has *two* decimals, and the divisor but *one*, we must point off *one* in the quotient. In like manner it may be shown universally, that

329. *The quotient must have as many decimal figures, as the decimal places in the dividend exceed those in the divisor; that is, the decimal places in the divisor and quotient together, must be equal in number to those in the dividend.*

2. What is the quotient of 3.775 divided by .2.5? *Ans.* 1.51.

3. What is the quotient of .0072 divided by 2.4.

Operation.
2.4).0072(.003 *Ans.*
72

Since the dividend has three decimals more than the divisor, the quotient must have three decimals. But as it has but one figure, we prefix two ciphers to it to make up the deficiency.

OBS. It will be noticed that 3, the first figure of the quotient, denotes *thousandths;* also the product of 2, the units figure of the divisor, into the first quotient figure, is written under the *thousandths* in the dividend. Hence,

The first figure of the quotient is of the same order, as that figure of the dividend under which is placed the product of the units of the divisor into the first quotient figure.

330. From the preceding illustrations we deduce the following general

RULE FOR DIVISION OF DECIMALS.

Divide as in whole numbers, and point off as many figures for decimals in the quotient, as the decimal places in the dividend exceed those in the divisor. If the quotient does not contain figures enough, supply the deficiency by prefixing ciphers.

PROOF.—*Division of Decimals is proved in the same manner as Simple Division.* (Art. 121.)

OBS. 1. When the number of decimals in the divisor is the *same* as that in the dividend, the quotient will be a *whole* number.

QUEST.—330. How are decimals divided? How point off the quotient? How is division of decimals proved?

2. When there are *more* decimals in the divisor than in the dividend, annex as many ciphers to the dividend as are necessary to make its decimal places *equal* to those in the divisor. The quotient thence arising will be a whole number. (Obs. 1.)

3. After all the figures of the dividend are divided, if there is a remainder, ciphers may be annexed to it and the division continued at pleasure. The ciphers annexed must be regarded as decimal places belonging to the dividend.

Note.—1. For ordinary purposes, it will be sufficiently exact to carry the quotient to three or four places of decimals; but when great accuracy is required, it must be carried farther.

2. When there is a remainder, the sign +, should be annexed to the quotient, to show that it is not complete.

EXAMPLES.

4. How many boxes will it require to pack 71.5 lbs. of butter, if you put 5.5 lbs. in a box?

5. How many suits of clothes will 29.6 yds. of cloth make, allowing 3.7 yds. to a suit?

6. If a man can walk 30.25 miles per day, how long will it take him to walk 150.75 miles?

7. How many loads will 134642.156 lbs. of hay make, allowing 1622.2 lbs. for a load?

8. If a team can plough 2.3 acres in a day, how long will it take to plough 63.75 acres?

9. How many bales of cotton in 56343.75 lbs., allowing 375 lbs. to a bale?

Divide the following decimals:

10. 46.84÷7.9.
11. 1.658÷.25.
12. 67234÷.85.
13. 4.00334÷6.31.
14. 73.8243÷.061.
15. 0.00033÷.011.
16. 236.041÷1.75.
17. 60.0001÷1.01.
18. 300.402÷12.1.
19. 4.32067÷.001.
20. 0.00006÷.003.
21. 167342÷.002.
22. 684234.6÷2682.
23. 0.000045÷9.
24. 7.231068÷.12.
25. 26.3845÷.125.
26. 4÷.00001.
27. 6÷.0000001.
28. 0.8÷.0000002.
29. 6541.234567÷21.

QUEST.—*Obs.* When the number of decimal places in the divisor is equal to that in the dividend, what is the quotient? When there are more decimals in the divisor than in the dividend, how proceed? When there is a remainder, what may be done?

CONTRACTIONS IN DIVISION OF DECIMALS.

CASE I.

331. When the divisor is 10, 100, 1000, &c., the division may be performed by simply *removing the decimal point* in the dividend as many places towards the *left*, as there are ciphers in the divisor, and it will be the quotient required. (Arts. 131, 330.)

1. Divide 4672.3 by 100. *Ans.* 46.723.
2. Divide 0.8 by 10000. *Ans.* 0.00008.
3. Divide 672345.67 by 10.
4. Divide 10342.306 by 100.
5. Divide 42643.621 by 100000.
6. Divide 6723000.45 by 1000000.
7. Divide 1.2300456 by 100000.
8. Divide 2.0076346 by 1000000.

CASE II

332. When the divisor contains a large number of decimal figures, the process of dividing may be very much abridged.

9. It is required to divide 3.2682 by 2.4736, and carry the quotient to four places of decimals.

Common Method.	*Contraction.*
2.4736)3.2682(1.3212	2.4736)3.2682(1.3212
2 4736	2 4736
7946 0	7946
7420 8	7421
525 20	525
494 72	495
30 480	30
24 736	25
5 7440	5
4 9472	5
7968	

Explanation.—We perceive the first figure of the quotient will be a whole number; for the number of decimals in the divisor is

QUEST.—331. When the divisor is 10, 100, 1000, &c., how may the division be performed?

equal to that of the dividend. (Art. 330. Obs. 1.) Now to obtain the decimals required, instead of annexing a cipher to the several remainders, which multiplies them respectively by 10, (Art. 98,) we may cut off a figure on the right of the divisor at each division, which is the same as dividing it successively by 10. (Art. 130.) When we multiply the divisor by 3, the second quotient figure, we carry 2 to the product of 3 into 3, because the product of 3 into 6, the figure omitted in the divisor, is nearer 20 than 10. (Art. 327.) We carry on the same principle to the first figure of each product of the divisor into the respective quotient figures. Hence,

333. To divide decimals, carrying the quotient to any required number of decimal places.

For the first quotient figure divide as usual; then instead of bringing down the next figure, or annexing a cipher to the remainder, cut off a figure on the right of the divisor at each successive division, and divide by the other figures. In multiplying the divisor by the quotient figure, carry for the nearest number of tens that would arise from the product of the figure last cut off into the figure last placed in the quotient. (Art. 327.)

OBS. 1. The *reason* for this contraction may be seen from the principle, that a *tenth* of the given divisor is contained in a *tenth* of the dividend, just as many times as the *whole* divisor is contained in the *whole* dividend; (Art. 145;) for, cutting off a figure on the right of the divisor, and omitting to annex a cipher to the dividend or remainder, is dividing each by *ten*. (Art. 130.)

2. When the divisor has *more figures* than the quotient is required to have, including the whole number and decimals, we may take as many on the left of the divisor as are required in the quotient, and divide by them as above.

3. If the divisor does not contain so *many figures* as are required in the quotient, we must divide in the usual way, until we obtain enough figures to make up this deficiency, and then begin the contraction.

10. Divide .4134 by .3243, and carry the quotient to four places of decimals.

11. Divide .079085 by .83497, and carry the quotient to five places of decimals.

12. Divide 2.3748 by 1.4736, and carry the quotient to three places of decimals.

13. Divide .3412 by 8.4736, and carry the quotient to five places of decimals.

14. Divide 1 by 10.473654, and carry the quotient to seven places of decimals.

15. Divide .4312672143 by .2134123406, and carry the quotient to four places of decimals.

16. Divide .879454 by .897, and carry the quotient to six places of decimals.

REDUCTION OF DECIMALS.

CASE I.

334. *Decimals reduced to Common Fractions.*

Ex. 1. Change the decimal .75 to a common fraction.

Suggestion.—Supplying the denominator, $.75=\frac{75}{100}$. (Art. 311.) Now $\frac{75}{100}$ is expressed in the form of a common fraction, and, as such, may be reduced to lower terms, and be treated in the same manner as any other common fraction. Thus, $\frac{75}{100}=\frac{15}{20}$, or $\frac{3}{4}$.

335. Hence, To reduce a Decimal to a Common Fraction.

Erase the decimal point; then write the decimal denominator under the numerator, and it will form a common fraction, which may be treated in the same manner as other common fractions.

2. Change .225 to a common fraction, and reduce it to the lowest terms. *Ans.* $\frac{9}{40}$.
3. Reduce .125 to a common fraction, &c.
4. Reduce .95 to a common fraction, &c.
5. Reduce .435 to a common fraction, &c.
6. Reduce .575 to a common fraction, &c.
7. Reduce .656 to a common fraction, &c.
8. Reduce .204 to a common fraction, &c.
9. Reduce .075 to a common fraction, &c.
10. Reduce .012 to a common fraction, &c.
11. Change .0025 to a common fraction, &c.
12. Change .1001 to a common fraction, &c.

QUEST.—335. How are Decimals reduced to Common Fractions?

13. Change .1844 to a common fraction, &c.
14. Change .0556 to a common fraction, &c.
15. Change .1216 to a common fraction, &c.
16. Change .2005 to a common fraction, &c.
17. Change .0015 to a common fraction, &c.

CASE II.

336. *Common Fractions reduced to Decimals.*

Ex. 1. Change $\frac{4}{5}$ to a decimal.

Suggestion.—Multiplying both terms by 10, the fraction becomes $\frac{40}{50}$. Again, dividing both terms by 5, it becomes $\frac{8}{10}$. (Art. 191.) But $\frac{8}{10}$=.8, which is the decimal required. (Art. 311.)

Note.—Since we make no use of the denominator 10 after it is obtained, we may omit the process of getting it; for, if we annex a cipher to the numerator and divide it by 5, we shall obtain the same result.

Operation.

5)4.0 A decimal point is prefixed to the quotient to
.8 *Ans.* distinguish it from a whole number.

2. Reduce $\frac{5}{8}$ to a decimal. *Ans.* 0.625.

337. Hence, to reduce a Common Fraction to a Decimal.

Annex ciphers to the numerator and divide it by the denominator. Point off as many decimal figures in the quotient, as you have annexed ciphers to the numerator.

OBS. 1. If there are not as many figures in the quotient as you have annexed ciphers to the numerator, supply the deficiency by prefixing ciphers to the quotient.

2. The *reason* of this rule may be illustrated thus. Annexing a cipher to the numerator multiplies the fraction by 10. (Arts. 98, 186.) If, therefore, the numerator with a cipher annexed to it, is divided by the denominator, the quotient will obviously be *ten* times too large. (Art. 141.) Hence, in order to obtain the true quotient, or a decimal equal to the given fraction, the quotient thus obtained must be divided by 10, which is done by *pointing off one figure.* (Art. 131.) Annexing 2 ciphers to the numerator multiplies the fraction by 100; annexing 3 ciphers by 1000, &c., consequently, when 2 ciphers are annexed, the quotient will be 100 times too large, and must therefore be divided by 100; when three ciphers are annexed, the quotient will be 1000 times too large, and must be divided by 1000, &c. (Art. 131.)

QUEST.—337 How are Common Fractions reduced to Decimals?

3. Reduce $\frac{17}{16}$ to a mixed number. *Ans.* 1.0625.

Reduce the following fractions to decimals:

4. $\frac{1}{2}$.	8. $\frac{1}{5}$.	12. $\frac{3}{6}$.	16. $\frac{4}{8}$.
5. $\frac{1}{4}$.	9. $\frac{2}{5}$.	13. $\frac{1}{8}$.	17. $\frac{5}{8}$.
6. $\frac{2}{4}$.	10. $\frac{3}{5}$.	14. $\frac{2}{8}$.	18. $\frac{6}{8}$.
7. $\frac{3}{4}$.	11. $\frac{4}{5}$.	15. $\frac{3}{8}$.	19. $\frac{7}{8}$.

20. Reduce $\frac{1}{6}$ to a decimal. *Ans.* 0.166666, &c.
21. Reduce $\frac{41}{333}$ to a decimal. *Ans.* 0.123123123, &c.

338. It will be seen that the last two examples cannot be *exactly reduced* to decimals; for there will continue to be a remainder after each division, as long as we continue the operation. In the 20th, the remainder is always 4; in the 21st, after obtaining three figures in the quotient, the remainder is the same as the given numerator, and the next three figures in the quotient are the same as the first three, when the same remainder will recur again. The same remainders, and consequently the same figures in the quotient, will thus continue to recur, as long as the operation is continued.

339. Decimals which consist of the same figure or set of figures continually repeated, as in the last two examples, are called *Periodical* or *Circulating Decimals;* also, *Repeating Decimals,* or *Repetends.*

OBS. When only a *single* figure is repeated, it is more accurate to call them *repeating* decimals; but when *two* or *more* figures recur at regular intervals, they are very properly called *periodical,* or *circulating* decimals.

340. When a common fraction can be exactly expressed by a decimal, the decimal is said to be *terminate,* or *finite;* but when it cannot be exactly expressed by a decimal, it is said to be *interminate,* or *infinite.*

OBS. It seems to be incongruous to call a fraction *infinite.* (Art. 180.) The term infinite, however, does not refer to the *value* of the fraction, but to the *number* of decimal figures required to *express* its value.

QUEST.—*Obs.* When there are not so many figures in the quotient as you have annexed ciphers, what is to be done? 339. What are periodical or circulating decimals? 340. When is a decimal terminate? When interminate?

341. If the denominator of a common fraction when reduced to its lowest term, contains no *prime factors* but 2 and 5, its equivalent decimal will *terminate;* on the other hand, if it contains *any other* prime factor besides 2 and 5, it will *not terminate.*

Thus $\frac{3}{60}$ reduced to its lowest terms, becomes $\frac{1}{20}$, and the prime factors of 20 are 2, 2, and 5; that is, $20=2\times2\times5$. (Art. 165.) We also find that $\frac{3}{60}=.05$; it is therefore *terminate.* Again, $\frac{3}{45}=\frac{1}{15}$; and the prime factors of 15 are 3 and 5; that is, $15=3\times5$; and $\frac{3}{45}=.0666666$, &c.; it is therefore *interminate.* Hence,

342. To ascertain whether a common fraction can be exactly expressed by decimals.

Reduce the given fraction to its lowest terms, and then resolve its denominator into its prime factors. (Art. 341.)

OBS. The *truth* of this principle is evident from the consideration, that annexing ciphers to the numerator, multiplies it successively by 10; but 2 and 5 are the prime factors of 10, and are the only numbers that can divide it without a remainder. (Art. 165. Obs. 2.) But any number that measures another, must also measure its product into any whole number; (Art. 161. Prop. 14;) consequently, when the denominator contains no prime factors but 2 and 5, the division will *terminate;* but when it contains other factors, the division can not *terminate.*

22. Will $\frac{1}{50}$ produce a terminate or interminate decimal?
23. Will $\frac{25}{40}$ produce a terminate or interminate decimal?
24. Will $\frac{17}{30}$ produce a terminate or interminate decimal?
25. Will $\frac{1}{80}$ produce a terminate or interminate decimal?
26. Will $\frac{6}{240}$ produce a terminate or interminate decimal?
27. Will $\frac{24}{396}$ produce a terminate or interminate decimal?
28. Will $\frac{16}{500}$ produce a terminate or interminate decimal?

343. When the decimal is *terminate,* the number of figures it contains, must be equal to the *greatest number* of times that either of the prime factors 2 or 5, is repeated in its denominator, when the given fraction is reduced to its lowest terms.

OBS. The *truth* of this principle may be illustrated thus: $\frac{1}{2}=.5$; that is, the decimal terminates with *one place;* for, the denominator 2, is taken only *once* as a factor in 10, and therefore only *one* cipher is required to be annexed to the numerator to reduce it. Again, $\frac{1}{4}=.25$, which contains two decimal places. Now the denominator $4=2\times2$; and since 2 is contained only *once* as a factor in 10, it is evident that 10 must be repeated as many times as a factor in the numerator, as 2 is taken times as a factor in the denominator, in order to reduce the fraction.

For the same reason $\frac{1}{8}$ will terminate with *three places*, and is equal to .125, for, $8=2\times2\times2$. So $\frac{1}{5}=.2$; that is, the decimal terminates with *one* place; for, since its denominator 5, is taken only *once* as a factor in 10, it is necessary to add only one cipher to its numerator in order to reduce it. In like manner it may be shown that the number of figures contained in *any terminate* decimal, is equal to the greatest number of times that either of the prime factors 2 or 5, is repeated in the denominator of the given fraction.

The same reasoning will evidently hold true when the numerator is 2, 3, 4, 5, &c., or any number greater than 1. In this case the decimal will be as many times greater, as the numerator is greater than 1.

344. The *number* of figures in the period must always be *one less* than there are *units* in the denominator; for, the number of remainders different from each other which can arise from any operation in division, must necessarily be one less than the units in the divisor. For example, in dividing by 7, it is evident, the only possible remainders are 1, 2, 3, 4, 5, and 6; and since in reducing a common fraction to a decimal, a cipher is annexed to each remainder, there cannot be more than *six* different dividends; consequently, there cannot be more than *six* different figures in the quotient. Thus, $\frac{1}{7}$=.142857,142857, &c.

When the decimal is *periodical* or *circulating*, it is customary to write the period but *once*, and put a *dot*, or *accent* over the first and last figure of the period to denote its continuance. Thus, .46135135135, &c., is written $.46\dot{1}3\dot{5}$, and .633333, &c., $.6\dot{3}$.

Reduce the following fractions to circulating decimals:

31. $\frac{1}{3}$.	36. $\frac{5}{6}$.	41. $\frac{5}{7}$.	46. $\frac{4}{9}$.
32. $\frac{2}{3}$.	37. $\frac{1}{7}$.	42. $\frac{6}{7}$.	47. $\frac{5}{9}$.
33. $\frac{1}{6}$.	38. $\frac{2}{7}$.	43. $\frac{1}{9}$.	48. $\frac{6}{9}$.
34. $\frac{2}{6}$.	39. $\frac{3}{7}$.	44. $\frac{2}{9}$.	49. $\frac{7}{9}$.
35. $\frac{4}{6}$.	40. $\frac{4}{7}$.	45. $\frac{3}{9}$.	50. $\frac{8}{9}$.

51. How many decimal figures are required to express $\frac{3}{16}$?

52. How many decimal figures are required to express $\frac{5}{65}$?

53. How many decimal figures are required to express $\frac{3}{125}$?

54. How many decimal figures are necessary to express $\frac{7}{625}$?

55. How many decimal figures are necessary to express $\frac{11}{40}$?
56. How many decimal figures are necessary to express $\frac{5}{1024}$?
57. Reduce $\frac{7}{12}$ to a decimal.
58. Reduce $\frac{1}{13}$ to a decimal.
59. Reduce $\frac{5}{256}$ to a decimal.
60. Reduce $\frac{37}{79}$ to a decimal.

Note.—For the method of *finding the value* of periodical decimals, or of reducing them to Common Fractions, also of *adding*, *subtracting*, *multiplying*, and *dividing* them, see the next Section.

CASE III.

345. *Compound Numbers reduced to Decimals.*

Ex. 1. Reduce 13s. 6d. to the decimal of a pound.

Analysis.—13s. 6d.=162d., and £1=240d. (Art. 281.) Now 162d. is $\frac{162}{240}$ of 240d. The question therefore resolves itself into this: reduce the fraction $\frac{162}{240}$ to decimals. *Ans.* £.675. Hence,

346. To reduce a compound number to the decimal of a higher denomination.

First reduce the given compound number to a common fraction; (Art. 295;) *then reduce the common fraction to a decimal.* (Art. 337.)

2. Reduce 4s. 9d. to the decimal of £1. *Ans.* £.2375.
3. Reduce 10s. 9d. to the decimal of £1.
4. Reduce 16s. 6d. to the decimal of £1.
5. Reduce 17s. 7d. to the decimal of £1.
6. Reduce 5d. to the decimal of a shilling.
7. Reduce $6\frac{1}{2}$d. to the decimal of a shilling.
8. Reduce 37 rods to the decimal of a mile.
9. Reduce 2 fur. 2 rods to the decimal of a mile.
10. Reduce 15 min. 30 sec. to the decimal of an hour.
11. Reduce 3 hrs. 3 min. to the decimal of a day.
12. Reduce 5 lbs. 4 oz. to the decimal of a cwt.
13. Reduce 7 oz. 8 drams to the decimal of a pound.
14. Reduce 3 pks. 4 qts. to the decimal of a bushel.
15. Reduce 4 qts. 1 pt. to the decimal of a peck.
16. Reduce 4 qts. 1 pt. to the decimal of a gallon.

QUEST.—346. How is a compound number reduced to the decimal of a higher denomination?

CASE IV.

347. *Decimal Compound Numbers reduced to whole ones.*

1. Reduce £.387 to shillings, pence and farthings.

Operation.	
£.387	
20	
shil. 7.740	
12	
pence 8.880	
4	
far. 3.520	
Ans. 7s. 8d. 3 far.	

Multiply the given decimal by 20, because 20s. make £1, and point off as many figures for decimals, as there are decimal places in the multiplier and multiplicand. (Art. 324.) The product is in shillings and a decimal of a shilling. Then multiply the decimal of a shilling by 12, and point off as before, &c. The numbers on the left of the decimal points, viz: 7s. 8d. 3 far., form the answer. Hence,

348. To reduce a decimal compound number to whole numbers of lower denominations.

Multiply the given decimal by that number which it takes of the next lower denomination to make ONE *of this higher, as in reduction, and point off the product, as in multiplication of decimal fractions.* (Art. 324.) *Proceed in this manner with the decimal figures of each succeeding product, and the numbers on the left of the decimal point of the several products, will be the whole number required.*

2. Reduce £.725 to shillings, pence and farthings.
3. Reduce £.1325 to shillings, &c.
4. Reduce .125s. to pence and farthings.
5. Reduce .825s. to pence and farthings.
6. Reduce .125 cwt. to pounds, &c.
7. Reduce .435 lbs. to ounces and drams.
8. Reduce .275 miles to rods, &c.
9. Reduce .4245 rods to feet, &c,
10. Reduce .1824 hhds. to gallons, &c.
11. Reduce .4826 gal. to qts., &c.
12. Reduce .4258 day to hours, &c.
13. Reduce .845 hr. to minutes and seconds.

QUEST.—348. How are decimal compound numbers reduced to whole ones of a lower denomination?

SECTION X.

PERIODICAL OR CIRCULATING DECIMALS.*

ART. **349.** Decimals which consist of the *same figures or set of figures repeated, are called* PERIODICAL, OR CIRCULATING DECIMALS. (Art. 339.)

350. The repeating figures are called *periods*, or *repetends.* If one figure only repeats, it is called a *single period*, or *repetend ;* as .11111, &c.; .33333, &c.

When the same set of figures recurs at equal intervals, it is called a *compound period*, or *repetend;* as .01010101, &c.; .123123123, &c.

351. If other figures arise before the period commences, the decimal is said to be a *mixed periodical ;* all others are called *pure*, or *simple periodicals.* Thus .42631631, &c., is a *mixed periodical ;* and .33333, &c., is a *pure periodical* decimal.

OBS. 1. When a pure circulating decimal contains as many figures as there are units in the denominator less *one*, it is sometimes called a *perfect period*, or *repetend.* (Art. 344.) Thus, $\frac{1}{7}$=.142857, &c., and is a perfect period.

2. The decimal figures which precede the period, are often called the *terminate* part of the fraction.

352. Circulating decimals are expressed by writing the period *once* with a *dot* over its first and last figure when compound; and when *single* by writing the repeating figure only once with a dot over it. Thus .46135135, &c., is written $.46\dot{1}3\dot{5}$ and .33, &c., $.\dot{3}$.

353. *Similar* periods are such as begin at the same place before or after the decimal point; as $.\dot{1}$ and $.\dot{3}$, or $\dot{2}.3\dot{4}$ and $\dot{3}.7\dot{6}$, &c.

Dissimilar periods are such as begin at different places; as $.\dot{1}2\dot{3}$ and $.42\dot{3}2\dot{5}$.

Similar and conterminous periods are such as begin and end in the same places; as $.2\dot{3}2\dot{1}$ and $1\dot{6}3\dot{4}$.

* Should Periodical Decimals be deemed too intricate for younger classes, they can be omitted till review.

REDUCTION OF CIRCULATING DECIMALS.

CASE I.—*To reduce pure circulating decimals to common fractions.*

354. To investigate this problem, let us recur to the origin of circulating decimals, or the manner of obtaining them. For example, $\frac{1}{9}=.11111$, &c., or $.\dot{1}$; therefore the true value of .11111, &c., or $.\dot{1}$, must be $\frac{1}{9}$ from which it arose. For the same reason .22222, &c., or $.\dot{2}$, must be twice as much or $\frac{2}{9}$; (Art. 186;) .33333, &c., or $.\dot{3}=\frac{3}{9}$; $.\dot{4}=\frac{4}{9}$; $.\dot{5}=\frac{5}{9}$, &c.

Again, $\frac{1}{99}=.010101$, &c., or $.\dot{0}\dot{1}$; consequently .010101, &c., or $.\dot{0}\dot{1}=\frac{1}{99}$; .020202, &c., or $.\dot{0}\dot{2}=\frac{2}{99}$; .030303, &c., or $.\dot{0}\dot{3}=\frac{3}{99}$; .070707, &c., or $.\dot{0}\dot{7}=\frac{7}{99}$, &c. So also $\frac{1}{999}=.001001001$, &c., or $.\dot{0}0\dot{1}$; therefore .001001, &c., or $.\dot{0}0\dot{1}=\frac{1}{999}$; $.\dot{0}0\dot{2}=\frac{2}{999}$; &c.

In like manner $\frac{1}{7}=.\dot{1}4285\dot{7}$; (Art. 337;) and $\dot{1}4285\dot{7}=\frac{142857}{999999}$; for, multiplying the numerator and denominator of $\frac{1}{7}$ by 142857, we have $\frac{142857}{999999}$. (Art. 191.) So $\frac{2}{7}$ is twice as much as $\frac{1}{7}$; $\frac{3}{7}$, three times as much, &c. Thus it will be seen that the value of a pure periodical decimal is expressed by the common fraction whose numerator is the given period, and whose denominator is as many 9s as there are figures in the period. Hence,

355. To reduce a pure circulating decimal to a common fraction.

Make the given period the numerator, and the denominator will be as many 9s as there are figures in the period.

Ex. 1. Reduce $.\dot{3}$ to a common fraction. *Ans.* $\frac{3}{9}$, or $\frac{1}{3}$.

2. Reduce $.\dot{6}$ to a common fraction. *Ans.* $\frac{6}{9}$, or $\frac{2}{3}$.

3. Reduce $.\dot{1}\dot{8}$ to a common fraction.

4. Reduce $.\dot{1}2\dot{3}$ to a common fraction.

5. Reduce $.\dot{2}9\dot{7}$ to a common fraction.

6. Reduce $.\dot{7}\dot{2}$ to a common fraction.

7. Reduce $.\dot{0}\dot{9}$ to a common fraction.

8. Reduce $.\dot{0}4\dot{5}$ to a common fraction.

9. Reduce $.\dot{1}4285\dot{7}$ to a common fraction.

10. Reduce $.\dot{0}7692\dot{3}$ to a common fraction.

CASE II.—*To reduce mixed circulating decimals to common fractions.*

356. 11. Reduce $.1\dot{6}$ to a common fraction.

Analysis.—Separating the mixed decimal into its terminate and periodical part, we have $.1\dot{6}=.1+.0\dot{6}$. (Art. 320.) Now $.1=\frac{1}{10}$; (Art. 312;) and $.0\dot{6}=\frac{6}{90}$; for, the pure period $.\dot{6}=\frac{6}{9}$, (Art. 351,) and since the mixed period $.0\dot{6}$, begins in hundredths' place, its value is evidently only $\frac{1}{10}$ as much; but $\frac{6}{9}\div10=\frac{6}{90}$. (Art. 227.) Therefore $.1\dot{6}=\frac{1}{10}+\frac{6}{90}$. Now $\frac{1}{10}$ and $\frac{6}{90}$, reduced to a common denominator and added together, make $\frac{15}{90}$, or $\frac{1}{6}$. *Ans.*

OBS. In mixed circulating decimals, if the period begins in *hundredths'* place it is evident from the preceding analysis that the value of the periodical part is only $\frac{1}{10}$ as much as it would be, if the period were pure or begun in *tenths'* place; when the period begins in *thousandths'* place, its value is only $\frac{1}{100}$ part as much, &c. Thus $.\dot{6}=\frac{6}{9}$; $.0\dot{6}=\frac{6}{9}\div10=\frac{6}{90}$; $.00\dot{6}=\frac{6}{9}\div100=\frac{6}{900}$, &c.

357. Hence, the *denominator* of the *periodical* part of a mixed circulating decimal, is always as many 9s as there are figures in the *period* with as many *ciphers annexed* as there are decimals in the *terminate part.*

12. Reduce $.85\dot{6}792\dot{3}$ to a common fraction.

Solution.—Reasoning as before $.85\dot{6}792\dot{3}=\frac{85}{100}+\frac{67923}{9999900}$. Reducing these two fractions to the least common denominator, (Art. 261.) $\frac{85}{100}\times99999=\frac{8499915}{9999900}$ whose denominator is the same as that of the other. Now $\frac{8499915}{9999900}+\frac{67923}{9999900}=\frac{8567838}{9999900}$. *Ans.*

Contraction.	
8500000	
85	
8499915	1st Nu.
67923	2d Nu.
8567838	
9999900	*Ans.*

To multiply by 99999, annex as many ciphers to the multiplicand as there are 9s in the multiplier, &c. (Art. 105.) This gives the numerator of the first fraction or terminate part, to which add the numerator of the second or periodical part, and the sum will be the numerator of the answer. The denominator is the same as that of the second or periodical part.

Second Method.

8567923	the given circulating decimal.
85	the terminate part which is subtracted.
8567838	the numerator of the answer.
$\frac{8567838}{9999900}$	*Ans.*

Note.—1. The *reason* of this operation may be shown thus: 8567923= 8500000+67923. Now 8500000—85+67923 is equal to 8567923—85.

2. It is evident that the required denominator is the same as that of the periodical part; (Art. 357;) for, the denominator of the periodical part is the least common multiple of the two denominators. Hence,

358. To reduce a mixed circulating decimal to a common fraction.

Change both the terminate and periodical part to common fractions separately, and their sum will be the answer required.

Or, from the given mixed periodical, subtract the terminate part, and the remainder will be the numerator required. The denominator is always as many 9s *as there are figures in the period with as many ciphers annexed as there are decimals in the terminate part.*

PROOF.—*Change the common fraction back to a decimal, and if the result is the same as the given circulating decimal, the work is right.*

13. Reduce $.13\dot{8}$ to a common fraction. *Ans.* $\frac{125}{900}$, or $\frac{5}{36}$.
14. Reduce $.5\dot{3}$ to a common fraction.
15. Reduce $.5\dot{9}2\dot{5}$ to a common fraction.
16. Reduce $.58\dot{3}$ to a common fraction.
17. Reduce $.02\dot{2}\dot{7}$ to a common fraction.
18. Reduce $.47\dot{4}\dot{5}$ to a common fraction.
19. Reduce $.59\dot{2}\dot{5}$ to a common fraction.
20. Reduce $.008\dot{4}9713\dot{3}$ to a common fraction.

CASE III.—*Dissimilar periodicals reduced to similar and conterminous ones.*

359. In changing dissimilar periods, or repetends, to similar and conterminous ones, the following particulars require attention.

1. Any *terminate* decimal may be considered as *interminate* by annexing ciphers continually to the numerator. Thus .46= .460000, &c.=$.46\dot{0}$.

2. Any *pure* periodical may be considered as *mixed*, by taking the given period for the *terminate part*, and making the given period the interminate part. Thus $.\dot{4}\dot{6}=.46+.00\dot{4}\dot{6}$, &c.

3. A *single* period may be regarded as a *compound* periodical. Thus $.\dot{3}$ may become $.\dot{3}\dot{3}$, or $.\dot{3}3\dot{3}$; so $.6\dot{3}$ may be made $.6\dot{3}3\dot{3}$, or $.6\dot{3}33\dot{3}$, &c.

4. A *single* period may also be made to begin at a lower order, regarding its higher orders as terminate decimals. Thus $.\dot{3}$ may be made $.3\dot{3}$, or $.333\dot{3}$, &c.

5. *Compound* periods may also be made to begin at a lower order. Thus $.\dot{3}\dot{6}$ may be changed to $.3\dot{6}\dot{3}$, or $.3\dot{6}36\dot{3}$, &c.; or by extending the number of places $.4\dot{7}\dot{9}$ may be made $.4\dot{7}97\dot{9}$, or $.4\dot{7}9797\dot{9}$, &c.; or making both changes at once $.\dot{5}3\dot{2}$ may be changed to $.5\dot{3}2532\dot{5}$, &c. Hence,

360. To make any number of dissimilar periodical decimals similar.

Move the points, so that each period shall begin at the same order as the period which has the most figures in its terminate part.

21. Change $6.\dot{8}1\dot{4}$, $3.\dot{2}\dot{6}$, and $.08\dot{3}$ to similar and conterminous periods.

Operation.

$6.\dot{8}1\dot{4}=6.81\dot{4}8148\dot{1}$

$3.\dot{2}\dot{6}=3.26\dot{2}6262\dot{6}$

$.08\dot{3}=0.08\dot{3}3333\dot{3}$

Having made the given periods similar, the next step is to make them conterminous. Now as one of the given periods contains 3 figures, another 2, and the other 1, it is evident the new periodical must contain a number of figures which is some multiple of the *number* of figures in the different periods; viz: 3, 2, and 1. But the least common multiple of 3, 2, and 1 is 6; therefore the new periods must at least contain 6 figures. Hence,

361. To make any number of *dissimilar* periodical decimals, *similar* and *conterminous.*

First make the periods similar; (Art. 360;) *then extend the figures of each to as many places, as there are units in the least common multiple of the* NUMBER *of periodical figures contained in each of the given decimals.* (Art. 176.)

27. Change $46.\dot{1}6\dot{2}$, $5.\dot{2}\dot{6}$, $63.\dot{4}2\dot{3}$, $.48\dot{6}$, and 12.5, to similar and conterminous periodicals.

Operation.

$46.\dot{1}6\dot{2} = 46.16\dot{2}1621\dot{6}$
$5.\dot{2}\dot{6} = 5.26\dot{2}6262\dot{6}$
$63.\dot{4}2\dot{3} = 63.42\dot{3}4234\dot{2}$
$.48\dot{6} = 0.48\dot{6}6666\dot{6}$
$12.5 = 12.50\dot{0}0000\dot{0}$

The numbers of periodical figures in the given decimals are 3, 2, 3, and 1; and the least common multiple of them is 6. Therefore the new periods must each have 6 figures.

23. Make $.\dot{2}\dot{7}$, $.\dot{3}$, and $.04\dot{5}$ similar and conterminous.
24. Make $4.\dot{3}2\dot{1}$, $6.4\dot{2}6\dot{3}$, and .6 similar and conterminous.

ADDITION OF CIRCULATING DECIMALS.

Ex. 1. What is the sum of $17.\dot{2}\dot{3} + 41.2\dot{4}7\dot{6} + 8.\dot{6}\dot{1} + 1.5 + 35.4\dot{2}\dot{3}$?

Operation.

Dissimilar.	Sim. & Conterminous.
$17.\dot{2}\dot{3}$	$=17.2\dot{3}2323\dot{2}$
$41.2\dot{4}7\dot{6}$	$=41.2\dot{4}7647\dot{6}$
$8.\dot{6}\dot{1}$	$=8.6\dot{1}6161\dot{6}$
1.5	$=1.5\dot{0}0000\dot{0}$
$35.4\dot{2}\dot{3}$	$=35.4\dot{2}3232\dot{3}$
Ans.	$104.0\dot{1}9364\dot{8}$

First make the given decimals similar and conterminous. (Art. 361.) Then add the periodical parts as in simple addition, and since there are 6 figures in the period, divide their sum by 999999; for this would be its denominator, if the sum of the periodicals were expressed by a common fraction. (Art. 355.) Setting down the remainder for the repeating decimals, carry the quotient 1 to the next column, and proceed as in addition of whole numbers. Hence,

362. We derive the following general

RULE FOR ADDING CIRCULATING DECIMALS.

First make the periods similar and conterminous, and find their sum as in Simple Addition. Divide this sum by as many 9s as there are figures in the period, set the remainder under the figures added for the period of the sum, carry the quotient to the next column, and proceed with the rest as in Simple Addition.

OBS. If the remainder has not so many figures as the period, ciphers must be prefixed to make up the deficiency.

2. What is the sum of $24.\dot{1}3\dot{2}+2.2\dot{3}+85.\dot{2}\dot{4}+67.\dot{6}$?

3. What is the sum of $328.12\dot{6}+81.\dot{2}\dot{3}+5.\dot{6}2\dot{4}+61.\dot{6}$?

4. What is the sum of $31.6\dot{2}+7.\dot{8}2\dot{4}+8.\dot{3}9\dot{2}+.02\dot{7}$?

5. What is the sum of $462.\dot{3}\dot{4}+60.\dot{8}\dot{2}+71.\dot{1}6\dot{4}+.3\dot{5}$?

6. What is the sum of $60.25+.3\dot{4}+6.43\dot{5}+.45+45.\dot{2}\dot{4}$?

7. What is the sum of $9.\dot{8}1\dot{4}+1.5+87.\dot{2}\dot{6}+0.8\dot{3}+124.0\dot{9}$?

8. What is the sum of $3.\dot{6}+78.3\dot{4}7\dot{6}+735.\dot{3}+375+.\dot{2}\dot{7}+187.\dot{4}$?

9. What is the sum of $5391.35\dot{7}+72.3\dot{8}+187.2\dot{1}+4.296\dot{5}+217.849\dot{6}+42.17\dot{6}+.52\dot{3}+58.3004\dot{8}$?

10. What is the sum of $.\dot{1}6\dot{2}+134.\dot{0}\dot{9}+2.\dot{9}\dot{3}+97.2\dot{6}+3.\dot{7}6923\dot{0}+99.083+1.5+.\dot{8}1\dot{4}$?

SUBTRACTION OF CIRCULATING DECIMALS.

Ex. 1. From $52.\dot{8}\dot{6}$ take $8.3\dot{7}23\dot{5}$.

Operation.

$$\begin{array}{rr} 52.\dot{8}\dot{6}= & 52.\dot{8}686\dot{8} \\ 8.3\dot{7}23\dot{5}= & 8.3\dot{7}23\dot{5} \\ \hline & 44.4\dot{9}63\dot{2} \end{array}$$

We first make the given decimals similar and conterminous, then subtract as in whole numbers. But since the period in the lower line is larger than that above it, we must borrow 1 from the next higher order. This will make the right hand figure of the remainder one less than if it was a terminate decimal. Hence,

363. We derive the following general

RULE FOR SUBTRACTING CIRCULATING DECIMALS.

Make the periods similar and conterminous, and subtract as in whole numbers. If the period in the lower line is larger than that above it, diminish the right hand figure of the remainder by 1.

OBS. The *reason* for diminishing the right hand figure of the remainder by 1, if the period in the lower line is larger than that above it, may be explained thus:

When the period in the lower line is larger than that above it, we must evidently borrow 1 from the next higher order. Now if the given decimals were extended to a second period, in this period the lower number would also be larger than that above it, and therefore we must borrow 1. But having borrowed 1 in the second period, we must also carry one to the next figure in the lower line, or, what is the same in effect, diminish the right hand figure of the remainder by 1.

2. From $85.\dot{6}\dot{2}$ take $13.7\dot{6}43\dot{2}$. *Ans.* $71.8\dot{6}19\dot{3}$.
3. From $476.3\dot{2}$ take $84.769\dot{7}$.
4. From $3.8\dot{5}6\dot{4}$ take $.038\dot{2}$.
5. From $46.\dot{1}2\dot{3}$ take $41.\dot{3}$.
6. From $801.\dot{6}$ take 400.75.
7. From $4.78\dot{2}\dot{4}$ take $.8\dot{7}$.
8. From 1419.6 take $1200.\dot{9}$.
9. From $.634\dot{8}5\dot{2}$ take $.0\dot{2}\dot{1}$.
10. From $8482.\dot{4}2\dot{1}$ take $6031.03\dot{5}$.

MULTIPLICATION OF CIRCULATING DECIMALS.

Ex. 1. What is the product of $.\dot{3}\dot{6}$ into $.2\dot{5}$?

Operation.

$.\dot{3}\dot{6}=\frac{36}{99}=\frac{4}{11}$

$.2\dot{5}=\frac{2}{10}+\frac{5}{90}=\frac{23}{90}$.

Now $\frac{4}{11}\times\frac{23}{90}=\frac{92}{990}$

But $\frac{92}{990}=.0\dot{9}\dot{2}$ *Ans.*

We first reduce the given periodicals to common fractions; (Art. 358;) then multiply them together. (Art. 219.) Finally, we reduce the product to a periodical decimal. Hence,

364. We derive the following general

RULE FOR MULTIPLYING CIRCULATING DECIMALS.

First reduce the given periodicals to common fractions, and multiply them together as usual. (Art. 219.) *Finally, reduce the product to decimals and it will be the answer required.*

OBS. If the numerators and denominators have *common* factors, the operation may be *contracted* by *canceling* those factors before the multiplication is performed. (Art. 221.)

2. What is the product of $37.2\dot{3}$ into $.2\dot{6}$? *Ans.* $9.92\dot{8}$.
3. What is the product of $.\dot{1}2\dot{3}$ into $.\dot{6}$?
4. What is the product of $.\dot{2}4\dot{5}$ into $7.\dot{3}$?
5. What is the product of $24.\dot{6}$ into $15.\dot{7}$?
6. What is the product of $48.\dot{2}\dot{3}$ into $16.1\dot{3}$?

7. What is the product of $8574.\dot{3}$ into $87.\dot{5}$?
8. What is the product of $3.\dot{9}7\dot{3}$ into 8?
9. What is the product of 49640.54 into $.\dot{7}0503$?
10. What is the product of $7.\dot{7}\dot{2}$ into $.\dot{2}9\dot{7}$?

DIVISION OF CIRCULATING DECIMALS.

Ex. 1. Divide $234.\dot{6}$ by $.\dot{7}$.

Operation.

$234.\dot{6}=234\frac{2}{3}=\frac{704}{3}$

$.\dot{7}=\frac{7}{9}$

Now $\frac{704}{3}\div\frac{7}{9}=\frac{704}{3}\times\frac{9}{7}=\frac{6336}{21}$

And $\frac{6336}{21}=301.\dot{7}14285\dot{}$ *Ans.*

We first reduce the divisor and dividend to common fractions; (Art. 358;) and divide one by the other; (Art. 229;) then reduce the quotient to a decimal. (Art. 337.) Hence,

365. We derive the following general

RULE FOR DIVIDING CIRCULATING DECIMALS.

Reduce the divisor and dividend to common fractions; divide one fraction by the other, and reduce the quotient to decimals.

OBS. After the divisor is inverted, if the numerators and denominators have factors common to both, the operation may be *contracted* by *canceling* those factors. (Art. 232.)

2. Divide $319.28007\dot{1}1\dot{2}$ by 764.5. *Ans.* $0.417632\dot{5}$.
3. Divide $18.\dot{5}\dot{6}$ by $.\dot{3}$.
4. Divide $.\dot{6}$ by $.\dot{1}2\dot{3}$.
5. Divide $2.\dot{2}9\dot{7}$ by $.\dot{2}97$.
6. Divide $750730.\dot{5}1\dot{8}$ by 87.5.
7. Divide $\dot{5}\dot{4}$ by $.1\dot{5}$.
8. Divide $13.5\dot{1}6953\dot{3}$ by $4.\dot{2}9\dot{7}$.
9. Divide $24.\dot{0}8\dot{1}$ by $.\dot{3}8\dot{6}$.
10. Divide $.\dot{3}\dot{6}$ by $.2\dot{5}$.

SECTION XI.

FEDERAL MONEY.

ART. **366.** FEDERAL MONEY is the currency of the *United States.* Its denominations, we have seen, are *Eagles, Dollars, Dimes, Cents,* and *Mills.* (Art. 244.)

367. All *accounts* in the United States are required by law to be kept in *dollars, cents,* and *mills.* Eagles are expressed in dollars, and dimes in cents. Thus, instead of 8 eagles, we say, 80 dollars; instead of 6 dimes and 7 cents, we say, 67 cents, &c.

368. Federal Money is based upon the *Decimal* system of Notation. Its denominations *increase* and *decrease* from right to left and left to right in a *tenfold ratio,* like whole numbers and decimals. (Art. 244. Obs. 1.)

369. The *dollar* is regarded as the *unit; cents* and *mills* are fractional parts of the dollar, and are distinguished from it by a *decimal point* or *separatrix* (.) in the same manner as common decimals are distinguished from whole numbers. (Art. 311.) *Dollars* therefore occupy *units'* place of simple numbers; *eagles,* or tens of dollars, *tens'* place, &c. *Dimes,* or tenths of a dollar, occupy the place of *tenths* in decimals; *cents,* or hundredths of a dollar, the place of *hundredths; mills,* or thousandths of a dollar, the place of *thousandths; tenths* of a mill, or ten thousandths of a dollar, the place of *ten thousandths,* &c.

OBS. 1. Since *dimes* in business transactions are expressed in *cents, two places* of decimals are assigned to cents. If therefore the number of cents is *less* than 10, a *cipher must always be placed on the left hand of them;* for cents are *hundredths* of a dollar, and hundredths occupy the second decimal place. (Art. 313.) For example, 4 cents are written thus .04; 7 cents thus .07; &c.

2. Mills occupy the *third* place of decimals; for they are *thousandths* of a dollar. Consequently, when there are no cents in the given sum, *two ciphers* must be placed before the mills. Hence,

QUEST.—366. What is Federal Money? 367. In what are accounts kept in the U. S.? How would you express 8 eagles? How express 6 dimes and 7 cents? 368. Upon what is Federal Money based? 369. What is regarded as the unit in Federal Money? What are cents and mills? How are they distinguished from dollars?

370. To read any sum of Federal Money.

Call all the figures on the left of the decimal point dollars; the first two figures after the point, are cents; the third figure denotes mills; the other places on the right are decimals of a mill. Thus, $3.25232 is read, 3 dollars, 25 cents, 2 mills, and 32 hundredths of a mill.

OBS. Sometimes all the figures after the point are read as decimals of a dollar. Thus, $5.356 is read, "5 and 356 thousandths dollars."

Write the following sums in Federal money:

1. 70 dollars, and 8 cents. *Ans.* $70.08.
2. 150 dollars, 3 cents, and 5 mills.
3. 400 dollars, 40 cents, and 3 mills.
4. 200 dollars, 5 cents, and 2 mills.
5. 4050 dollars, 65 cents, and 3 mills.

Note.—In business transactions, when dollars and cents are expressed together, the cents are frequently written in the form of a common fraction Thus, the sum of $75.45, is written $75\frac{45}{100}$ dollars.

REDUCTION OF FEDERAL MONEY.

CASE I.

Ex. 1. How many cents are there in 95 dollars?

Solution.—Since in 1 dollar there are 100 cents, in 95 dollars there are 95 times as many. And $95\times100=9500$.

Ans. 9500 cents.

2. In 20 cents how many mills? *Ans.* 200 mills.

Note.—To multiply by 10, 100, &c., we simply annex as many ciphers to the multiplicand, as there are ciphers in the multiplier. (Art. 99.) Hence,

371. *To reduce dollars to cents, annex two ciphers.*

To reduce dollars to mills, annex three ciphers.

To reduce cents to mills, annex one cipher.

OBS. To reduce dollars, cents, and mills, to mills, *erase the sign of dollars and the separatrix.* Thus, $25.36 reduced to cents, becomes 2536 cents.

QUEST.—370. How do you read Federal Money? *Obs.* What other mode of reading Federal Money is mentioned? 371. How are dollars reduced to cents? Dollars to mills? Cents to mills? *Obs.* Dollars, cents, and mills, to mills?

3. In $12 how many cents? *Ans.* 1200 cents.
4. In $460 how many cents?
5. In $95 how many mills?
6. In 90 cents how many mills?
7. Reduce $25.15 to cents.
8. Reduce $864.08 to cents.
9. Reduce $1265.05 to mills.
10. Reduce $4580.10 to mills.
11. Reduce $6886.258 to mills.
12. Reduce $85625.40 to mills.

CASE II.

13. In 6400 cents, how many dollars?

Suggestion.—Since 100 cents make 1 dollar, 6400 cents will make as many dollars as 100 is contained times in 6400. And 6400÷100=64. *Ans.* $64.

14. In 260 mills, how many cents? *Ans.* 26 cents.

Note.—To divide by 10, 100, &c., we simply cut off as many figures from the right of the dividend as there are ciphers in the divisor. (Art. 131.) Hence,

372. *To reduce cents to dollars, cut off two figures on the right.*

To reduce mills to dollars, cut off three figures on the right.

To reduce mills to cents, cut off one figure on the right.

OBS. The figures cut off are cents and mills.

15. In 626 cents, how many dollars? *Ans.* $6.26.
16. In 1516 cents, how many dollars?
17. In 162 mills, how many cents?
18. In 1000 mills, how many dollars?
19. In 2360 mills, how many cents?
20. In 3280 mills, how many dollars?
21. Reduce 8500 cents to dollars.
22. Reduce 2345 cents to dollars, &c.
23. Reduce 92355 mills to dollars, &c.

QUEST.—372 How are cents reduced to dollars? Mills to dollars? Mills to cents? *Obs.* What are the figures cut off?

24. Reduce 150233 mills to dollars, &c.

25. Reduce 450341 cents to dollars, &c.

373. Since *Federal Money* is expressed according to the decimal system of notation, it is evident that it may be subjected to the same operations and treated in the same manner as *Decimal Fractions.*

ADDITION OF FEDERAL MONEY.

Ex. 1. A man bought a cloak for $35.375, a hat for $4.875, a pair of boots for $6.50, and a coat for $23.625: what did he pay for all?

Operation.

$35.375
4.875
6.50
23.625
$70.375 *Ans.*

We write the dollars under dollars, cents under cents, &c. Then add each column separately, and point off as many figures for cents and mills, in the amount, as there are places of cents and mills in either of the given numbers. Hence,

374. We derive the following general

RULE FOR ADDING FEDERAL MONEY.

Write dollars under dollars, cents under cents, &c., so that the same orders or denominations may stand under each other. Add each column separately, and point off the amount as in addition of decimal fractions. (Art. 320.)

Obs. If either of the given numbers have no cents expressed, it is customary to supply their place by ciphers.

2. What is the sum of $48.25, $95.60, $40.09, and $81.10?

3. What is the sum of $103.40, $68.253, $89.455, $140.02, and $180?

4. What is the sum of $136.255, $10.30, $248.50, $65.38, and $100.125?

Quest.—374. How is Federal Money added? How point off the amount? Obs. When any of the given numbers have no cents expressed, how is their place supplied?

5. What is the sum of $170, $400.02, $130, $250.10, and $845.22?

6. What is the sum of $268.45, $800.05, $192.125, $80.625, and $90.25?

7. What is the sum of $1500.20, $1050.07, $100.70, $95.025, $360.437, and $425?

8. What is the sum of $2600, $1927.404, $1603.40, $3304.17, $165.47, and $2600.08?

9. A man bought a load of hay for $19.675, a horse for $73.25, a yoke of oxen for $69.56, a cow for $17, and a calf for $5.80: what did he pay for all?

10. A lady gave $21.50 for a dress, $9.25 for a bonnet, $28.33 for a shawl, and $15.25 for a muff: what was her bill?

11. A jockey bought a span of horses for $276.87, and sold them so as to gain $73.45: how much did he sell them for?

12. A man gave $4925.68 for a farm, and sold it so as to gain $1565.37: how much did he sell it for?

13. A man sold a sloop for $7623.87, which was $1141.25 less than cost: how much did it cost?

14. A man bought a block of stores for $15268, which was $1721 less than cost: what was the cost?

15. What is the sum of 134 dolls. 3 cts. 7 mills, 108 dolls. 6 cts. 8 mills, 90 dolls. 9 cts. 4 mills, and 46 dolls. 18 cts. 4 mills?

16. What is the sum of 61 dolls. 1 ct. 2 mills, 19 dolls. 11 cts. 4 mills, 140 dolls. and 80 dolls. 4 cts.?

17. What is the sum of 140 dolls. 10 cts., 69 dolls. 3 cts. 8 mills, 18 dolls. 7 cts., and 29 dolls. 5 mills?

18. What is the sum of 860 dolls. 8 cts., 298 dolls. 4 cts. 8 mills, 416 dolls., 280 dolls. 13 cts., and 91 dolls.?

19. What is the sum of 14209 dolls., 65241 dolls., 1050 dolls., 610 dolls. 7 cts., and 1000 dolls. 10 cts.?

20. What is the sum of 1625 dolls., 4025 dolls., 1863 dolls. 75 cts., 16000 dolls., and 48261 dolls.?

21. What is the sum of 8 thousand dolls., 2 hundred and 60 dolls. 5 cts., 19 thousand dolls. 60 cts., 6 hundred dolls. 9 cts.?

22. What is the sum of 19 thousand dolls. 50 cts., 61 thousand dolls. 10 cts., 18 hundred dolls. 3 cts.?

SUBTRACTION OF FEDERAL MONEY.

Ex. 1. A merchant bought a quantity of molasses for $75.40, and a box of sugar for $42.63: how much more did he pay for one than the other?

Operation.	
$75.40	We write the less number under the greater, placing dollars under dollars, &c., then subtract
42.63	and point off the answer, as in subtraction of
$32.77	decimals. Hence,

375. We derive the following general

RULE FOR SUBTRACTING FEDERAL MONEY.

Write the less number under the greater, with dollars under dollars, cents under cents, &c.; *then subtract and point off the remainder as in subtraction of decimal fractions.* (Art. 322.)

OBS. If either of the given numbers have no cents expressed, it is customary to supply their place by ciphers.

2. A man bought a horse for $75.50, and sold it for $87.63: how much did he make by his bargain?

3. If a man deposits $204.65 in a bank, and afterwards checks out $119.83, how much will he have left?

4. A man owing $682.40, paid $435.25: how much does he still owe?

5. A man owing $982.68, paid all but $64.20: how much did he pay?

6. A merchant bought a quantity of goods for $833.63, and retailed them for $1016.85: how much did he make by the bargain?

7. A merchant bought a lot of goods for $1265.82, and sold them for $942.35: how much did he lose?

8. A grocer sold a lot of sugar for $635.20, and made thereby $261.38: how much did he pay for the sugar?

9. A man sold his farm for $12250.62, which was $1379.87 more than it cost: how much did it cost?

QUEST.—375. How is Federal Money subtracted? How point off the remainder? Obs. When either of the given numbers have no cents, how is their place supplied?

10. From $10600.75 take $8901.26
11. From $20206.85 take $10261.062.
12. From $61219.40 take $100.036.
13. From $19 take 1 cent and 9 mills.
14. From 89 dollars take 89 cents.
15. From 506 dolls. take 316 dolls. and 8 cts.
16. From 5 dolls. 7 mills take 2 dolls. 7 cts.
17. From 61 dolls. 6 cts. take 29 dolls. 4 mills.
18. From 11000 dolls. 10 cts. take 110 dolls. 3 cts.
19. From 100100 dolls. take 10110 dolls. 10 cts.

MULTIPLICATION OF FEDERAL MONEY.

376. In multiplication of Federal Money, as well as in simple numbers, the multiplier must always be considered an *abstract number*. (Art. 82. Obs. 2.)

Ex. 1. What will 8 bbls. of flour cost, at $5.62 per bbl.?

Analysis.—Since 1 bbl. costs $5.62, 8 bbls. will cost 8 times as much; and $5.62×8=$44.96 *Ans.*

2. What cost 21.7 bushels of apples, at 15 cts. per bushel?

Operation.

```
  21.7
   .15
  ----
  1085
  217
  ----
$3.255 Ans.
```

Reasoning as before, 21.7 bushels will cost 21.7 times 15 cents. But in performing the multiplication, it is more convenient to make the .15 the multiplier, and the result will be the same as if it was placed for the multiplicand. (Art. 83.) Point off the product as before. Hence,

377. When the *price* of *one* article, *one* pound, *one* yard, &c., is given, to find the *cost* of any *number* of articles, pounds, yards, &c.

Multiply the price of one article and the number of articles together, and point off the product as in multiplication of decimals. (Art. 324.)

QUEST.—376. In Multiplication of Federal Money, what must one of the given factors be considered? 377. When the price of one article, one pound, &c., is given, how is the cost of any number of articles found?

3. What cost 17.6 yards of cloth, at $4.75 per yard?
4. Multiply $25.625 by 20.2.

378. From the preceding illustrations we derive the following general

RULE FOR MULTIPLYING FEDERAL MONEY.

Multiply as in simple numbers, and point off the product as in multiplication of decimal fractions. (Art. 324.)

OBS. 1. When the price, or the quantity contains a common fraction, the fraction may be changed to a decimal. (Art. 337.)

2. In business operations, when the mills in the answer are 5, or over, it is customary to call them a cent; when under 5, they are disregarded.

5. What cost 12½ yards of cotton, at 9¼ cts. per yard?

Solution.—12½ yards=12.5, and 9¼ cts.=.0925; now .0925× 12.5=$1.15625. *Ans.*

6. What cost 45¼ yards of satin, at 87½ cts. per yard?
7. What cost 169½ bbls. of pork, at $8½ per barrel?
8. What cost 324¾ lbs. of sugar, at 12½ cts. a pound?
9. What cost 97 gals. of oil, at 87½ cts. per gallon?
10. What cost 310 lbs. of tea, at 62½ cts. a pound?
11. What cost 23½ tons of hay, at $8¾ per ton?
12. What cost 45 bbls. of flour, at $7¼ per barrel?
13. At 15½ cts. per doz., what cost 13½ dozen of eggs?
14. At 8¾ cts. per pound, what will 32½ lbs. of pork come to?
15. At $6¼ per bbl., what will 145½ bbls. of flour cost?
16. At 22½ cts. per doz., what will a gross of buttons cost?
17. At 31¼ cts. per doz., what cost 45 doz. skeins of silk?
18. At 17¼ cts. per yard, what cost 91½ yards of calico?
19. What cost 45 doz. plates, at 62½ cts. per doz.?
20. What cost 63 doz. pen-knives, at $3½ per doz.?
21. What cost 19 doz. silver spoons, at $7½ per dozen?
22. What cost 1865½ bushels of wheat, at $1¼ per bushel?
23. What cost 2560¾ yds. of broadcloth, at $5½ per yard?

QUEST.—378. What is the rule for Multiplication of Federal Money? *Obs.* When the price or quantity contains a common fraction, what should be done with it?

DIVISION OF FEDERAL MONEY.

Ex. 1. A man bought 8 sheep for $42.24: what did he give apiece?

Analysis.—If 8 sheep cost $42.24, 1 sheep will cost $\frac{1}{8}$ of $42.24; and $42.24 ÷ 8 = $5.28. *Ans.* $5.28.

PROOF.—If 1 sheep costs $5.28, 8 sheep will cost 8 times as much; and $5.28 × 8 = $42.24. Hence,

379. When the number of articles, pounds, yards, &c., and the *cost* of the *whole* are given, to find the price of *one* article, *one* pound, &c.

Divide the whole cost by the whole number of articles, and point off the quotient as in division of decimal fractions. (Art. 330.)

2. A shoemaker sold 15 pair of boots for $67.50: how much did he get a pair?

3. A merchant sold 65½ lbs. of sugar for $3.93: how much was that a pound?

4. A man bought 6.5 yards of cloth for $20.345: how much was that per yard?

5. How many bbls. of flour, at $5.38 per bbl., can be bought for $34.97?

Analysis.—Since $5.38 will buy 1 bbl., $34.97 will buy as many bbls. as $5.38 are contained times in $34.97. We divide as in simple numbers, and point off one decimal figure in the quotient.

Operation.

```
5.38)34.97(6.5 Ans.
     32 28
     -----
      2 690
      2 690
      -----
```

PROOF.—$5.38 × 6.5 = $34.97, the given amount.

380. Hence, when the *price* of *one* article, pound, yard, &c., and the *cost* of the *whole* are given, to find the number of articles, &c.

Divide the whole cost by the price of one, and point off the quotient as in division of decimals.

QUEST.—379. When the number of articles, pounds, &c., and the cost of the whole are given, how is the cost of one article found? 380. When the price of one article, one pound &c., and the cost of the whole are given, how is the number of articles found?

6. How many coats, at $12.56, can be bought for $103.085?

7. How many times is $11.13 contained in $87.606?

8. A gentleman distributed $68 equally among 32 poor persons: how much did each receive?

Operation.

```
32)$68(2.125 Ans.
   64
    4000
    32
     80
     64
     160
     160
```

After dividing the $68 by 32, there is a remainder of 4 dollars, which should be reduced to cents and mills, and then be divided as before. (Art. 371.) The ciphers thus annexed must be regarded as decimals; consequently there will be three decimal figures in the quotient.

381. From the preceding illustrations we derive the following general

RULE FOR DIVIDING FEDERAL MONEY.

Divide as in simple numbers, and point off the quotient as in division of decimal fractions. (Art. 330.)

OBS. 1. In dividing Federal Money, if the number of decimals in the divisor is the *same* as that in the dividend, the quotient will be a *whole* number. (Art. 330. Obs. 1.)

2. When there are *more* decimals in the divisor than in the dividend, annex as many ciphers to the dividend as are necessary to make its decimal places *equal* to those in the divisor. The quotient thence arising will be a whole number. (Obs. 1.)

3. After all the figures of the dividend are divided, if there is a remainder, ciphers may be annexed to it, and the operation may be continued as in division of decimals. (Art. 330. Obs. 3.) The ciphers thus annexed must be regarded as decimal places of the dividend.

9. How many gallons of molasses, at 28 cts. per gallon, can you buy for $86.25?

QUEST.—381. What is the rule for Division of Federal Money? *Obs.* When there is a remainder after all the figures of the dividend are divided, how proceed? When there are more decimals in the divisor than in the dividend, how proceed?

10. How many yards of calico, at 13½ cts. per yard, can be bought for $73.37½?

11. How many doz. of eggs, at 9½ cts. per doz., can be bought for $94.185?

12. At 18¾ cts. per doz., how many skeins of sewing silk can be bought for $67.50?

13. A man paid $72.25 for 20.5 yards of cloth: how much did he pay per yard?

14. A man paid $76.50 for 51 sheep: what was the price per head?

15. A man paid $150 for 24 pair of boots: how much was that a pair?

16. If you give $56.25 for 28½ bbls. of flour, how much do you pay per barrel?

17. If a man gives $316.375 for 87½ yards of cloth, what is that per yard?

18. A grocer sold 965½ lbs. of sugar for $81.25: what did he get a pound?

19. The fare from Albany to Buffalo, a distance of 326 miles, is $13.20: how much is it per mile?

20. The fare from Boston to Albany, a distance of 203 miles, is $5.50: how much is it per mile?

21. If a clerk's salary is $650 per annum, how much does he receive per day?

22. If a man spends $563.38 a year, how much are his average expenses per day?

23. At 87½ cts. per bushel, how many bushels of wheat can you buy for $1500?

24. How many tons of coal, at $6.625 per ton, can you buy for $752.36?

25. If a man's income is $100 per week, how much is it per hour?

26. At $14.50 per acre, how many acres of land can you buy for $3560?

27. At $15½ apiece, how many cows can you buy for $7750?

28. At $375.75 apiece, how many carriages can be bought for $56362.50?

COUNTING ROOM EXERCISES.

EX. 1. What cost 320 yards of satinet, at \$1.12½ per yard?

Analysis.—If the price were \$1 per yard, the cloth would evidently cost as many dollars as there are yards. But \$1.12½ is equal to 1 and $\frac{1}{8}$ dollars; hence, the cloth will cost $\frac{1}{8}$ more dollars than there are yards; consequently, if we add to the number of yards $\frac{1}{8}$ of itself, it will give the cost. Now $\frac{1}{8}$ of 320=40, and 320+40=360. *Ans.* \$360.

PROOF.—\$1.12½×320=\$360, the same as before. Hence,

382. When the price of 1 article, 1 pound, &c., is \$1.12½; \$1.25; \$1.37½; &c., to find the cost of any number of articles.

To the given number of articles, add $\frac{1}{8}$, $\frac{1}{4}$, $\frac{3}{8}$, &c., *of itself, as the case may be, and the sum will be the cost required.*

OBS. When the price of 1 article, &c., is \$2.12½, \$2.25, \$3.37½, &c., the operation may be contracted by multiplying the given number of articles by $2\frac{1}{8}$, $2\frac{1}{4}$, $3\frac{3}{8}$, &c., as the case may be.

2. What cost 640 bushels of wheat, at \$1.25 per bushel?
3. What cost 372 pair of shoes, at \$1.37½ a pair?
4. What cost 480 bbls. of cider, at \$1.62½ a barrel?
5. What cost 520 yards of silk, at \$1.50 per yard?
6. What cost 720 drums of figs, at \$1.87½ per drum?
7. At \$2.12½ apiece, what will 480 sheep cost?
8. At \$2.37½ apiece, what will 364 vests cost?
9. At \$3.25 per yard, what cost 744 yards of cloth?
10. At \$4.62½ apiece, what cost 960 hats?
11. At \$5.12½ a pair, what cost 278 pair of boots?
12. At \$7.37½ per lb., what will 365 lbs. of opium cost?
13. A collier sold 856 tons of coal, at \$6.87½ per ton: how much did it amount to?
14. At \$19.62½ per acre, what will 537 acres of land cost?
15. What cost 72 lbs. of flax, at \$8.25 per hundred?

Analysis.—72 pounds are $\frac{72}{100}$ of 100 pounds; therefore 72 pounds will cost $\frac{72}{100}$ of \$8.25; and $\frac{72}{100}$ of $\$8.25=\frac{8.25\times72}{100}$.

Operation.

```
 $8.25
    72
  1650
 5775
$5.9400 Ans.
```

We multiply the price of 100 lbs. ($8.25) by 72, the given number of pounds, and the product $594.00, is the cost of 72 lbs. at $8.25 per *pound.* But the price is $8.25 per *hundred;* consequently the product $594.00 is 100 times too large, and must therefore be divided by 100, to give the true answer. But to divide by 100, we simply remove the decimal point two places towards the left. (Art. 331.)

16. What cost 367 bricks, at $4.45 per 1000?

Operation.

```
  4.45
  3 67
$1.63315 Ans.
```

Reasoning as before, 367 bricks will cost $\frac{367}{1000}$ of $4.45. We multiply the price of 1000 bricks by the given number of bricks, and divide the product by 1000. (Art. 331.) Hence,

383. To find the cost of articles sold by the 100, or 1000.

Multiply the given price by the given number of articles; then if the price is for 100, *divide the product by* 100; *but if the price is for* 1000, *divide it by* 1000. (Art. 331.)

17. A farmer sold 563 lbs. of hay, at $1.12½ per hundred: how much did it come to?

18. What cost 1640 lbs. of beef, at $6.37½ per hundred?

19. What cost 2719 lbs. of fish, at $4.20 per hundred?

20. What is the freight on 3568 lbs. from New York to Buffalo, at $1.67 per hundred?

21. What cost 6521 lbs. of cheese, at $7¾ per hundred?

22. What cost 15214 lbs. of butter, at $12½ per hundred?

23. At $6.25 per 1000, what cost 865 feet of spruce boards?

24. At $19.45 per 1000, what cost 2680 feet of pine boards?

25. At $67.33 per 1000, what cost 6500 feet of mahogany?

26. When ginger is $16.53 per cwt., what is it per pound?

Analysis.—Since 100 lbs. cost $16.53, 1 lb. will cost $\frac{1}{100}$ of $16.53. But to divide by 100, we remove the decimal point two places to the left. (Art. 331.) *Ans.* $0.1653.

QUEST.—383. How do you find the cost of articles sold, by the 100, or 1000?

27. When pine boards are $21.63 per 1000, what are they per foot?

Solution.—Reasoning as before, 1 foot will cost $\frac{1}{1000}$ of $21.63. Now to divide by 1000, we remove the decimal point three places to the left. (Art. 331.) *Ans.* $.02163. Hence,

384. When the cost of 100, or 1000 articles, pounds, &c., is given, the price of *one* is found by simply *removing the decimal point in the given cost or dividend, as many places to the left as there are ciphers in the divisor.* (Art. 331.)

28. Bought 1000 bricks for $7.20: what is that apiece?

29. If 1000 feet of hemlock boards cost $6.40, what will one foot cost?

30. Bought 42 cwt. of tobacco for $565.82: what is that per cwt.; and what per pound?

31. Bought 75 cwt. of butter for $966.38: what is that per cwt.; and what per pound?

BILLS, ACCOUNTS, &C.

385. A *Bill*, in mercantile operations, is a paper containing a written statement of the items, and the price or amount of goods sold.

32. What is the cost of the several articles, and what the amount, of the following bill?

NEW YORK, May 21st, 1847.

G. B. Grannis, Esq.,

Bought of Mark H. Newman & Co.,

75	Thomson's Mental Arithmetic,	at	$ .12½	- -
50	" Practical Arithmetic,	"	.31¼	- -
36	Porter's Rhetorical Reader,	"	.62½	- -
25	Willson's School History,	"	.46	- -
30	M'Elligott's Young Analyzer,	"	.31¼	- -
75	Thomson's Day's Algebra,	"	.50	- -
50	" Legendre's Geometry,		.47½	- -

Received Payment,

Mark H. Newman & Co.

(33.)

PHILADELPHIA, June 10th, 1847.

Hon. Horace Binney,

Bought of Leverette & Griggs,

163 lbs. Butter,	at	$ .14½	- - -
235 lbs. Coffee,	"	.08¼	- - -
86 lbs. Chocolate,	"	.11	- - -
685 lbs. Sugar,	"	.10½	- - -
21 doz. Eggs,	"	.13	- -
860 lbs. Lard,	"	.09½	- - -

What was the cost of the several articles, and what the amount of his bill?

(34.)

ALBANY, July 1st, 1847

Messrs. Collins & Brothers,

To G. W. Bunker, Dr.

For 320 yds. Silk,	at	$1.12½	- - -
" 256 " Broadcloth,	"	3.62½	- - -
" 175 pair Cotton Hose,	"	0.12½	- - -
" 100 " Silk "	"	0.87½	- - -
" 15 doz. Gloves,	"	0.62½	- - -
" 120 Straw Hats,	"	1.87½	- - -

What was the cost of the several articles, and what the amount of his bill?

(35.)

ST. LOUIS, Aug. 25th, 1847.

James Henry, Esq.

To J. L. Hoffman & Co., Dr.

For 15260 lbs. Pork,	at	$0.05½	- - -
" 7265 lbs. Cheese,	"	0.08½	- - -
" 11521 bu. Corn,	"	0.50	- - -
" 1560 bbls. Flour,	"	6.12½	- - -
	CREDIT.		
By 1150 lbs. Cotton,	at	$0.06¼	- - -
" 8256 lbs. Sugar,	"	0.07	- - -
" 6450 gals. Molasses,	"	0.37½	- - -
" Cash to balance account,	-	-	- - -

What is the amount of cash requisite to balance the account?

SECTION XII.

PERCENTAGE.

ART. **386.** The terms *Percentage* and *Per Cent.* signify a certain *allowance on a hundred;* that is, a certain *part* of a hundred, or simply *hundredths.* Thus, the expression 6 per cent. signifies 6 hundredths, ($\frac{6}{100}$,) 7 per cent., 7 hundredths, ($\frac{7}{100}$,) &c., of the number, or sum of money under consideration.

Note.—The terms *Percentage* and *Per Cent.* are derived from the Latin *per* and *centum*, signifying *by the hundred.*

387. We have seen that *hundredths* are *decimal* expressions, occupying the first two places of figures on the right of the decimal point. (Arts. 311, 314.) Now, since *percentage* and *per cent.* signify *hundredths,* it is manifest that they can be expressed by decimals, as in the following

PERCENTAGE TABLE.

1 per cent.	is written thus:	.01
2 per cent. . . .	" " "	.02
3 per cent. . . .	" " "	.03
6 per cent. . . .	" " "	.06
7 per cent. . . .	" " "	.07
10 per cent. . . .	" " "	.10
12 per cent. . . .	" " "	.12
50 per cent. . . .	" " "	.50
100 per cent. . . .	" " "	1.00
103 per cent. . . .	" " "	1.03
125 per cent., &c. . .	" " "	1.25
$\frac{1}{2}$ per cent., that is, $\frac{1}{2}$ of 1 per cent.	" " "	.005
$\frac{1}{4}$ per cent., that is, $\frac{1}{4}$ of 1 per cent.	" " "	.0025
$\frac{3}{4}$ per cent., that is, $\frac{3}{4}$ of 1 per cent.	" " "	.0075
$13\frac{1}{8}$ per cent. . . .	" " "	.13125
$25\frac{3}{8}$ per cent. . . .	" " "	.25375

OBS. 1. It will be seen from the preceding Table, that when the given per cent. is *less* than 10, a cipher must be prefixed to the figure expressing it, in the same manner as when the number of cents is less than 10. (Art. 369. Obs. 1.)

QUEST.—386. What do the terms percentage and per cent. signify? What is meant by 6 per cent., 7 per cent., &c., of any number, or sum?

When the given per cent. is *more* than 100, it must plainly require a mixed number to express it. (Art. 315. Obs. 2.)

2. Parts of 1 per cent. may be expressed either by a *common* fraction, or by *decimals*. Thus, the expression $17\frac{5}{8}$ per cent., is equivalent to .17625 per cent.

3. The *first two* decimal figures properly denote the *per cent.*, for they are *hundredths*; the other decimals being *parts of hundredths*, express *parts* of 1 per cent.

EXAMPLES.

1. Write 1 per cent., 2 per cent., 4 per cent., 6 per cent., 7 per cent., 8 per cent., in decimals.

2. Write 11 per cent.; 12; 14; 15; 16; 23; 65; 93.

3. Write $\frac{1}{2}$ per ct.; $\frac{1}{4}$; $\frac{1}{5}$; $\frac{2}{5}$; $\frac{4}{5}$; $\frac{3}{4}$; $\frac{1}{6}$; $\frac{1}{8}$; $\frac{2}{8}$; $\frac{5}{8}$; $\frac{1}{3}$; $\frac{2}{3}$; $\frac{1}{7}$; $\frac{1}{9}$

4. Write $4\frac{1}{2}$ per ct.; $6\frac{1}{4}$; $7\frac{1}{8}$; $9\frac{1}{5}$; $12\frac{1}{2}$; $16\frac{1}{4}$; 115; $400\frac{1}{4}$.

5. An agent collected $700 for a merchant, and received 5 per cent. for his services: how much did he receive?

Analysis.—Since 5 per cent. is the same as $\frac{5}{100}$, the agent must have received $\frac{5}{100}$ of $700. Now $\frac{1}{100}$ of $700 is $\$\frac{700}{100}$, which is equal to $7; and 5 hundredths is 5 times $7, or $35.

Operation.	
$700	Since $\frac{5}{100}$=.05, we multiply the given number of dollars by .05, and it gives the answer in cents, which we reduce to dollars by pointing off 2 decimals. (Art. 372.) Hence,
.05	
$35.00 *Ans.*	

388. To calculate percentage on any number, or sum of money.

Multiply the given number or sum by the given per cent. expressed decimally; and point off the product as in multiplication of decimal fractions. (Art. 324.)

OBS. 1. It is important for the learner to observe, that the *amount* of money *collected*, is made the *basis* upon which the percentage is computed. That is, the agent is entitled to 3 dollars, as often as he *collects* 100 dollars, and *not* as often as he *pays over* 100 dollars, as is frequently supposed. For in the latter case he would receive only $\frac{3}{103}$, instead of $\frac{3}{100}$ of the sum in question. This distinction is important, especially in calculating percentage on large sums.

QUEST.—387. How may per centage or per cent. be expressed? *Obs.* When the given per cent. is less than 10, how is it written? When more than 100, how? 388. How is percentage calculated? *Obs.* In collecting money, upon what basis is the per cent. calculated? If the per cent. contains a common fraction which cannot be expressed decimally, how proceed?

2. Hence, if the per cent. contains a common fraction which cannot be expressed decimally, first multiply by the decimal, then by the common fraction of the given per cent., and point off the sum of their products as above.

6. What is $4\frac{1}{5}$ per cent. of $300 ?

Solution.—Expressed decimally, $4\frac{1}{5}$ per cent.=.042 ; (Art. 384 Obs. 2 ;) and $300×.042=$12.60. *Ans.*

7. What is 3 per cent. of $256.25 ?

8. What is 2 per cent. of $437.63 ?

9. What is $2\frac{1}{2}$ per cent. of $138.432 ?

10. What is 6 per cent. of $145.13 ?

11. What is 7 per cent. of $1630.10 ?

12. A man borrowed $150, and paid 7 per cent. for the use of it: how much did he pay ?

13. A merchant bought goods amounting to $1825, and sold them so that he gained 12 per cent.: how much did he gain ?

14. A constable collected $862.56, and charged 5 per cent. for his services: how much did he receive; and how much did he pay over ?

15. What is 10 per cent. of $4020.50 ?

16. What is 8 per cent. of $1675 ?

17. What is $4\frac{1}{2}$ per cent. of $725 ?

18. What is $5\frac{1}{4}$ per cent. of $648.30 ?

19. What is $6\frac{1}{4}$ per cent. of $1000 ?

20. What is $7\frac{1}{3}$ per cent. of $2000 ?

21. What is $8\frac{3}{4}$ per cent. of $100.25 ?

22. A farmer having 1500 sheep, lost 25 per cent. of them: how many did he lose ?

23. A merchant having $1960 on deposit, drew out 20 per cent. of it: how much had he left in the bank ?

24. A merchant imported 1500 boxes of oranges, and $12\frac{1}{4}$ per cent. of them decayed: how many boxes did he lose; and how many had he left ?

25. What is $\frac{1}{2}$ per cent. of $1625 ?

26. What is $\frac{1}{4}$ per cent. of $2526.40 ?

27. What is $\frac{1}{5}$ per cent. of $42260.08 ?

28. What is $\frac{1}{3}$ per cent. of $75000 ?

29. What is $\frac{3}{4}$ per cent. of $100000 ?

30. What is $\frac{1}{5}$ per cent. of \$45241.20?

31. What is $\frac{1}{8}$ per cent. of \$675264?

32. A merchant bought a stock of goods amounting to \$4565, and paid $3\frac{1}{2}$ per cent. for freight: what was the whole cost of his goods?

33. A man's salary is \$2000 a year, and he lays up $37\frac{1}{2}$ per cent. of it: how much does he spend?

34. A youth who inherited \$20000, spent 40 per cent. of it in dissipation: how much had he left?

35. Two merchants embarked in business with \$18250 capital apiece; one gained 20 per cent. and the other lost 20 per cent. the first year: what was then the amount of each man's property?

36. Two men invested \$10000 apiece in stocks; one lost 8 per cent., the other 6 per cent.: what was the difference of their loss?

37. What is the difference between 6 per cent. of \$1040, and 7 per cent. of \$905?

APPLICATIONS OF PERCENTAGE.

389. PERCENTAGE, *or the method of reckoning by hundredths*, is applied to various calculations in the practical concerns of life. Among the most important of these are Commission, Brokerage, the Rise and Fall of Stocks, Interest, Discount, Insurance, Profit and Loss, Duties, and Taxes. Its principles, therefore, should be thoroughly understood by every scholar.

COMMISSION, BROKERAGE, AND STOCKS.

390. *Commission* is the *per cent.* or *sum charged* by agents for their services in buying and selling goods, or transacting other business.

OBS. An *Agent* who buys and sells goods for another, is called a *Commission Merchant, a Factor, or Correspondent.*

391. *Brokerage* is the *per cent.* or *sum charged* by money dealers, called *Brokers,* for negotiating *Bills of Exchange,* and other monetary operations, and is of the same nature as Commission.

QUEST.—390. What is commission? *Obs.* What is an agent who buys and sells goods for another usually called?

392. By the term *Stocks*, is meant the *Capital* of moneyed institutions, as incorporated Banks, Manufactories, Railroad and Insurance Companies; also, Government and State Bonds, &c.

OBS. 1. Stocks are usually divided into portions of $100 each, called *shares;* and the owners of these shares are called *Stockholders*.

2. The association or company thus formed, is called a *corporation;* the instrument specifying the powers, rights, and privileges invested in the corporation, is called a *charter*.

393. The original *cost* or *valuation* of a share is called its *nominal*, or *par value;* the *sum* for which it can be sold, is its *real value*.

OBS. 1. The *rise* or *fall* of Stocks is reckoned at a certain per cent. of its *par value*. The term *par* is a Latin word, which signifies *equal*, or a *state of equality*.

2. When Stocks sell for their original cost or valuation, they are said to be *at par;* when they sell for more than cost, they are said to be *above par*, *at a premium*, or *an advance;* when they do not sell at cost, they are said to be *below par*, or *at a discount*.

3. Persons who deal in Stocks are usually called *Stock Brokers*, or *Stock Jobbers*.

394. The *commission* or *allowance* made to factors and brokers, also the *rise* and *fall* of stocks, are usually reckoned at a *certain percentage* on the *amount* of money employed in the transaction, or on the *par value* of the given shares. Hence,

395. To compute commission, brokerage, and the premium or discount on stocks.

Multiply the given sum by the given per cent. expressed in decimals, and point off the product as in Percentage. (Art. 388.)

OBS. The commission for the collection of bills, taxes, &c., also for the sale or purchase of goods, varies from $2\frac{1}{2}$ to 12 or 15 per cent., and should always be reckoned on the *amount* of money *collected*, or paid out, or employed in the transaction.

The brokerage for the sale or purchase of stocks, varies from $\frac{1}{8}$ to $\frac{3}{4}$ per cent., reckoned on the par value of the stock.

QUEST.—391. What is brokerage? 392. What is meant by the term stocks? *Obs.* How are stocks usually divided? 393. What is the par value of stocks? What the real value? *Obs.* What is the meaning of the term par? When are stocks at par? When above par? When below? 395. How do you compute commission, brokerage, &c.?

EXAMPLES.

1. An auctioneer sold goods amounting to \$463, at 3 per cent. commission: how much did he receive? *Ans.* \$13.89.

2. An agent bought goods amounting to \$625.375: what is his commission, at 2 per cent.?

3. What is the commission on \$1682.25, at $3\frac{1}{2}$ per cent.?

4. What is the commission on \$1463.18, at 5 per cent.?

5. What is the commission on \$2560.07, at $4\frac{1}{2}$ per cent.?

6. What is the commission on \$10250, at 6 per cent.?

7. What is the commission on \$8340.60, at 7 per cent.?

8. What is the commission on \$960.625, at $5\frac{1}{2}$ per cent.?

9. A commission merchant sold goods to the amount of \$6235, at $2\frac{1}{2}$ per cent.: what was his commission?

10. An attorney collected a debt of \$8265.17, and charged $7\frac{1}{2}$ per cent. for his services: how much did he receive?

11. Bought \$1108 worth of books, at 4 per cent. commission: what was the amount of commission?

12. A tax-gatherer collected \$12250, for which he was entitled to $5\frac{1}{2}$ per cent. commission: how much did he receive?

13. Sold goods amounting to \$1432.26: how much was the commission, at 4 per cent.?

14. A commission merchant sold a quantity of hardware amounting to \$9240.71: how much would he receive, allowing $2\frac{1}{2}$ per cent. for selling, and 2 per cent. more for guaranteeing the payment?

15. An auctioneer sold carpeting amounting to \$2136.63, and charged $2\frac{1}{4}$ per cent. for selling, and $2\frac{3}{4}$ per cent. for guaranteeing the payment: how much did the auctioneer receive; and how much did he remit the owner?

396. Commission merchants, agents, &c., generally keep an account with their employers, and as they make investments or sales of goods, charge their commission on the amount invested, or the sum employed in the transaction.

Sometimes, however, a specific amount is sent to an agent or broker, requesting him, after deducting his commission, to lay out the balance in a certain manner.

16. A gentleman sent his agent \$1500 to purchase a library: how much had he to lay out after deducting his commission at 5 per cent.; and what was his commission?

Note.—The money actually laid out by the agent in books, is manifestly the proper basis on which to calculate his commission, for it would be unjust to charge commission on the sum he retains. (Art. 395. Obs.)

Analysis.—The money laid out is $\frac{100}{100}$ of itself, and the commission is $\frac{5}{100}$ of this sum; consequently the money laid out added to the commission, must be $\frac{105}{100}$ the whole amount. The question therefore resolves itself into this: \$1500 is $\frac{105}{100}$ of what sum? If \$1500 is $\frac{105}{100}$, $\frac{1}{100}$ must be $1500 \div 105 = \frac{1500}{105}$, and $\frac{100}{100} = \frac{1500}{105} \times 100 = \$1428.57$, the sum laid out. Now $\$1500 - \$1428.57 = \$71.43$, the commission.

PROOF. — $\$1428.57 \times .05 = \$71.43$; and $\$1428.57 + \$71.43 = \$1500$, the amount sent. Hence,

397. To compute commission when it is to be deducted in advance from a given amount, and the balance is to be invested.

Divide the given amount by \$1 increased by the per cent. commission, and the quotient will be the part to be invested. Subtract the part invested from the given amount, and the remainder will be the commission.

OBS. The commission may also be found by multiplying the sum invested by the given per cent. according to the preceding rule. (Art. 395.)

17. An agent received \$21500 to lay out in provisions, after deducting 2 per cent. commission: what sum did he lay out?

18. A country merchant sent \$3560 to his agent in the city, to purchase goods: after taking out his commission, at 3½ per cent., how much remained to lay out?

19. Baring, Brothers & Co. sent their agents \$800000 to buy flour: after deducting 5 per cent. commission, how much would be left to invest?

20. A broker negotiated a bill of exchange of \$82531, at 5 per cent.: how much did he receive for his services?

21. What is the brokerage on \$94265, at 1½ per cent.?

22. What is the brokerage on \$6200, at ¾ per cent.?

23. What is the brokerage on $8845.50, at $\frac{1}{4}$ per cent.?

24. What is the brokerage on $2500, at $\frac{3}{8}$ per cent.?

25. A broker made an investment of $21265, and charged $1\frac{1}{2}$ per cent.: what was the amount of his brokerage?

26. If you buy 20 shares of Western Railroad stock, at 7 per cent. advance, how much will your stock cost you? *Ans.* $2140.

Note.—The stock evidently cost its par value, which is $2000 and 7 per cent. besides. Now $\$2000 \times .07 = \$140.00$; and $\$2000 + \$140 = \$2140$.

27. What cost 20 shares of bank stock, at 7 per cent. discount?
Ans. $\$2000 - \$140 = \$1860$.

28. What cost 35 shares of New York and Erie Railroad stock, at $5\frac{1}{2}$ per cent. premium?

29. A merchant bought 45 shares of Commercial Bank stock at par, and afterwards sold them, at 50 per cent. discount: how much did he lose?

30. A man invested $8460 in the New England Manufacturing Co., and afterwards sold out at $4\frac{1}{2}$ per cent. advance: how much did he sell his stock for?

31. Sold 64 shares of Hudson River Railroad stock, at $10\frac{1}{2}$ per cent. premium: how much did they come to?

32. A man bought 35 shares of Utica and Syracuse Railroad stock at par, and afterwards sold them at $1\frac{1}{2}$ per cent. advance: how much did he get for them?

33. A man bought 15 shares of Albany and Schenectady Railroad stock, at 2 per cent. advance, and sold them at 10 per cent. disc.: how much did he sell them for; and how much did he lose?

34. Bought 71 shares in the Albany Gas Co. at $5\frac{1}{2}$ per cent. premium: how much did they amount to?

35. A broker bought 48 shares of Michigan Railroad stock, at 14 per cent. discount, and sold them at 6 per cent. advance: how much did he make by the operation?

36. If I employ a broker to buy me 55 shares of Railroad stock, which is 20 per cent. below par, and pay him $\frac{1}{2}$ per cent. brokerage, how much will my stock cost me?

37. If my agent buys 78 shares of New York and Philadelphia Railroad stock, at 15 per cent. advance, and charges me $\frac{3}{4}$ per cent. brokerage, how much will my stock cost?

INTEREST.

398. Interest is the sum paid for the *use of money* by the borrower to the lender. It is reckoned at a given *per cent. per annum;* that is, so many dollars are paid for the use of $100 for *one year;* so many cents for 100 cents; so many pounds for £100; &c.

Obs. The student should be careful to notice the distinction between *Commission* and *Interest*. The *former* is reckoned at a certain *per cent.* without regard to time; (Art. 395;) the latter is reckoned at a certain *per cent.* for *one year;* consequently, for *longer* or *shorter* periods than one year, *like proportions* of the percentage for one year are taken.

The term *per annum*, signifies *for a year.*

399. The *money lent,* or that for which interest is paid, is called the *principal.*

The *per cent.* paid per annum, is called the *rate.*

The *sum* of the principal and interest, is called the *amount.* Thus, if I borrow $100 for 1 year, and agree to pay 5 per cent. for the use of it, at the end of the year I must pay the lender $100, the sum which I borrowed, and $5 interest, making $105. The principal in this case, is $100; the interest $5; the rate 5 per cent.; and the amount $105.

Obs. The term *per annum,* is seldom expressed in connection with the *rate per cent.*, but it is always understood; for the *rate* is the *per cent.* paid *per annum.* (Art. 399.)

400. The *rate* of interest is usually established by law. It varies in different countries and in different parts of our own country.

Obs. When no rate is mentioned, the rate established by the laws of the State in which the transaction takes place, is always understood to be the one intended by the parties.

401. Any rate of interest *higher* than the legal rate, is called *usury,* and the person exacting it is liable to a heavy penalty.

Any rate *less* than the legal rate may be taken, if the parties concerned so agree.

Quest.—398. What is Interest? How is it reckoned? *Obs.* What is the difference between Commission and Interest? What is meant by the term per annum? 399. What is meant by the principal? The rate? The amount? 400. How is the rate usually determined? Is it the same everywhere? *Obs.* When no rate is mentioned, what rate is understood? 401. What is any rate higher than the legal rate called?

402. The *legal rates* of interest, and the *penalty* for *usury* in the several States of the Union, are as follows:

States.	*Legal rates.*	*Penalty for Usury.*
Maine,	6 per cent.	Forfeit of the whole debt.
N. Hampshire,	6 per cent.	Forfeit of three times the usury.
Vermont,	6 per cent.	Recovery in action with costs.
Massachusetts,	6 per cent.	Forfeit of three times the usury.
Rhode Island,	6 per cent.	Forfeit of the usury and int. on the debt.
Connecticut,	6 per cent.	Forfeit of the whole debt.
New York,	7 per cent.	Forfeit of the whole debt.
New Jersey,	6 per cent.	Forfeit of the whole debt.
Pennsylvania,	6 per cent.	Forfeit of the whole debt.
Delaware,	6 per cent.	Forfeit of the whole debt.
Maryland,	6 per cent. *a*	Usurious contracts void.
Virginia,	6 per cent.	Forfeit of double the usury.
N. Carolina,	6 per cent.	Forfeit of double the usury.
S. Carolina,	7 per cent.	Forfeit of interest and usury with costs.
Georgia,	8 per cent.	Forfeit of three times the usury.
Alabama,	8 per cent.	Forfeit of interest and usury.
Mississippi,	8 per cent. *b*	Forfeit of usury and costs.
Louisiana,	5 per cent. *c*	Usurious contracts void.
Tennessee,	6 per cent.	Usurious contracts void.
Kentucky,	6 per cent.	Usury may be recovered with costs.
Ohio,	6 per cent.	Usurious contracts void.
Indiana,	6 per cent.	Forfeit of double the excess.
Illinois,	6 per cent. *d*	Forfeit of three times the usury, and int. due.
Missouri,	6 per cent. *e*	Forfeit of the usury, and the interest due.
Michigan,	7 per cent.	Forfeit of the usury, and one fourth the debt.
Arkansas,	6 per cent. *f*	Forfeit of usury.
Florida,	8 per cent.	Forfeit of interest and usury.
Wisconsin,	7 per cent. *g*	Forfeit of three times the usury.
Iowa,	7 per cent. *h*	Forfeit of three times the usury.
Texas,	10 per cent.	Usurious contracts void.
Dist. Columbia,	6 per cent.	Usurious contracts void.

OBS. 1. On debts and judgments in favor of the *United States*, interest is computed at 6 per cent.

2. In *Canada* and *Nova Scotia*, the legal rate of interest is 6 per cent. In *England* and *France* it is 5 per cent.; in *Ireland* 6 per cent. In *Italy*, about the commencement of the 13th century, it varied from 20 to 30 per cent.

a On tobacco contracts 8 per cent. *b* By contract as high as 10 per cent. *c* Bank interest 6 per cent.; conventional as high as 10 per cent. *d* By agreement as high as 12 per cent. *e* By agreement as high as 10 per cent. *f* By agreement, any rate not exceeding 10 per cent. *g* By contract as high as 12 per cent. *h* By agreement as high as 12 per cent.

403. Ex. 1. What is the interest of $30 for 1 year, at 6 per cent.?

Analysis.—We have seen that 6 per cent. is $\frac{6}{100}$; that is, $6 for $100, 6 cents for 100 cents, &c. (Art. 386.) Since therefore the interest of $1 (100 cents) for 1 year is 6 cents, the interest of $30 for the same time must be 30 times as much; and $30×.06 =$1.80. *Ans.*

Operation.

$30	Prin.
.06	Rate.
$1.80	Int. 1 yr.

We first multiply the principal by the given rate per cent. expressed in decimals, as in percentage, and point off as many decimals in the product as there are decimal places in both factors.

Ex. 2. What is the interest of $140.25 for 1 year, 1 month, and 10 days, at 7 per cent.? What is the amount?

Operation.

	$140.25	Prin.
	.07	Rate.
12)	$9.8175	Int. 1 yr.
3)	8181	" 1 mo.
	2727	" 10 d.
	$10.9083	Interest.
	$140.25	Prin. added.
	$151.1583	Amount.

1 month is $\frac{1}{12}$ of a year; therefore the interest for 1 month is $\frac{1}{12}$ of 1 year's interest. 10 days are $\frac{1}{3}$ of 1 month, consequently the interest for 10 days, is $\frac{1}{3}$ of 1 month's interest. The amount is found by adding the principal and interest together.

Note.—1. In adding the principal and interest, care must be taken to add dollars to dollars, cents to cents, &c. (Art. 374.)

2. When the rate per cent. is *less* than 10, a cipher must always be prefixed to the figure denoting it. (Art. 387. Obs. 1.) It is highly important that the principal and the rate should both be *written* correctly, in order to prevent mistakes in pointing off the product.

Ex. 3. What is the interest of $250.80 for 4 years, at 5 per cent.? What is the amount?

Solution.—$250.80×.05=$12.54, the interest for 1 year.
Now $12.54× 4=$50.16, " " 4 years.
And $250.80+$50.16=$300.96, the amount required.

11*

404. From the foregoing illustrations and principles we deduce the following general

RULE FOR COMPUTING INTEREST.

I. FOR ONE YEAR. *Multiply the principal by the given rate, and from the product point off as many figures for decimals, as there are decimal places in both factors.* (Art. 324.)

II. FOR TWO OR MORE YEARS. *Multiply the interest of* 1 *year by the given number of years.*

III. FOR MONTHS. *Take such a fractional part of* 1 *year's interest, as is denoted by the given number of months.*

IV. FOR DAYS. *Take such a fractional part of one month's interest, as is denoted by the given number of days.*

The amount is found by adding the principal and interest together.

OBS. 1. The *reason* of this rule is evident from the consideration that the given *rate per cent. per annum* denotes *hundredths.* (Arts. 386, 398.) Now when the rate is 6 per cent. we multiply by .06, when 7 per cent. by .07, &c., and point off two figures in the product; consequently the result will be the same as to multiply by $\frac{6}{100}$, $\frac{7}{100}$, &c.

2. In calculating interest, a month, whether it contains 30 or 31 days, or even but 28 or 29, as in the case of February, is assumed to be *one twelfth* of a year. Therefore, for 1 month we take $\frac{1}{12}$ of 1 year's interest; for 2 months, $\frac{1}{6}$; for 3 months, $\frac{1}{4}$; for 4 months, $\frac{1}{3}$; for 6 months, $\frac{1}{2}$; for 8 months, $\frac{2}{3}$, &c.

Again, 30 days are commonly considered a month; consequently the interest for 1 day, or any number of days under 30, is so many *thirtieths* of a month's interest. (Art. 303. Obs. 2.) Therefore, for 1 day we take $\frac{1}{30}$ of 1 month's interest; for 2 days, $\frac{1}{15}$; for 3 days, $\frac{1}{10}$; for 5 days, $\frac{1}{6}$; for 10 days, $\frac{1}{3}$, &c.

This practice seems to have been originally adopted on account of its convenience. Though not strictly accurate, it is sanctioned by general usage.

3. Allowing 30 days to a month, and 12 months to a year, a year would contain only 360 days, which in point of fact is $\frac{5}{365}$, or $\frac{1}{73}$ less than an ordinary year. Hence,

To find the interest for any number of days with *entire accuracy,* we must take so many 365ths of 1 year's interest, as is denoted by the given number of days; or, find the interest for the days as above; from this subtract $\frac{1}{73}$ of

QUEST.—404. How is interest computed for a year? How for any number of years? How for months? How for days? How find the amount? *Obs.* In reckoning interest, what part of a year is a month considered? How many days are commonly considered a month? Is this practice accurate?

itself, and the remainder will be the exact interest. The laws of New York, and several other states, require this deduction to be made.

In business, when the mills in the result are 5, or over, it is customary to add 1 to the cents; if under 5, to disregard them.

EXAMPLES.

1. What is the interest of $423 for 1 yr., at 7 per cent.?
2. What is the interest of $240.31 for 3 yrs., at 6 per cent.?
3. What is the interest of $403.67 for 2 yrs., at 5 per cent.?
4. What is the interest of $640 for 1 yr., at 8 per cent.?
5. What is the interest of $430.45 for 2 yrs., at 7 per cent.?
6. What is the interest of $185.06 for 4 yrs., at 6 per cent.?
7. What is the interest of $864.80 for 5 yrs., at $4\frac{1}{2}$ per cent.?
8. What is the interest of $763 for 4 months, at 7 per cent.?
9. What is the interest of $940.20 for 6 mo., at 6 per cent.?
10. What is the interest of $243.10 for 5 mo., at 8 per cent.?
11. What is the interest of $195.82 for 7 mo., at 6 per cent.?
12. What is the interest of $425.35 for 9 mo., at 6 per cent.?
13. At 7 per cent., what is the int. of $738 for 1 yr. and 2 mo.?
14. At 6 per cent., what is the int. of $894 for 1 yr. and 8 mo.?
15. At 7 per cent., what is the amount of $926 for 6 mo.?
16. At 7 per cent., what is the amt. of $648 for 2 mo. 15 d.?
17. At 6 per cent., what is the amt. of $1000 for 1 mo. 11 d.?
18. At 5 per cent., what is the amt. of $1565.45 for 3 mo.?
19. At 6 per cent., what is the amt. of $872 for 4 mo.?
20. What is the int. of $681 for 10 days, at 6 per cent.?
21. What is the int. of $483.26 for 15 d., at 7 per cent.?
22. What is the int. of $569.40 for 20 d., at 6 per cent.?
23. What is the amt. of $95 for 1 yr. and 6 mo., at 5 per cent.?
24. What is the amt. of $148 for 8 mo. 12 d., at 6 per cent.?
25. What is the amt. of $700 for 30 d., at 7 per cent.?
26. What is the int. of $340 for 60 d., at $5\frac{1}{4}$ per cent.?
27. What is the int. of $4685 for 90 d., at $6\frac{1}{2}$ per cent.?
28. What is the amt. of $3293 for 30 d., at 7 per cent.?
29. What is the amt. of $5265 for 15 d., at 6 per cent.?
30. What is the int. of $8310 for 10 d., at 7 per cent.?
31. What is the int. of $50625 for 21 d., at 7 per cent.?
32. What is the amt. of $65256 for 4 mo., at 7 per cent?

SECOND METHOD OF COMPUTING INTEREST.

405. There is another method of computing interest, which is very simple and convenient in its application, particularly when the interest is required for *months* and *days*, at 6 *per cent.*

406. We have seen that for 1 year, the interest of $1 at 6 per cent. is 6 cents., or $.06; (Art. 404;) therefore,

For 1 month, the interest of $1 is $\frac{1}{12}$ of 6 cents, which is $.005;
" 2 months, " " is $\frac{2}{12}$, or $\frac{1}{6}$ of 6 cents, " " .01;
" 3 months, " " is $\frac{3}{12}$, or $\frac{1}{4}$ of 6 cents, " " .015;
" 4 months, " " is $\frac{4}{12}$, or $\frac{1}{3}$ of 6 cents, " " .02;
" 5 months, " " is $\frac{5}{12}$, of 6 cents, " " .025;
" 6 months, " " is $\frac{6}{12}$, or $\frac{1}{2}$ of 6 cents, " " .03;

Hence, *The interest of $1 for 1 month, at 6 per-cent., is 5 mills, for every 2 months, it is 1 cent; and for any number of months, it is as many cents, or hundredths of a dollar, as 2 is contained times in the given number of months.*

407. Since the interest of $1 for 1 month (30 days) is 5 mills, or $.005, (Art. 406,)

For 6 days ($\frac{1}{5}$ of 30 days) the interest of $1 is $\frac{1}{5}$ of 5 mills, or $.001;
" 12 days ($\frac{2}{5}$ of 30 days) " " is $\frac{2}{5}$ of 5 mills, or .002;
" 18 days ($\frac{3}{5}$ of 30 days) " " is $\frac{3}{5}$ of 5 mills, or .003;
" 3 days ($\frac{1}{10}$ of 30 days) " " is $\frac{1}{10}$ of 5 mills, or .0005;

That is, the interest of $1 for *every* 6 *days*, is 1 *mill*, or $.001; and for any number of days, it is as many *mills*, or *thousandths of a dollar, as* 6 *is contained times* in the given number of days.

408. Hence, to find the interest of $1 for any number of *days*, at 6 per cent.

Divide the given number of days by 6, and set the first quotient figure in thousandths' place, when the days are 6, or more than 6; but in ten thousandths' place, when they are less than 6.

OBS. For 60 days (2 mo.) the interest of $1 is 1 cent; (Art. 406;) when, therefore, the number of days is 60 or over, the first quotient figure must occupy *hundredths'* place.

QUEST.—408. How find the interest of $1 for any number of days, at 6 per cent?

Ex. 1. What is the interest of $185 for 1 year, 6 months and 18 days, at 6 per cent.?

Analysis.—The interest of $1 for 1 year is 6 cents; for 6 months it is 3 cents; and for 18 days it is 3 mills. (Arts. 406, 407.) Now .06+.03+.003=$.093. Since therefore the interest of $1 for the given time is $.093, the interest of $185 must be 185 times as much.

Operation.
$185 Prin.
.093 Int. $1.
555
1665
$17.205 *Ans.*

409. From these principles we may derive a

SECOND RULE FOR COMPUTING INTEREST.

I. To compute the interest on any sum, at 6 per cent.

Multiply the principal by the interest of $1 *for the given time, at* 6 *per cent., and point off the product as in multiplication of decimals.* (Art. 324.)

II. To compute int. at any rate, *greater* or *less* than 6 per cent.

First find the interest on the given sum at 6 *per cent.; then add to this interest, or subtract from it, such a fractional part of itself, as the required rate exceeds or falls short of* 6 *per cent.*

The amount is found by adding the principal and interest together as in the former method. (Art. 404.)

OBS. 1. The *amount* may also be found by multiplying the given principal by the *amount* of one dollar for the time.

2. The *reason* of the first part of this rule, is manifest from the principle that the interest of 2 dollars for any given time and rate, must be *twice* as much as the interest of 1 dollar for the same time and rate; the interest of 50 dollars, 50 times as much as that of 1 dollar, &c.

3. When the required rate is 7 per cent., we first find the interest at 6 per cent., then add $\frac{1}{6}$ of it to itself; if 5 per cent., subtract $\frac{1}{6}$ of it from itself, &c., for the obvious reason, that 7 per cent. is *once* and 1 *sixth*, or $\frac{7}{6}$ of 6 per cent.; 5 per cent. is only $\frac{5}{6}$ of 6 per cent., &c.

4. When the decimal denoting the int. of $1 for the days, is *long*, or is a *repetend*, it is more accurate to retain the common fraction. (Art. 387. Obs. 2.)

2. What is the interest of $746 for 4 months and 18 days, at 6 per cent.? *Ans.* $17.158.

QUEST.—409. What is the second method of computing interest, at 6 per cent? When the rate per cent. is greater or less than 6 per cent., how proceed?

3. What is the interest of $240 for 6 months and 12 days, at 7 per cent.?

Operation.

$240 Prin.	The interest of $1 for 6 mo. at 6 per ct., is .03; for 12 d. it is .002; and .03+.002=$.032.
.032 Int. of $1.	
480	
720	The required rate is 1 per cent. more than 6 per cent.; we therefore find the interest at 6 per cent., and add $\frac{1}{6}$ of it to itself.
6)$7.680=Int. at 6 per ct.	
1.280=$\frac{1}{6}$ of 6 per ct.	
Ans. $8.960 Int. at 7 per ct.	

4. What is the interest of $680 for 3 mo., at 5 per cent.?
5. What is the interest of $213.08 for 1 mo., at 6 per cent.?
6. What is the interest of $859 for 1 yr. 2 mo., at 7 per cent.?
7. What is the interest of $768 for 1 yr. 7 mo., at 8 per cent.?
8. What is the interest of $684 for 9 mo., at 6 per cent.?
9. At 7 per cent., what is the amount of $387 for 5 mo.?
10. At 4 per cent., what is the amt. of $1125 for 1 yr. 2 mo.?
11. At 6 per cent., what is the amt. of $1056 for 10 mo. 24 d.?
12. At 6 per cent., what is the int. of $1340 for 1 mo. 15 d.?
13. At 6 per cent., what is the int. of $815 for 2 mo. 21 d.?
14. At 8 per cent., what is the amt. of $961 for 4 mo. 10 d.?
15. What is the int. of $2345.10 for 6 mo., at 7 per cent.?
16. What is the int. of $1567.18 for 4 mo., at 7½ per cent.?
17. What is the int. of $3500 for 11 mo., at 10 per cent.?
18. What is the int. of $39.375 for 2 yrs., at 12½ per cent.?
19. What is the int. of $113.61 for 5 yrs., at 15 per cent.?
20. What is the int. of $1000 for 2 yrs., at 20 per cent.?
21. What is the int. of $1260.34 for 10 yrs., at 13 per cent.?
22. At 16 per cent., what is the int. of $150 for 6 years.?
23. At 30 per cent., what is the int. of $300 for 1 year.?
24. What is the amt. of $12645 for 10 d., at 6 per cent.?
25. What is the amt. of $16285 for 24 d., at 7 per cent.?
26. At 4½ per cent., what is the int. of $10255 for 8 months?
27. At 5¼ per cent., what is the int. of $17371 for 3 months?
28. What is the amt. of $1 for 100 yrs., at 7 per cent.?
29. What is the amt. of 1 cent for 100 yrs., at 6 per cent.?

410. Since the interest of \$1 at 6 per cent. for 12 mo. is 6 cents, (Art. 406,) for 6 mo. it must be 3 cents; for 3 mo., $1\frac{1}{2}$ cents; for 2 mo., 1 cent; for 1 mo. or 30 d. $\frac{1}{2}$ cent; for 15 d., $\frac{1}{4}$ cent; for 20 d. $\frac{1}{3}$ cent, &c. That is, the interest of \$1 at 6 per cent. is as many cents as are equal to *half* the given number of months.

411. Hence, to compute interest at 6 per cent. by *months*.

Multiply the principal by half the number of months, and point off two more figures for decimals in the product than there are decimal places in the multiplicand.

OBS. 1. When there are years and days, reduce the years to months, and the days to a common fraction of a month.

Or, divide the days by 3, and annex the quotient to the months considered as *hundredths; half of the number thus produced will be the decimal multiplier.*

2. The latter method is the same as dividing the days by 6, and setting the first quotient figure in *thousandth's* place; for, we divide the days by 3 and 2, and $3\times2=6$. (Arts. 407, 408.)

30. What is the int. of \$460.384 for 8 mos. and 15 d., at 6 per ct.?

Operation.

\$460.384	We multiply by $4\frac{1}{4}$, for, 8 months+15 days=$8\frac{1}{2}$ months, and $8\frac{1}{2}\div2=4\frac{1}{4}$. And since there are three decimals in the multiplicand, we point off 5 in the product.
$4\frac{1}{4}$	
1841536	
115096	
\$19.56632 *Ans.*	

31. What is the interest of \$780 for 4 months, at 6 per cent.?
32. What is the interest of \$1406 for 3 mo., at 6 per cent.?
33. What is the interest of \$109 for 2 mo., at 7 per cent.?
34. What is the interest of \$119.45 for 8 mo., at 6 per cent.?
35. What is the interest of \$618 for 1 yr. 3 mo., at 6 per cent.?
36. What is the interest of \$861 for 2 yrs. 6 mo., at 6 per cent.?
37. What is the interest of \$936.40 for 3 yrs., at 6 per cent.?
38. What is the interest of \$4526 for 6 mo. 2 d., at 6 per cent.?
39. What is the interest of \$8246 for 10 mo., at 7 per cent.?
40. What is the interest of \$31285 for 3 mo., at 5 per cent.?
41. What is the interest of \$17500 for 1 yr. 3 mo., at 7 per ct.?
42. What is the amount of \$3286 for 8 mo. 15 d., at 6 per ct.?
43. What is the amount of \$15876 for 5 mo. 18 d., at 6 per ct.?

412. We have seen that the interest of $1 at 6 per cent. for any number of days is equal to as many mills, as 6 is contained times in the given days. (Art. 407.) Hence,

413. To compute interest at 6 per cent. by *days*.

Multiply the principal by one sixth of the given number of days, and point off three more figures for decimals in the product than there are decimal places in the principal. (Art. 411. Obs. 2.)

Or, multiply the principal by the given number of days, divide the product by 6, and point off the quotient as above.

OBS. The product is in mills and parts of a mill. The object, therefore, of pointing off three more places for decimals in the product than there are decimals in the principal, is to reduce it to dollars. (Art. 372.)

44. What is the interest of $976.22 for 33 days, at 6 per cent.?

Solution.—$\frac{1}{6}$ of 33 d.$=5\frac{1}{2}$; and 976.22\times 5\frac{1}{2}=5369.21$ mills. Pointing off 3 more decimals, we have $5.36921. *Ans.*

45. What is the interest of $536.30 for 24 days, at 6 per cent.?

46. What is the interest of $7085 for 63 d., at 6 per cent.?

47. What is the interest of $8126.21 for 8 d., at 6 per cent.?

48. What is the interest of $25681 for 93 d., at 6 per cent.?

49. What is the interest of $764.85 for 114 d., at 6 per cent.?

APPLICATIONS OF INTEREST.

414. In the application of interest to business transactions the following particulars deserve attention.

1. A *promissory* note is a writing which contains a promise of the payment of money or other property to another, at or before a time specified, in consideration of value received by the *promiser* or *maker* of the note.

Unless a note contains the words "value received," by some authorities it is deemed *invalid;* consequently these words should always be inserted.

2. The person who signs a note is called the *maker*, *drawer*, or *giver* of the note. The person to whom a note is made payable, is called the *payee;* the person who has the legal possession of a note, is called the *holder* of it.

3. A note which is made payable "*to order*," "*or bearer*," is said to be *negotiable;* that is, the holder may *sell* or *transfer* it to whom he pleases, and it can be collected by any one who has lawful possession of it. Notes without these words are not negotiable. (See Nos. 1, 2.)

4. If the holder of a negotiable note which is made payable *to order* wishes to sell or transfer it, the law requires him to *endorse* it, or write his name on the back of it. The person to whom it is transferred, or the holder of it, is

then empowered to collect it of the drawer; if the drawer is *unable*, or *refuses* to pay it, then the endorser is responsible for its payment. (See No. 1.)

5. When a note is made payable to the *bearer*, the holder can sell or transfer it without endorsing it, or incurring the liability for its payment. Bank notes or bills are of this description. (See No. 2.)

6. When a note is made payable to any particular person without the words *order* or *bearer* it is *not negotiable;* for, it cannot be collected or sued except in the name of the person to whom it is made payable. (See No. 3.)

7. A note should always specify the time at which it is to be paid; but if no time is mentioned, the presumption is that it is intended to be paid on *demand*, and the giver must pay it when demanded.

8. According to custom and the statutes of most of the States, a note or draft is not presented for collection until *three* days after the time specified for its payment. These three days are called *days of grace*. Interest is therefore reckoned for *three* days more than the time specified in the note. When the last day of grace comes on Sunday, or a national holiday, as the 4th of July, &c., it is customary to pay a note on the day previous.

9. If a note is not paid at *maturity* or the *time specified*, it is necessary for the holder to *notify the endorser* of the fact in a legal manner, as soon as circumstances will admit; otherwise the responsibility of the endorser ceases.

10. Notes do not draw interest unless they contain the words "with interest." But if a note is not paid when it becomes due, it then draws legal interest till paid, though no mention is made of interest. (Art. 400. Obs.)

11. Notes which contain the words "*with interest*," though the *rate* is not mentioned, are entitled to the legal rate established by the State in which the note is made. In writing notes therefore it is unnecessary to specify the rate, unless by agreement it is to be less than the legal rate.

12. When a note is made payable on a given day, and in a specified article of merchandise, as grain, stock, &c., if the article specified is not *tendered* at the given time and place, the holder can demand payment in money. Such notes, are *not negotiable;* nor is the drawer entitled to the days of grace.

13. When *two* or *more* persons jointly and severally give their note, it may be collected of either of them. (See No. 4.)

14. The *sum* for which a note is given, is called the *principal*, or *face of the note;* and should always be written out in words.

415. When it is required to compute the interest on a *note*, we must first find the *time* for which the note has been on interest, by subtracting the *earlier* from the *later* date; (Art. 303;) then cast the interest on the face of the note for the time, by either of the preceding methods. (Arts. 404, 409.)

OBS. In determining the time, the day on which a note is dated, and that on which it becomes due should not both be reckoned; it is customary to exclude the former.

Ex. 1. What is the interest due on a note of $625 from Feb. 2d, 1846, to June 20th, 1847, at 6 per cent.?

Operation.

	Yrs.	mo.	ds.
	1847	6	20
	1846	2	2
Time	1	4	18

$625 Prin.
.083 Int. of $1.
1875
5000
$51.875 *Ans.*

Compute the interest on the following notes:

(No. 1.)

$450. NEW YORK, June 3d, 1847.

2. Sixty days after date, I promise to pay George Baker, or order, Four Hundred and Fifty Dollars, with interest, value received. ALEXANDER HAMILTON.

(No. 2.)

$630. BOSTON, Aug. 5th, 1847.

3. Thirty days after date, I promise to pay Messrs. Holmes & Homer, or bearer, Six Hundred and Thirty Dollars, with interest, value received. JAMES UNDERWOOD

(No. 3.)

$850. PHILADELPHIA, Sept. 16th, 1847.

4. Four months after date, I promise to pay Horace Williams, Eight Hundred and Fifty Dollars, with interest, value received.
JOHN C. ALLEN.

(No. 4.)

$1000. CINCINNATI, Oct. 3d, 1847.

5. For value received, we jointly and severally promise to pay to the order of Wm. D. Moore & Co., One Thousand Dollars, in one year from date, with interest. JOSEPH HENRY,
SANDFORD ATWATER.

6. What is the interest on a note of $634 from Jan. 1st, 1846 to March 7th, 1847, at 6 per cent.?

7. What is the interest on a note of $820 from April 16th, 1846, to Jan. 10th, 1847, at 6 per cent.?

8. What is the interest on a note of $615.44 from Oct. 1st, 1836, to June 13th, 1840, at 4 per cent.?

9. What is the interest on a note of $1830.63 from Aug. 16th, 1841, to June 19th, 1842, at 7 per cent.?

10. What is the amount due on a note of $520 from Sept. 2d, 1846, to March 14th, 1847, at 5 per cent.?

11. What is the amount due on a note of $25000 from Aug. 17th, 1845, to Jan. 17th, 1846, at 7 per cent.?

12. What is the amount due on a note of $6200 from Feb. 3d, 1846, to Jan. 9th, 1847, at 6 per cent.?

PARTIAL PAYMENTS.

416. When *partial payments* are made and endorsed upon Notes and Bonds, the rule for computing the interest adopted by the *Supreme Court of the United States,* is the following.

I. "*The rule for casting interest, when partial payments have been made, is to apply the payment, in the first place, to the discharge of the interest then due.*

II. "*If the payment exceeds the interest, the surplus goes towards discharging the principal, and the subsequent interest is to be computed on the balance of principal remaining due.*

III. "*If the payment be less than the interest, the surplus of interest must not be taken to augment the principal; but interest continues on the former principal until the period when the payments, taken together, exceed the interest due, and then the surplus is to be applied towards discharging the principal; and interest is to be computed on the balance as aforesaid.*"

Note.—The above rule is adopted by *New York*, *Massachusetts*, and most of the other States of the Union. It is given in the language of the distinguished Chancellor Kent.—*Johnson's Chancery Reports*, Vol. I. p. 17.

QUEST.—416. What is the general method of casting interest on Notes and Bonds, when partial payment have been made?

$965. NEW YORK, March 8th, 1843.

13. For value received, I promise to pay George B. Granniss, or order, Nine Hundred and Sixty-five Dollars, on demand, with interest at 7 per cent. HENRY BROWN.

The following payments were endorsed on this note:

Sept. 8th, 1843, received $75.30.
June 18th, 1844, received $20.38.
March 24th, 1845, received $80.

What was due on taking up the note, Feb. 9th, 1846?

Operation.

Principal, - - - - - - -		$965.00
Interest to first payment, Sept. 8th. (6 months,)		33.775
Amount due on note Sept. 8th, - - - -		$998.775
1st payment, (to be deducted from amount,) -		75.30
Balance due after 1st pay't., Sept. 8th, 1843, -		$923.475
Interest on Balance to 2d pay't., June 18th, (9 mo. 10 d.,)	$50.278	
2d pay't., (being less than int. then due,)	20.38	
Surplus int. unpaid June 18th, 1844,	$29.898	
Int. continued on Bal. from June 18th, to March 24th, 1845, (9 mo. 6 d.,)	49.559	79.457
Amount due March 24th, 1845, - - -		$1002.932
3d pay't., (being greater than the int. now due,) is to be deducted from the amount,		80.00
Balance due March 24th, 1845, - - -		$922.932
Int. on Bal. to Feb. 9th, (10 mo. 15 d.,) - -		56.529
Bal. due on taking up the note, Feb. 9th, 1846,		$979.461

$650. BOSTON, Jan. 1st, 1842.

14. For value received, I promise to pay John Lincoln, or order, Six Hundred and Fifty Dollars on demand, with interest at 6 per cent. GEORGE LEWIS.

Endorsed, Aug. 13th, 1842, $100.
Endorsed, April 13th, 1843, $120.

What was due on the note, Jan. 20th, 1844?

$2460. PHILADELPHIA, April 10th, 1844.

15. Four months after date, I promise to pay James Buchanan, or order, Two Thousand Four Hundred and Sixty Dollars, with interest, at 6 per cent., value received.

GEORGE WILLIAMS.

Endorsed, Aug. 20th, 1845, $840.
" Dec. 26th, 1845, $400.
" May 2d, 1846, $1000.

How much was due Aug. 20th, 1846 ?

$5000. NEW ORLEANS, May 1st, 1845.

16. Six months after date, I promise to pay John Fairfield, or order, Five Thousand Dollars, with interest at 5 per cent., value received.

WILLIAM ADAMS.

Endorsed, Oct. 1st, 1845, $700.
" Feb. 7th, 1846, $45.
" Sept. 13th, 1846, $480.

What was the balance due Jan. 1st, 1847 ?

CONNECTICUT RULE.

417. I. " Compute the interest on the principal to the time of the first payment; if that be one year or more from the time the interest commenced, add it to the principal, and deduct the payment from the sum total. If there be after payments made, compute the interest on the balance due to the next payment, and then deduct the payment as above; and in like manner, from one payment to another, till all the payments are absorbed; provided the time between one payment and another be one year or more."

II. " If any payments be made before one year's interest has accrued, then compute the interest on the principal sum due on the obligation, for one year, add it to the principal, and compute the interest on the sum paid, from the time it was paid up to the end of the year; add it to the sum paid, and deduct that sum from the principal and interest added as above."

III. " If a year extends beyond the time of payment, then find the amount of the principal remaining unpaid up to the time of settlement, likewise the amount of the endorsements from the time they were paid to the time of settlement, and deduct the sum of these several amounts from the amount of the principal."

" If any payments be made of a less sum than the interest arisen at the time of such payment, no interest is to be computed, but only on the principal sum for any period."—*Kirby's Reports.*

THIRD RULE.

418. *First find the amount of the given principal for the whole time; then find the amount of each payment from the time it was endorsed to the time of settlement. Finally, subtract the amount of the several payments from the amount of the principal, and the remainder will be the sum due.*

Note.—It will be an excellent exercise for the pupil to cast the interest on the preceding notes by each of the above rules.

419. *To compute Interest on Sterling Money.*

17. What is the interest of £241, 10s. 6d. for 1 year, at 6 per cent.?

Operation.

```
     £241.525 Prin.
          .06 Rate.
    £14.49150 Int. 1 yr.
         20s.=£1.
    s. 9.83000
         12d.=1s.
    d. 9.96000
          4f.=1d.
  far. 3.84000
```

Ans. £14, 9s. 9¾d.

We first reduce the 10s. 6d. to the decimal of a pound, (Art. 346,) then multiply the principal by the rate, and point off the product as in Art. 404. The 14 on the left of the decimal point, denotes pounds; the figures on the right are decimals of a pound, and must be reduced to shillings, pence, and farthings. (Art. 348.) Hence,

419. *a.* To compute the interest on pounds, shillings, pence, and farthings.

Reduce the given shillings, pence, and farthings to the decimal of a pound; (Art. 346;) *then find the interest as on dollars and cents; finally, reduce the decimal figures in the answer to shillings, pence, and farthings.* (Art. 348.)

18. What is the amount of £156, 15s. for 1 year and 4 months, at 5 per cent.? *Ans.* £167, 4s.

19. What is the int. of £275, 12s. 6d. for 1 yr., at 7 per cent.?

20. What is the int. of £89, 7s. 6½d. for 2 yrs., at 5 per cent.?

21. What is the int. of £500 for 6 mo., at 5 per cent.?

22. What is the amt. of £1825, 10s. for 8 mo., at 6 per cent.?

23. What is the amt. of £2000 for 10 yrs., at 4½ per cent.?

QUEST.—419. *a.* How is interest computed on pounds, shillings, and pence?

PROBLEMS IN INTEREST.

420. It will be observed that there are *four parts* or *terms* connected with each of the preceding operations, viz: *the principal, the rate per cent., the time, and the interest, or the amount.* These parts or terms have such a relation to each other, that if any *three* of them are given, the *other* may be found. The questions, therefore, which may arise in interest, are numerous; but they may be reduced to a few *general principles*, or *Problems.*

OBS. A number or quantity is said to be *given*, when its value is stated, or may be easily inferred from the conditions of the question under consideration. Thus, when the principal and interest are known, the *amount* may be said to be *given*, because it is merely the *sum* of the principal and interest. So, if the principal and the amount are known, the *interest* may be said to be *given*, because it is the *difference* between the amount and the principal.

PROBLEM I.

421. *To find the* INTEREST, *the principal, rate per cent., and the time being given.*

This problem embraces all the preceding examples pertaining to Interest, and has already been illustrated.

PROBLEM II.

To find the RATE PER CENT., *the principal, the interest, and the time being given.*

Ex. 1. A man borrowed $80 for 5 years, and paid $36 for the use of it: what was the rate per cent.?

Analysis.—The interest of $80 at 1 per cent. for 1 year is 80 cents; (Art. 404;) consequently for 5 years it is 5 times as much, and $.80×5=$4. Now since $4 is 1 per cent. on the principal for the given time, $36 must be $\frac{36}{4}$ of 1 per cent., which is equal to 9 per cent. (Art. 196.)

Or, we may reason thus: Since $4 is 1 per cent. on the principal for the given time, $36 must be as many per cent. as $4 is contained times in $36; and $36÷$4=9. *Ans.* 9 per cent.

QUEST.—420. How many terms are connected with each of the preceding examples? What are they? When three are given, can the fourth be found? *Obs.* When is a number or quantity said to be given?

PROOF.—$80×.09=$7.20, the interest of $80 for 1 year at 9 per cent., and $7.20×5=$36.00, the interest for 5 years, which is equal to the sum paid. Hence,

422. To find the *rate per cent.* when the principal, interest and time are given.

Divide the given interest by the interest of the principal at 1 per cent. for the given time, and the quotient will be the required per cent.

Or, find the interest of the principal at 1 per cent. for the given time; then make the interest thus found the denominator and the given interest the numerator of a common fraction; reduce this fraction to a whole or mixed number, and the result will be the per cent. required. (Art. 196.)

2. If I loan $500 for 2 years, and receive $50 interest, what is the rate per cent.? *Ans.* 5 per cent.

3. A man borrowed $620 for 8 months, and paid $24.80 for the use of it: what per cent. interest did he pay?

4. At what per cent. interest must $2350 be loaned, to gain $47 in 4 months?

5. At what per cent. interest must $1925 be loaned, to gain $154 in 1 year?

6. A man has $12000 from which he receives $900 interest annually: what per cent. is that?

7. A man deposited $2600 in a savings bank, and received $143 interest annually: what per cent. was that?

8. A man invested $4500 in the Bank of New York, and received a semi-annual dividend of $157.50: what per cent. was the dividend?

9. A man paid $16250 for a house, and rented it for $975 a year: what per cent. did it pay?

10. A hotel which cost $250000, was rented for $12500 a year: what per cent. did it pay on the cost?

11. A capitalist invested $500000 in manufacturing, and received a semi-annual dividend of $12500: what per cent. was his dividend?

QUEST.—422. When the principal, interest and time are given, how is the rate per ct. found?

PROBLEM III.

To find the PRINCIPAL, *the interest, the rate per cent., and the time being given.*

12. What sum must be put at interest, at 6 per cent., to gain $75 in 2 years?

Analysis.—The interest of $1 for 2 years at 6 per cent., (the given time and rate,) is 12 cents. Now 12 cents interest is $\frac{12}{100}$ of its principal $1; consequently, $75, the given interest, must be $\frac{12}{100}$ of the principal required. The question therefore resolves itself into this: $75 is $\frac{12}{100}$ of what number of dollars? If $75 is $\frac{12}{100}$, $\frac{1}{100}$ is $\frac{1}{12}$ of $75, which is 6\frac{1}{4}$; and $\frac{100}{100}$=6\frac{1}{4}\times$100, which is $625, the principal required.

Or, we may reason thus: Since 12 cents is the interest of 1 dollar for the given time and rate, 75 dollars must be the interest of as many dollars for the same time and rate, as 12 cents is contained times in 75 dollars. And $75÷.12=625. *Ans.* $625.

PROOF.—$625×.06=$37.50, the interest for 1 year at the given per cent., and $37.50×2=$75, the given interest. Hence,

423. To find the *principal*, when the interest, rate per cent., and time are given.

Divide the given interest by the interest of $1 *for the given time and rate, expressed in decimals; and the quotient will be the principal required.*

Or, make the interest of $1 *for the given time and rate, the numerator, and* 100 *the denominator of a common fraction; then divide the given interest by this fraction, and the quotient will be the principal required.* (Art. 234.)

13. What sum must be put at 7 per cent. interest, to gain $63 in 6 months?

14. What sum must be put at 5 per cent. interest, to gain $90 in 4 months?

15. What sum must be invested in 6 per cent. stock, to gain $300 in 6 months?

QUEST.—423. When the interest, rate per cent., and time are given, how is the principal found?

16. What sum must be invested in 7 per cent. stock, to gain $560 in one year?

17. A man founded a professorship with a salary of $1000 a year: what sum must be invested at 7 per cent. to produce it?

18. What sum must be put at 6 per cent. interest to pay a salary of $1200 a year?

19. What sum must be invested in 5 per cent. stock to make a semi-annual dividend of $750?

20. A man bequeathed his wife $1250 a year: what sum must be invested at 6 per cent. interest to pay it?

PROBLEM IV.

To find the TIME, *the principal, the interest, and the rate per cent. being given.*

21. A man loaned $200 at 6 per cent., and received $42 interest: how long was it loaned?

Analysis.—The interest of $200 at 6 per cent. for 1 year is $12, (Art. 404.) Now, since $12 interest requires the principal 1 year at the given per cent., $42 interest will require the same principal $\frac{42}{12}$ of 1 year, which is equal to $3\frac{1}{2}$ years. (Art. 196.)

Or, we may reason thus: If $12 interest requires the use of the given principal 1 year, $42 interest will require the same principal as many years as $12 is contained times in $42. And $42÷$12=3.5. *Ans.* 3.5 years. Hence,

424. To find the *time*, when the principal, interest, and rate per cent. are given.

Divide the given interest by the interest of the principal at the given rate for 1 *year, and the quotient will be the time required.*

Or, make the given interest the numerator, and the interest of the principal for 1 *year at the given rate the denominator of a common fraction; reduce this fraction to a whole or mixed number, and it will be the time required.*

OBS. If the quotient contains a decimal of a year, it should be reduced to months and days. (Art. 348.)

QUEST.—424. When the principal, interest, and rate per cent. are given, how is the time found? *Obs.* When the quotient contains a decimal of a year, what should be done with it?

22. A man loaned $765.50, at 6 per cent., and received $183.72 interest: how long was it loaned?

23. In what time will $850 gain $29.75, at 7 per cent. per annum?

24. A man received $136.75 for the use of $1820, which was 6 per cent. interest for the time: what was the time?

25. In what time will $6280 gain $471, at 5 per cent. interest?

26. How long will it take $100, at 5 per cent., to gain $100 interest; that is, to double itself?

Operation. The interest of $100 for 1 year, at 5 per cent., is $5. (Art. 404.)

$$\$5)\underline{\$100}$$
$$20 \quad \textit{Ans. } 20 \text{ years.}$$

PROOF.—$\$100 \times .05 \times 20 = \$100$, the given principal. (Art. 404.)

TABLE,

Showing in what time any given principal will double itself at any rate, from 1 to 20 per cent. Simple Interest.

Per cent.	Years.	Per cent.	Years.	Per cent.	Years.	Per cent.	Years.
1	100	6	$16\frac{2}{3}$	11	$9\frac{1}{11}$	16	$6\frac{1}{4}$
2	50	7	$14\frac{2}{7}$	12	$8\frac{1}{3}$	17	$5\frac{15}{17}$
3	$33\frac{1}{3}$	8	$12\frac{1}{2}$	13	$7\frac{9}{13}$	18	$5\frac{5}{9}$
4	25	9	$11\frac{1}{9}$	14	$7\frac{1}{7}$	19	$5\frac{5}{19}$
5	20	10	10	15	$6\frac{2}{3}$	20	5

27. How long will it take $365 to double itself, at 6 per cent.?

28. How long will it take $1181 to double itself, at 7 per cent.?

29. In what time will $2365.24 double itself at 7 per cent.?

30. In what time will $5640 double itself, at 10 per cent.?

31. How long will it take $10000 to gain $5000, at 6 per cent. interest?

32. A man hired $15000, at 7 per cent., and retained it till the principal and interest amounted to $25000: how long did he have it?

33. A man loaned his clerk $25000 to go into business, and agreed to let him have it, at 5 per ct., till it amounted to $60000: how long did he have it?

COMPOUND INTEREST.

425. *Compound Interest* is the interest arising not only from the principal, but also from the *interest itself*, after it becomes due.

OBS. Compound Interest is often called *interest upon interest.* When interest is paid on the *principal only*, it is called *Simple Interest.*

Ex. 1. What is the compound interest of $842 for 4 years, at 6 per cent.?

Operation.

	$842.00	Principal.
$842×.06=	50.52	Int. for 1st year.
	892.52	Amt. for 1 year.
$892.52×.06=	53.55	Int. for 2d year.
	946.07	Amt. for 2 years.
$946.07×.06=	56.76	Int. for 3d year.
	1002.83	Amt. for 3 years.
$1002.83×.06=	60.17	Int. for 4th year.
	1063.00	Amt. for 4 years.
	842.00	Prin. deducted.
Ans.	$221.00	Compound int. for 4 years.

426. Hence, to calculate compound interest.

Cast the interest on the given principal for 1 *year, or the specified time, and add it to the principal; then cast the interest on this amount for the next year, or specified time, and add it to the principal as before. Proceed in this manner with each successive year of the proposed time. Finally, subtract the given principal from the last amount, and the remainder will be the compound interest.*

2. What is the compound interest of $600 for 5 years, at 7 per cent.? *Ans.* $241.53.

3. What is the compound int. of $1260 for 5 yrs., at 7 per cent.?

4. What is the amount of $1535 for 6 yrs., at 6 per cent. compound interest?

5. What is the amount of $4000 for 2 yrs., at 7 per cent., payable semi-annually?

QUEST.—426. How is compound interest calculated?

TABLE,

Showing the amount of \$1, *or* £1, *at* 3, 4, 5, 6, *and* 7 *per cent. compound interest, for any number of years, from* 1 *to* 40.

Yrs.	3 per cent.	4 per cent.	5 per cent.	6 per cent.	7 per cent.
1.	1.030,000	1.040,000	1.050,000	1.060,000	1.07,000
2.	1.060,900	1.081,600	1.102,500	1.123,600	1.14,490
3.	1.092,727	1.124,864	1.157,625	1.191,016	1.22,504
4.	1.125,509	1.169,859	1.215,506	1.262,477	1.31,079
5.	1.159,274	1.216,653	1.276,282	1.338,226	1.40,255
6.	1.194,052	1.265,319	1.340,096	1.418,519	1.50,073
7.	1.229,874	1.315,932	1.407,100	1.503,630	1.60,578
8	1.266,770	1.368,569	1.477,455	1.593,848	1.71,818
9.	1.304,773	1.423,312	1.551,328	1.689,479	1.83,845
10.	1.343,916	1.480,244	1.628,895	1.790,848	1.96,715
11.	1.384,234	1.539,451	1.710,339	1.898,299	2.10,485
12.	1.425,761	1.601,032	1.795,856	2.012,196	2.25,219
13.	1.468,534	1.665,074	1.885,649	2.132,928	2.40,984
14.	1.512,590	1.731,676	1.979,932	2.260,904	2.57,853
15.	1.557,967	1.800,944	2.078,928	2.396,558	2.75,903
16.	1.604,706	1.872,981	2.182,875	2.540,352	2.95,216
17.	1.652,848	1.947,900	2.292,018	2.692,773	3.15,881
18.	1.702,433	2.025,817	2.406,619	2.854,339	3.37,293
19.	1.753,506	2.106,849	2.526,950	3.025,600	3.61,652
20.	1.806,111	2.191,123	2.653,298	3.207,135	3.86,968
21.	1.860,295	2.278,768	2.785,963	3.399,564	4.14,056
22.	1.916,103	2.369,919	2.925,261	3.603,537	4.43,040
23.	1.973,587	2.464,716	3.071,524	3.819,750	4.74,052
24.	2.032,794	2.563,304	3.225,100	4.048,935	5.07,236
25.	2.093,778	2.665,836	3.386,355	4.291,871	5.42,743
26.	2.156,592	2.772,470	3.555,673	4.549,383	5.80,735
27.	2.221,289	2.883,369	3.733,456	4.822,346	6.21,386
28.	2.287,928	2.998,703	3.920,129	5.111,687	6.64,883
29.	2.356,566	3.118,651	4.115 136	5.418,388	7.11,425
30.	2.427,262	3.243,398	4.321,942	5.743,491	7.61,225
31.	2.500,080	3.373,133	4.538,039	6.088,101	8.14,571
32.	2.575,083	3.508,059	4.764,941	6.453,386	8.71,527
33.	2.652,335	3.648,381	5.003,189	6.840,590	9.32.533
34.	2.731,905	3.794,316	5.253,348	7.251,025	9.97,811
35.	2.813,862	3.946,089	5.516,015	7.686,087	10.6,765
36.	2.898,278	4.103,933	5.791,816	8.147,252	11.4,239
37.	2.985,227	4.268,090	6.081,407	8.636,087	12.2,236
38.	3.074,783	4.438,813	6.385,477	9.154,252	13.0,792
39.	3.167,027	4.616,366	6.704,751	9.703,507	13.9,948
40.	3.262,038	4.801,021	7.039,989	10.285,72	14.9,744

427. To calculate compound interest by the preceding table.

Find the amount of $1 *or* £1 *for the given number of years by the table, multiply it by the given principal, and the product will be the amount required. Subtract the principal from the amount thus found, and the remainder will be the compound interest.*

6. What is the compound interest of $500 for 15 years, at 6 per cent.? What is the amount?

Operation.

$2.396558	Amt. of $1 for 15 yrs. by Table.
500	The given principal.
$1198.279000	Amt. required.
$500	Principal to be subtracted.
$698.279	Interest required.

7. What is the amount of $960 for 10 yrs., at 7 per ct.?
8. What is the amount of $1000 for 9 yrs., at 5 per ct.?
9. What is the compound int. of $1460 for 12 yrs., at 4 per ct.?
10. What is the compound int. of $2500 for 15 yrs., at 6 per ct.?
11. What is the amount of $5000 for 20 yrs., at 6 per ct.?
12. What is the amount of $10000 for 40 yrs., at 7 per ct.?

DISCOUNT.

428. DISCOUNT is the *abatement* or *deduction* made for the payment of money before it is *due*. For example, if I owe a man $100, payable in one year without interest, the *present worth* of the note is less than $100; for, if $100 were put at interest for 1 year, at 6 per cent., it would amount to $106; at 7 per cent., to $107, &c. In consideration, therefore, of the *present payment* of the note, justice requires that he should make some *abatement* from it. This abatement is called *Discount.*

429. The *present worth* of a debt payable at some future time without interest, is that sum which, being put at legal interest, *will amount to the debt*, at the time it becomes due.

QUEST.—428. What is discount? 429. What is the present worth of a debt, payable at some future time, without interest?

Ex. 1. What is the present worth of $756, payable in 1 year and 4 months, without interest, when money is worth 6 per cent. per annum?

Analysis.—The *amount*, we have seen, is the sum of the principal and interest. (Art. 399.) Now the amount of $1 for 1 year and 4 months, at 6 per cent., is $1.08; (Art. 404;) that is, the amount is $\frac{108}{100}$ of the principal $1. The question then resolves itself into this: $756 is $\frac{108}{100}$ of what principal? If $756 is $\frac{108}{100}$ of a certain sum, $\frac{1}{100}$ is $\frac{1}{108}$ of $756; now $756÷108=$7, and $\frac{100}{100}$=$7×100, which is $700.

Or, we may reason thus: Since $1.08 (amount) requires $1 principal for the given time, $756 (amount) will require as many dollars as $1.08 is contained times in $756; and $756÷$1.08= $700, the same as before.

Proof.—$700×.08=$56, interest for 1 year and 4 months; and $700+56=$756, the sum whose present worth is required. Hence,

430. To find the *present worth* of any sum, payable at a future time without interest.

First find the amount of $1 for the time, at the given rate, as in simple interest; then divide the given sum by this amount, and the quotient will be the present worth. (Art. 404.)

The present worth subtracted from the debt, will give the true discount.

Obs. This process is often classed among the Problems of Interest, in which the amount, (which answers to the given sum or debt,) the rate per cent., and the time are given, to find the *principal*, which answers to the *present worth.*

2. What is the present worth of $424.83, payable in 4 months, when money is worth 6 per cent.? What is the discount?

Solution.—$424.83÷$1.02=$416.50, Present worth.
And $424.83—$416.50=$8.33, Discount.

3. What is the present worth of $1000, payable in 1 year, when the rate of interest is 7 per cent.?

4. What is the present worth of $1645, payable in 1 year and 8 months, when the rate of interest is 7 per cent.?

Quest.—430. How do you find the present worth of a debt? How find the discount?

5. What is the discount on a note for $2300, payable in 6 months, when the rate of interest is 8 per cent.?

6. What is the discount, at 6 per cent., on $4260, payable in 4 months?

7. What is the present worth of a note for $4800, due in 3 months, when the rate of interest is 6 per cent.?

8. What is the present worth of a draft for $6240, payable in 1 month, when the rate of interest is 6 per cent.?

9. A man sold his farm for $3915, payable in $2\frac{1}{2}$ years: what is the present worth of the debt, at 6 per cent. discount?

10. What is the present worth of a draft of $10000, payable at 30 days sight, when interest is 6 per cent. per annum?

11. What is the difference between the discount of $8000 for 1 year, and the interest of $8000 for 1 year, at 7 per cent.?

BANK DISCOUNT.

431. A *Bank*, in commerce, is an institution established for the safe keeping and issue of money, for discounting notes, dealing in exchange, &c.

OBS. 1. There are *three* kinds of banks, viz: banks of *deposit*, *discount*, and *circulation*.

A *bank of deposit* receives money to keep, subject to the order of the depositor. This was the primary object of these institutions.

A *bank of discount* is one which loans money, or discounts notes, drafts, and bills of exchange.

A *bank of circulation* issues *bills*, or *notes* of its own, which are redeemable in specie, at its place of business, and thus become a circulating medium of exchange. Banks of this country generally perform the three-fold office of deposit, discount, and circulation.

2. The affairs of a bank are managed by a *board of directors*, chosen annually by the stockholders. (Art. 392. Obs.) The directors appoint a *president* and *cashier*, who sign the bills. and transact the ordinary business of the bank.

A *teller* is a clerk in a bank, who receives and pays the money on checks.

A *check* is an order for money, drawn on a banker, or the cashier, by a depositor, payable to the bearer.

3. Banks originated in Italy. The first one was established in Venice, in 1171, called the Bank of Venice.

QUEST.—431. What is a bank? *Obs.* Of how many kinds are banks?

432. It is customary for *Banks*, in discounting a note or draft, to deduct in advance the *legal interest* on the given sum from the time it is discounted to the time it becomes due. Hence,

Bank discount is the same as simple interest paid *in advance.* Thus, the *bank discount* on a note of $106, payable in 1 year, at 6 per cent., is $6.36, while the *true discount* is but $6. (Art. 430.)

OBS. 1. The difference between *bank discount* and *true discount*, is the interest of the true discount for the given time. On small sums for a short period this difference is trifling, but when the sum is large, and the time for which it is discounted is long, the difference is worthy of notice.

2. Taking *legal interest in advance*, according to the general rule of law, is *usury*. An exception is generally allowed, however, in favor of notes, drafts, &c., which are payable in *less* than a *year*.

The Safety Fund Banks of the State of New York, though the legal rate of interest is 7 per cent., are not allowed by their charters to take over 6 per cent. discount in advance on notes and drafts which mature within 63 days from the time they are discounted.*

Banks charge interest for the three days grace.

CASE I.

12. What is the bank discount on a note for $850.20 for 6 months, at 6 per cent.? What is the present worth of the note?

Operation.

$850.20 Principal.
.03 05 Int. $1 for 6 mo. 3 ds. grace.
4251 00
25 5060
$25.9311 00 Bank discount.

And $850.20—$25.93=$824.27, Present worth. Hence,

433. To find the bank discount on a note or draft.

Cast the interest on the face of the note or draft for three days more than the specified time, and the result will be the discount.

The discount subtracted from the face of the note, will give the present worth or proceeds of a note discounted at a bank.

QUEST.—432. How do banks usually reckon discount? What then is bank discount? *Obs.* What is the difference between bank discount and true discount? Is this difference worth noticing? How is taking interest in advance generally regarded in law? What exception to this rule is allowed?

* Revised Statutes of New York, (3d edition,) Vol. I. p. 741.

Note.—Interest should be computed for the *three days grace* in each of the following examples.

14. What is the bank discount on a note for $465, payable in 6 months, at 6 per cent.?

15. What is the bank discount on a note for $972, payable in 4 months, at 5 per cent.?

16. What is the bank discount on a note for $1492, payable in 3 months, at 7 per cent.?

17. What is the bank discount on a draft of $628, payable at 60 days sight, at 5 per cent.?

18. What is the present worth of $21·35, payable in 8 months, at 7 per cent.?

19. What is the present worth of a note for $2790, payable in 1 month, discounted at 6 per cent. at a bank?

20. What is the bank discount, at 5½ per cent., on a draft of $1747, payable at 90 days sight?

21. What is the bank discount, at 4½ per cent., on a draft of $3143, payable in 4 months?

22. What is the bank discount on $5126.63, payable in 30 days, at 8 per cent.?

23. What is the bank discount on $3841.27, payable in 60 days, at 6½ per cent.?

24. What is the present worth of a note for $6721, payable in 10 months, discounted at 6 per cent. at a bank?

25. What is the present worth of a note for $1500, payable in 12 days, at 7 per cent. discount?

26. What is the bank discount on $10000, payable in 45 days, at 6 per cent.?

27. What is the bank discount on $25260, payable in 90 days, at 7 per cent.?

28. What is the difference between the true discount and bank discount on $5000 for 10 years, at 6 per cent.?

CASE II.

29. A man wishes to make a note payable in 1 year, at 6 per cent., the present worth of which, if discounted at a bank, shall be just $200: for what sum must the note be made?

Analysis.—The present worth of $1, payable in 1 year, at 6 per cent. discount, is 100 cts.—6 cts.=94 cts.; that is, the present worth is $\frac{94}{100}$ of the principal or sum discounted. The question then resolves itself into this: $200 (present worth) is $\frac{94}{100}$ of what sum? Now, if $200 is $\frac{94}{100}$ of a certain sum, $\frac{1}{100}$ is $\frac{1}{94}$ of $200; and $200÷94=$2.12766, and $\frac{100}{100}$=$2.12766×100, which is $212.766. *Ans.*

Or, we may reason thus: Since 94 cents present worth requires $1, (100 cents) principal, or sum to be discounted for the given time, $200 present worth will require as many dollars, as 94 cents is contained times in $200; and $200÷$.94=$212.766.

PROOF.—$212.766×.06=$12.7659, the bank discount for 1 year; and $212.766—$12.7659=$200, the given sum. Hence,

434. To find what sum, payable in a specified time, will produce a given amount, when discounted at a bank, at a given per cent.

Divide the given amount to be raised by the present worth of $1, *for the time, at the given rate of bank discount, and the quotient will be the sum required to be discounted.*

30. How large must I make a note payable in 6 months, to raise $400, when discounted at 7 per cent. bank discount?

31. What sum payable in 4 months must be discounted at a bank, at 5 per cent., to produce $950?

32. What sum payable in 60 days, will produce $1236, if discounted at a bank, at 8 per cent.?

33. For what sum must a note be drawn, payable in 34 days, the avails of which, at 6 per cent., bank discount, will be $2500?

34. For what sum must a note be drawn, payable in 90 days, so that the avails, at 7 per cent. bank discount, shall be $3755?

35. A man bought a farm for $4268 cash: how large a note, payable in 4 months, must he take to a bank to raise the money at 6 per cent. discount?

QUEST.—434. How find what sum, payable in a given time, will produce a given amount, at a given per cent., bank discount?

36. A man wishes to obtain $63240 from a bank at 6 per cent. discount: how large must he make his note, payable in 1 month and 15 days?

37. What sum payable in 8 months, if discounted at a bank, at 6 per cent., will produce $10000?

38. What sum payable in 4 months, will produce $50000, if discounted at 7 per cent. at a bank?

39. A man received $46250 as the avails of a note, payable in 60 days, discounted at a bank at 5 per cent.: what was the face of the note?

40. A merchant wished to pay a debt of $8246 at a bank, by getting a note payable in 30 days discounted, at 8 per cent.: how large must he make the note?

INSURANCE.

435. Insurance is *security* against *loss* or *damage* of property by fire, storms at sea, and other casualties. This security is usually effected by contract with Insurance Companies, who, for a stipulated sum, agree to restore to the owners the amount insured on their houses, ships, and other property, if destroyed or injured during the specified time of insurance.

Obs. 1. Insurance on ships and other property at sea is sometimes effected by contract with individuals. It is then called *out-door insurance*.

2. The insurers, whether an incorporated company or individuals, are often termed *Underwriters*.

436. The *written instrument* or *contract* is called the *Policy*.

The *sum* paid for insurance is called the *Premium*.

The premium paid is a *certain per cent.* on the amount of property insured for 1 year, or during a voyage at sea, or other specified time of risk.

Obs. 1. Rates of insurance on dwelling-houses and furniture, stores and goods, shops, manufactories, &c., vary from $\frac{3}{5}$ to 2 per cent. per annum on the sum insured, according to the exposure of the property and the difficulty of moving the goods in case of casualty. It is a rule with most Insurance

---

Quest.—435. What is Insurance? *Obs.* When insurance is effected with individuals what is it called? What are the insurers sometimes called? 436. What is meant by the policy? The premium?

Companies not to insure more than *two thirds* of the value of a building, or goods on land.

2. Coasting vessels are commonly insured by the season or year. In time of peace, the rate varies from 4 to $7\frac{1}{2}$ per cent. per annum; in time of war it is much higher. Whale ships are generally insured for the voyage, at a rate varying from 5 to 8 per cent. on the sum insured.

3. When the general average of loss is *less* than 5 per cent., the underwriters are not liable for its payment.

CASE I.

437. To compute Insurance for 1 year, or a specified time.

Multiply the sum insured by the given rate per cent., as in interest. (Art. 404.)

Ex. 1. A man effected an insurance on his house for $500, at $1\frac{1}{4}$ per cent. per annum: how much premium did he pay?

Solution.—$1500×.0125 (the rate)=$18.75. *Ans.*

2. What is the premium for insuring a store to the amount of $2760, at $\frac{3}{4}$ per cent.?

3. What premium must I pay for insuring a quantity of goods, worth $6280, from New York to Liverpool, at $1\frac{1}{2}$ per cent.?

4. What is the annual premium for insuring a stock of goods, worth $10200, at $\frac{5}{8}$ per cent.?

5. What is the annual premium for insuring a coasting vessel, worth $1600, at $6\frac{1}{2}$ per cent.?

6. A bookseller shipped a quantity of books, valued at $4700, from Boston to New Orleans, at $1\frac{1}{2}$ per cent. insurance: what amount of premium did he pay?

7. A merchant shipped a cargo of flour, worth $45000, from New York to Liverpool, at 2 per cent.: how much premium did he pay?

8. What is the insurance on a cargo of teas, worth $75000, from Canton to Philadelphia, at $2\frac{1}{2}$ per cent.?

9. What is the annual insurance on a factory, worth $65000, at $\frac{3}{4}$ per cent.?

10. A powder mill was insured for $1945, at $12\frac{1}{2}$ per cent.: what was the annual premium?

QUEST.—437. How is insurance computed for 1 year or a specified time?

11. A ship embarking on an exploring expedition, was insured for $45360, at $8\frac{1}{2}$ per cent. per annum: what did the insurance amount to in 5 years?

12. A policy of insurance for $45000 was obtained on a whale ship, at $7\frac{1}{2}$ per cent. for the voyage: what was the amount paid for insurance?

CASE II.

13. If a man pays $16 annually for insuring $800 on his shop, what per cent. does he pay?

Analysis.—If $800, the amount insured, costs $16 premium, $1 will cost $\frac{1}{800}$ of $16; and $16÷800=.02; which is 2 per cent.

PROOF.—$800×.02=$16, the premium paid. Hence,

438. To find the *rate per cent.* when the sum insured and the annual premium are given.

Divide the given premium by the sum insured, and the quotient will be the rate per cent. required.

Note.—This case is similar in principle to Problem II. in Interest.

14. If a man pays $60 annually for insuring $2400 on his house, what per cent. does it cost him?

15. A merchant pays $200 per annum for insuring $8000 on his goods: what per cent. does he pay?

16. A grocer paid $122.50 premium on a cargo of flour, worth $12250, from Charleston to Portland: what per cent. did he pay?

17. An importer paid $350 insurance on a quantity of cloths, worth $28000, from Havre to New York: what per cent. did he pay?

CASE III.

18. A man pays $45 annually for insuring his library, which is 3 per cent. on the amount of his policy: what is the sum insured?

Analysis.—Since 3 cents will insure $1 at the given rate, for a year, $45 will insure as many dollars as 3 cents are contained times in $45; and $45÷.03=$1500. *Ans.*

PROOF.—$1500×.03=$45, the given premium. Hence,

439. To find the *sum insured* when the premium and the rate per cent. are given.

Divide the given premium by the rate per cent., expressed in decimals, and the quotient will be the sum insured.

Note.—This case is similar in principle to Problem III. in Interest.

19. An importer paid $650 premium on goods from Hamburgh to New York, which was $1\frac{1}{4}$ per cent. on the amount insured: how much did he insure?

20. A merchant paid $1640 premium on goods from Philadelphia to Constantinople, which was $2\frac{1}{2}$ per cent. on the worth of the goods insured: how much did he insure?

21. A premium of $487.50 was paid on a cargo of cotton from New Orleans to Liverpool, which was $\frac{3}{4}$ per cent. on its value: what amount was insured on the cargo?

22. When the rate of insurance is $1\frac{1}{2}$ per cent., what sum can you get insured for $860 premium?

23. At $\frac{3}{5}$ per cent. per annum, what amount can a man get insured on his house and furniture for $20.50 per annum?

CASE IV.

To find what sum must be insured on any given property, so that, if destroyed, its value and the premium may both be recovered.

24. If a man owns a vessel worth $1920, what sum must he get insured on it, at 4 per cent., so that if wrecked, he may recover both the value of the vessel and the premium?

Analysis.—It is plain, when the rate of insurance is 4 per cent. on a policy of $1, or 100 cents, the owner would receive but 96 cents towards his loss; for, he has paid 4 cents for insurance. Since therefore the recovery of 96 cents requires $1 to be insured, the recovery of $1920 will require as many dollars to be insured as 96 cents is contained times in $1920; and $1920÷.96=$2000. *Ans.*

PROOF.—$2000×.04=$80, the premium paid, and $2000—$80=$1920, the value of the vessel.

440. Hence, to find what sum must be insured on a given amount of property, so that if destroyed, both the value of the property and the premium may be recovered.

Subtract the rate per cent. from $1, then divide the value of the property insured by the remainder, and the quotient will be the sum to be insured.

25. What sum must be insured on property worth $8240, at 1½ per cent., so that the owner may suffer no loss if the property is destroyed?

26. What sum must be insured on $13460, at 3 per cent., in order to cover both the premium and property insured?

27. If I send an adventure to the Sandwich Islands worth $25000, what sum must I get insured, at 7½ per cent., that I may sustain no loss in case of a total wreck?

LIFE INSURANCE.

441. A LIFE INSURANCE is a contract for the payment of a certain sum of money on the death of an individual, in consideration of a stipulated sum paid down, or, more commonly, of an annual premium, to be continued during the life of the assured.

The *average duration* of human life is often called the *Expectation of Life.** This is different in different countries, but it may be determined with great accuracy in any given country, by calculations founded on the register of births and deaths in that country.

OBS. At birth, the expectation of life, according to the Carlisle Table, is 38.72 y.; at 5, it is 51.25 y.; at 10, it is 48.82 y.; at 15, it is 45 y.; at 20, it is 41.46 y.; at 25, it is 37.86 y.; at 30, it is 34.34 y.; at 35, it is 31 y.; at 40, it is 27.61 y.; at 45, it is 24.46 y.; at 50, it is 21.11 y.; at 55, it is 17.58 y.; at 60, it is 14.34 y.; at 65, it is 11.79 y.; at 70, it is 9.19 y.; at 75, it is 7.01 y.; at 80, it is 5.51 y.; at 85, it is 4.12 y.; at 90, it is 3.28 y.; at 100, it is 2.28 y.

442. The *premium* paid for life insurance, like that for other insurance, is calculated at a certain per cent. on the amount insured. The per cent. varies according to the age and employment of the assured, and the time embraced in the policy.

QUEST.—441. What is Life Insurance? What is meant by the expectation of life? 442. How is Life insurance calculated?

* See Registers of London, Breslau, Northampton, &c.

OBS. 1. At the age of 21 years, the per cent. on a policy for life is from $1\frac{8}{10}$ to $2\frac{1}{4}$ per cent. per annum on the sum insured; for 7 years, it is from $\frac{19}{20}$ to $1\frac{1}{2}$ per cent. per annum; for 1 year, from $\frac{9}{10}$ to $1\frac{3}{8}$ per cent.

At 30, on a policy for life, it is from $2\frac{7}{20}$ to $2\frac{7}{10}$ per cent. per annum; for 7 years, from $1\frac{7}{20}$ to $1\frac{8}{10}$ per cent.; for 1 year, from $1\frac{3}{10}$ to $1\frac{7}{10}$ per cent.

At 40, on a policy for life, it is from $3\frac{2}{10}$ to $3\frac{4}{10}$ per cent.; for 7 years, from $1\frac{8}{10}$ to $2\frac{2}{10}$ per cent.; for 1 year, from $1\frac{2}{3}$ to $2\frac{1}{20}$ per cent.

At 50, on a policy for life, it is from $4\frac{1}{2}$ to $4\frac{6}{10}$ per cent.; for 7 years, from $2\frac{1}{10}$ to $3\frac{5}{20}$ per cent.; for 1 year, from $1\frac{9}{10}$ to $2\frac{8}{10}$ per cent.

At 60, on a policy for life, it is from $6\frac{3}{10}$ to 7 per cent.; for 7 years, from $4\frac{3}{10}$ to 5 per cent.; for 1 year, from $3\frac{9}{10}$ to $4\frac{4}{10}$ per cent.

28. A young man, at the age of 21 years, effected an insurance for $1500 for life, at $2\frac{1}{10}$ per cent.: what was the annual premium? *Ans.* $31.50.

29. A man, at the age of 30, effected a life insurance for $2700, for 7 years, at $1\frac{8}{10}$ per cent.: what was the annual premium?

30. At 60 years of age, a man effected a life insurance for 1 year for $5750, at $6\frac{1}{2}$ per cent.: how much premium did he pay?

31. At 40 years of age, a man effected an insurance for $10000 for life, at $3\frac{1}{2}$ per cent. per annum; he lived till he was 75 years old: which was the larger, the sum paid for insurance, or the sum insured?

PROFIT AND LOSS.

443. PROFIT and Loss in commerce, signify the sum *gained* or *lost* in ordinary business transactions. They are reckoned at a certain per cent. on the *purchase price,* or *sum paid* for the articles under consideration.

CASE I.

To find the AMOUNT *of profit or loss, the purchase price and rate per cent. being given.*

Ex. 1. A grocer bought a lot of flour for $84, and sold it for 7 per cent. profit: how much did he make by his bargain?

QUEST.—443. What is meant by profit and loss? How are they reckoned? **444.** How is the amount of profit or loss found, when the cost and rate per cent. are given?

Analysis.—Since he gained 7 per cent. on the cost of the flour, he must have gained $\frac{7}{100}$ of \$84. Now $\frac{1}{100}$ of \$84 is $\frac{84}{100}$, and $\frac{7}{100}$ is 7 times as much, which is $\frac{588}{100}$=\$5.88. *Ans.*

Or thus: If \$1 (100 cents) gain 7 cents, \$84 will gain 84 times as much; and \$84 × .07=\$5.88, the same as before. Hence,

444. To find the *amount of profit* or *loss*, when the purchase price and rate per cent. are given.

Multiply the purchase price by the given per cent. as in percentage; and the product will be the amount gained or lost by the transaction. (Art. 388.)

OBS. In order to obtain the *exact* profit and loss in mercantile operations, it is manifest that the *interest* on the cost or purchase price of the goods, during the time they have been on hand, also for the time before payment is received should be taken into consideration.

2. If I buy a piece of broadcloth for \$120, and after keeping it 6 months, sell it at 8 per cent. advance on 6 months credit, how much shall I gain if I pay 7 per cent. for the money invested? *Ans.* \$1.20.

3. If I buy a farm for \$1740, and sell it 8 per cent. less than cost, how much do I lose? *Ans.* \$139.20.

4. If you buy a house for \$2180, and sell it at 10 per cent. advance, how much will you gain by your bargain?

5. A merchant bought goods amounting to \$3400, and retailed them at 20 per cent. profit: how much did he make?

6. A grocer bought a lot of flour for \$6235, and sold it 15 per cent. less than cost: what was his loss?

7. A speculator bought a quantity of cotton for \$24850, and sold it at 5½ per cent. advance: how much did he make by the operation?

8. A man bought a block of stores for \$58246, and sold them at 118 per cent. advance: how much did he gain?

9. A man bought wild land amounting to \$125000, and after keeping it 10 years, sold it at 50 per cent. advance: allowing money to be worth 6 per cent., did he make or lose by the operation; and how much?

CASE II.

To find how an article must be sold to gain or lose a specified per cent., the cost being given.

10. A man bought a building lot for $625, and afterwards sold it so as to gain 10 per cent.: how much did he sell it for?

Operation.

$625 purchase price.
.10 per cent. profit.
$62.50 profit.
$687.50 selling price.

Since he gained 10 per cent., it is obvious he sold it for the purchase price together with 10 per cent. of that price. We therefore find 10 per cent. on the cost, and add it to itself. (Art. 388.)

11. A man bought a small house for $840, and afterwards sold it so as to lose 10 per cent.: how much did he get for it?

Operation.

$840 purchase price.
.10 per cent. loss.
$84.00 sum lost.
$756.00 selling price.

Having found the sum lost, (Art. 388,) subtract it from the cost, and the remainder is obviously the selling price. Hence,

445. To find how any article must be sold, in order to *gain* or *lose* a given rate per cent. when the cost is given.

First find the amount of profit or loss on the purchase price at the given rate, as in the last Case; then the amount thus found added to, or subtracted from the purchase price, as the case may be, will give the selling price required.

12. A grocer bought a quantity of cheese for $130.67: for how much must he sell it, to gain 20 per cent.?

13. Bought a stock of goods for $3460: for how much must they be sold, to gain $22\frac{1}{2}$ per cent.?

14. Bought a quantity of flour for $5245: for how much must it be sold, to gain 13 per cent.?

15. Bought 2500 bales of cotton for $30575, which were sold at a loss of $3\frac{1}{2}$ per cent.: what did they fetch?

QUEST.—445. What is the method of finding how any article must be sold, in order to gain or lose a given per cent.? 446. How is the rate per cent. of profit or loss found, when the cost and selling price are given?

CASE III.

To find the RATE PER CENT. *of profit or loss, the cost and selling price being given.*

16. If a merchant buys a quantity of butter for $75, and sells it for $90, what per cent. profit will he make?

Analysis.—Subtracting the cost from the selling price, shows that he gained $15. Now 15 dollars are $\frac{15}{75}$ of 75 dollars; therefore he gained $\frac{15}{75}$ of his *outlay*, or the purchase price of the goods. And $\frac{15}{75}$ reduced to a decimal, is equal to 20 *hundredths*, or 20 per cent. (Art. 387. Obs. 3.)

Or, we may reason thus: If $75 (outlay) gain 15 dollars, $1 will gain $\frac{1}{75}$ of $15. And $15÷75=.20, the same as before.

446. Hence, to find the *rate per cent.* of profit or loss, when the *cost* and *selling prices* are given.

First find the amount gained or lost by subtraction; then make the gain or loss the numerator and the purchase price the denominator of a common fraction; reduce this fraction to a decimal, and the result will be the per cent. required. (Art. 337.)

Or, simply annex ciphers to the profit or loss, and divide it by the cost; the quotient will be the per cent.

OBS. 1. As *per cent.* signifies *hundredths*, the *first two* decimal figures which occupy the place of hundredths, are properly the per cent.; the other decimals are *parts* of 1 per cent. After obtaining two decimal figures, there is sometimes an advantage in placing the remainder over the divisor, and annexing it to the decimals thus obtained. (Art. 387. Obs. 3.)

2. It should be remembered that the percentage which is *gained* or *lost*, is always calculated on the *purchase* price, or the *sum paid* for the article, and not on the *selling* price, or *sum received*, as it is often supposed.

17. Bought a quantity of cotton at $6\frac{1}{2}$ cents per yard, and sold it at 8 cents: what per cent. was the profit?

18. Bought a quantity of calico, at 12 cents per yard, and sold it at $12\frac{1}{2}$ cents: what per cent. was the profit?

19. Bought a lot of corn, at 45 cents per bushel, and sold it at 38 cents: what per cent. was the loss?

QUEST.—*Obs.* What figures properly signify the per cent.? What do the other decimal figures on the right of hundredths denote? On what is the per cent. gained or lost calculated?

20. A grocer bought a pipe of wine for $252, and retailed it at $12\frac{1}{2}$ cents per gill: what per cent. did he make?

21. A man bought a house for $4325, and sold it for $5216: what per cent. did he make?

22. A speculator invested $75000 in stocks, which he sold for $77225: what per cent. did he make by the operation?

CASE IV.

To find the COST, *the selling price and per cent. gained or lost being given.*

23. A man sold a lot of salt for $360, which was 20 per cent. more than cost: what did he pay for the salt?

Analysis.—The cost is $\frac{100}{100}$ of itself, and the gain is $\frac{20}{100}$ of the cost. (Art. 386.) Now $\frac{100}{100}+\frac{20}{100}=\frac{120}{100}$; hence, the selling price is $\frac{120}{100}$ of the cost. The question then is this: $360 is $\frac{120}{100}$ of what sum? If $360 is $\frac{120}{100}$ of a certain sum, $\frac{1}{100}$ of that sum is $\frac{1}{120}$ of $360. Now $\$360\div120=\$3$, and $\frac{100}{100}=\$3\times100$, which is $300. *Ans.*

Or, if we divide $360, the selling price, by the fraction $\frac{120}{100}$, the quotient $300, will be the cost. (Art. 234.)

PROOF.—$\$300\times.20=\$60.00$ the gain; (Art. 388;)
and $\$300+\$60=\$360$, the selling price.

24. A miller sold a lot of flour for $170, which was 15 per cent. less than cost: how much did the flour cost him?

Analysis.—Reasoning as before, the cost is $\frac{100}{100}$ of itself, and the loss is $\frac{15}{100}$ of the cost. Now $\frac{100}{100}-\frac{15}{100}=\frac{85}{100}$; consequently the selling price is $\frac{85}{100}$ of the cost. The question therefore is this: $170 is $\frac{85}{100}$ of what sum? If $170 is $\frac{85}{100}$ of a certain sum, $\frac{1}{100}$ is $\frac{1}{85}$ of $170. Now $\$170\div85=\$2$, and $\frac{100}{100}=\$2\times100$, which is $200. *Ans.*

Or, thus: Since he lost 15 per cent., he realized only 85 cents on $1 outlay. Therefore, if 85 cents, selling price, requires $1 outlay, $170, selling price, will require as many dollars outlay as 85 cents are contained times in $170; and $\$170\div.85=\$200$.

PROOF.—$\$200\times.15=\$30.00$, the loss; (Art. 388;)
and $\$200-\$30=\$170$, the selling price. Hence,

447. To find the *cost* when the *selling price* and the per cent. *gained* or *lost* are given.

Divide the selling price by $1, *increased or diminished by the per cent. gained or lost, as the case may be, and the quotient will be the cost required.*

Or, make the given per cent. added to or subtracted from 100, *as the case may be, the numerator, and* 100 *the denominator of a common fraction; then divide the selling price by this fraction, and the quotient will be the cost.*

OBS. It is not unfrequently supposed that if we find the percentage on the selling price at the given rate, and add the percentage thus found to, or subtract it from, the selling price, as the case may be, the sum or remainder will be the cost. This is a mistake, and leads to serious errors in the result. It will easily be avoided by remembering, that the basis on which *profit* and *loss* are calculated, is always the *purchase price* or *sum paid* for the articles under consideration. (Art. 446. Obs. 2.)

25. A grocer sold a quantity of cheese for $530, which was 15 per cent. more than cost: what was the cost?

26. A man sold a carriage for $175, which was 15 per cent. less than cost: what was the cost?

27. A man sold a farm for $2360, which was 10 per cent. less than cost: what did he give for it?

28. An importer sold a library for $3078, which was $12\frac{1}{2}$ per cent. advance on the cost: how much did it cost him?

29. A merchant sold a cargo of crockery for $12000, which was 8 per cent. less than cost: what was the cost?

30. A commission merchant sold a lot of cloths for $7265, which was 15 per cent. more than cost: how much did they cost?

31. A builder sold a house for $17450, which was 2 per cent. less than cost: what was the cost?

32. A broker sold stocks to the amount of $45000, which was $5\frac{1}{2}$ per cent. advance: what was the cost?

33. A manufacturer sold a quantity of carpeting for $63240, which was 50 per cent. more than the cost of the materials: what did the materials cost?

QUEST.—447. How is the cost found, when the selling price and the rate per cent. gained or lost, are given? *Obs.* What mistake is sometimes made in finding the cost? How may it be avoided?

DUTIES.

448. DUTIES, in commerce, signify a *sum of money* required by Government to be paid on *imported* goods.

OBS. 1. In every port of entry in the United States, the Government has an establishment, called a *Custom House*, at which the duties on all foreign goods entered at that port, are to be paid.

2. The persons appointed to inspect the cargoes of vessels engaged in foreign commerce, to examine the invoices of goods, collect the duties, &c., are called *custom house officers*.

449. Duties are of two kinds, *specific* and *ad valorem*. A *specific duty* is a certain sum imposed on a ton, hundred weight, hogshead, gallon, square yard, foot, &c., without regard to the value of the article.

Ad valorem duties are those which are imposed on goods, at a certain per cent. on their *value* or *purchase price.*

Note.—The term *ad valorem* is a Latin phrase, signifying *according to*, or *upon the value.*

450. Before specific duties are imposed, it is customary to make certain deductions called *tare, draft* or *tret, leakage,* &c.

Tare is an allowance of a certain number of pounds made for the box, cask, &c., which contains the article under consideration.

Draft or *Tret* is an allowance of a certain per cent. (usually 4 per cent.) on the weight of goods for waste, or refuse matter.

Leakage is an allowance of a certain per cent. (usually 2 per cent.) for the waste of liquors contained in casks, &c.

OBS. 1. All duties, both specific and ad valorem, are regulated by the Government, and have been different at different times and in different countries.

2. The allowances or deductions for draft, tare, leakage, &c., are different on different articles, and are also regulated by law.

3. In buying and selling groceries in large quantities, allowances are sometimes made for draft, tare, leakage, &c., similar to those in reckoning duties.

QUEST.—448. What are duties in commerce? 449. Of how many kinds are they? What are specific duties? Ad valorem duties? *Note.* What is the meaning of the term ad valorem? 450. What deductions are made before specific duties are imposed? What is tare? Draft or tret? Leakage? *Obs.* How are duties regulated? Are allowances for draft, &c., ever made in buying and selling groceries?

CASE I.—*Calculation of Specific Duties.*

Ex. 1. What is the specific duty on 15 hhds. of molasses, at 10 cents per gallon, allowing 2 per cent. for leakage?

Analysis.—Since there are 63 gallons in one hhd., in 15 hhds. there are 15 times as many, and 63 gals.$\times$15=945 gals. But 2 per cent. of 945 gals. is equal to 945$\times$.02, or 18.9 gals.; (Art. 388;) and 945 gals.—18.9 gals.=926.1 gals., the *net* gallons. Now if the duty on 1 gallon is 10 cents, on 926.1 gals. it is 926.1$\times$.10= $92.61, the duty required. Hence,

451. To find the *specific* duty on any given merchandise.

First deduct the legal draft, tare, leakage, &c., from the given quantity of goods; then multiply the remainder by the given duty per gallon, pound, yard, &c., and the product will be the duty required.

2. If the specific duty on tea is 12 cents a pound, how much will it be on 30 chests, each weighing 115 lbs., allowing 12 lbs. per chest for draft?

3. At 4 cents a pound, what is the specific duty on 160 drums of figs, weighing 28 lbs. apiece, allowing 2½ lbs. a drum for tare?

4. At 15 cents a pound, what is the specific duty on 63 chests of opium, each weighing 150 lbs., allowing 10 lbs. per chest for draft?

5. At 3½ cents a pound, what is the specific duty on 250 bags of coffee, weighing 65 lbs. apiece, allowing 4 per cent. for tret?

6. What is the specific duty, at 6 cents a pound, on 173 kegs of tobacco, each weighing 125 lbs., allowing 6 lbs. per keg for tare?

7. At 5½ cents a pound, what is the specific duty on 430 boxes of paints, weighing 175 lbs. a box, reckoning the tare at 15 lbs. per box?

8. At 8 cents per gallon, what is the specific duty on 140 bar. of olive oil, allowing 2 per cent. for leakage?

9. At 22 cents per gallon, what is the specific duty on 50 hhds of wine, allowing 2 per cent. for leakage?

10. At 7½ cents per pound, what is the duty on 345 sacks of almonds, weighing 75 lbs. apiece, allowing 3 per cent. for tare?

QUEST.—451. How are specific duties calculated?

CASE II.—*Calculation of Ad Valorem Duties.*

452. When duties are imposed upon the actual cost of merchandise, there are of course no deductions to be made; consequently we have only to find the legal per cent. on the amount of the given invoice, or cost of the goods, and it will be the duty required.

Ex. 11. What is the ad valorem duty, at 25 per cent., on a case of bombazines, invoiced at $450?

Solution.—$450×.25=$112.50, the ad valorem duty. Hence,

453. To find the *ad valorem* duty on any given merchandise.

Multiply the amount of the given invoice by the legal per cent., and the product will be the duty required. (Art. 324.)

OBS. 1. An *invoice* is a written statement of merchandise, with the value or prices of the articles annexed.

2. The law requires that the invoice shall be verified by the owner, or one of the owners of the goods, certifying that the invoice annexed contains a *true and faithful account of the actual costs* thereof, and of all charges thereon, and no other different discount, bounty, or drawback, but such as has been actually allowed on the same; which oath shall be administered by a consul, or commercial agent of the United States, or by some public officer duly authorized to administer oaths in the country where the goods were purchased, and the same shall be duly certified by the said consul, &c. Fraud on the part of the owners, or the consul, &c., who administers the oath, is visited with a heavy penalty.—*Laws of the United States.*

12. What is the ad valorem duty, at 20 per cent., on an invoice of broadcloths which cost $1240 in Manchester?

13. What is the ad valorem duty, at 34 per cent., on an invoice of silks, which cost $2110 in Italy?

14. What is the duty, at 25 per cent., on a quantity of indigo, the invoice of which is $1968?

15. What is the duty on a bale of Irish linens, which cost $3187, at 33 per cent.?

16. At 25 per cent., what is the duty on an invoice of hosiery, amounting to $2863?

QUEST.—453. How are ad valorem duties calculated? *Obs.* What is an invoice? What does the law require respecting the invoice of imported goods?

17. At 33⅓ per cent., what is the duty on an invoice of mousseline de laines, amounting to $3690?

18. At 35 per cent., what is the duty on an invoice of watches, amounting to $45385?

19. What is the duty, at 20 per cent., on an invoice of boots and shoes, amounting to $63212?

20. What is the duty, at 15 per cent., on a quantity of ready-made clothing, worth $18714?

21. What is the duty on $37241 worth of spices, at 30 per ct.?

22. What is the duty on $46210 worth of liquor, at 37½ per cent.?

23. At 22 per cent., what is the duty on $71685 worth of crockery?

ASSESSMENT OF TAXES.

454. A *Tax* is a sum of money assessed on individuals for the support of Government, Corporations, Parishes, Districts, &c. Taxes levied by the Government, are assessed either on the *person* or *property* of the *citizens*. When assessed on the *person*, they are called *poll taxes*, and are usually a *specific sum*. Those assessed on the *property* are usually apportioned at a *certain per cent.* on the amount of *real estate* and *personal property* of each citizen or taxable individual.

OBS. Property is divided into two kinds, viz: *real estate* and *personal* property. The *former* denotes possessions that are *fixed;* as houses, lands, &c. The *latter* comprehends *all other* property; as money, stocks, notes, mortgages, ships, furniture, carriages, cattle, tools, &c.

455. When a tax of any given amount is to be assessed, the first thing to be done is to obtain an inventory of the amount of *taxable* property, both personal and real, in the State, County, Corporation, or District, by which the tax is to be paid; also the amount of property of every citizen who is to be taxed, together with the number of Polls.

QUEST.—454. What are taxes? Upon what are they assessed? When assessed upon the person, what are they called? When assessed upon the property, how are they apportioned? *Obs.* How is property divided? What does real estate denote? What is personal property? 455. When a tax is to be assessed, what is the first step?

OBS. 1. By the *number of polls* is meant the number of *taxable individuals*, which usually includes every *native* or *naturalized freeman* over the age of 21 and under 70 years. In Massachusetts poll taxes are assessed upon every male inhabitant of the state, between the ages of 16 and 70 years, whether a citizen or an alien.*

2. When any part or the whole of a tax is assessed upon the polls, each citizen is taxed a *specific sum*, without regard to the amount of property he possesses.

Ex. 1. The tax assessed by a certain town is \$990; its property, both personal and real, is valued at \$28000, and it contains 300 polls, which are assessed 50 cents apiece. What per cent. is the tax; that is, how much is the tax on a dollar; and how much is a man's tax who pays for 3 polls, and whose property is valued at \$1500?

Solution.—Since 1 poll pays 50 cents, 300 polls must pay 300 times 50 cents, which is \$150. Now \$990—\$150=\$840, the sum to be assessed on the property. Now if \$28000 is to pay \$840, \$1 must pay $\frac{1}{28000}$ of \$840; and \$840÷\$28000=\$.03, or 3 per cent. Finally, the tax on \$1500, the amount of the man's property, at 3 per cent., is \$1500×.03=\$45; and \$45+\$1.50 (3 polls)=\$46.50, the man's tax. Hence,

456. To assess a State, County, or other tax.

I. *First find the amount of tax on all the polls, if any, at the given rate, and subtract this sum from the whole tax to be assessed. Then dividing the remainder by the whole amount of taxable property in the State, County, &c., the quotient will be the per cent. or tax on one dollar.*

II. *Multiply the amount of each man's property by the tax on one dollar, and the product will be the tax on his property.*

III. *Add each man's poll tax to the tax he pays on his property, and the amount will be his whole tax.*

PROOF.—*When a tax bill is made out, add together the taxes of all the individuals in the town, district, &c., and if the amount is equal to the whole tax assessed, the work is right.*

QUEST.—*Obs.* What is meant by the number of polls? 456. How are taxes assessed? When a tax bill is made out, how is its correctness proved?

* Revised Statutes of Massachusetts.

2. A certain corporation is taxed $537.50; the whole property of the corporation is valued at $35000, and there are 50 polls which are assessed 25 cents apiece. What per cent. is the tax; and how much is a man's tax, who pays for 2 polls, and whose property is valued at $4240.

Operation.

Multiply $.25 the tax on 1 poll,
By 50 the number of polls.
$12.50 Amount on polls.

But $537.50—12.50=$525, the sum assessed on the corporation; and $525÷$35000=.015, the per cent. or tax on $1.

Now $4240×.015=$63.60, the tax on the man's property,
And .25×2= .50, the tax for polls.
Ans. $64.10, whole tax.

3. What is B's tax, who pays for 3 polls, and whose property is valued at $3560?

4. What is C's tax, who is assessed for 1 poll and $5350?

5. The city of New York levied a tax of $1945600; its taxable property was rated at $243200000: what per cent. was the tax?

6. What was A's tax, whose property was valued at $10000?

7. What was B's tax, who was assessed for $15240?

8. What was C's tax, who was assessed for $35460?

457. Having ascertained the expenditures of a State, County, Town, &c., it is necessary in assessing the tax, to take into consideration the *expense of collecting it.* Collectors are paid a certain per cent. commission on the amount collected; (Art. 388. Obs. 1;) consequently, in determining the exact sum to be assessed, allowance must be made not only for the commission on the *net amount* to be raised, but also on the commission itself; for the commission is to be paid out of the money collected.

9. If the expenses of a town are $950, what sum must be assessed to raise this amount, with 5 per cent. commission for collecting it?

Analysis.—Since the commission is 5 per cent. the net value of $1 assessment is 95 cents. Therefore, if 95 cents *net*, require $1 assessment, $950 *net*, will require as many dollars assessment, as 95 cents are contained times in $950; and $950÷$.95=1000.

Ans. $1000.

PROOF.—$1000×.05=$50, the commission;
and $1000—$50=$950, the net sum required. Hence,

458. To find what sum must be assessed, to raise a given net amount.

Subtract the given per cent. commission from $1, *and the remainder will be the net value of* $1 *assessment.*

Divide the net amount to be raised by the net value of $1 *assessment, and the quotient will be the sum to be assessed.*

OBS. To meet the expense of collecting a tax, assessors not unfrequently calculate the commission at the given per cent. on the *net amount* to be raised, and add it to the tax bill. This method is wrong, and leads to erroneous results. Thus, on a tax of $1000, at 5 per cent. commission, the net amount is $2.50 too small; on $100000, the error is $250; on $1000000, it is $2500.

10. What sum must be assessed to raise a net amount of $8500, with 4 per cent. commission for collection?

11. What sum must be assessed to raise $15400 net, allowing 4½ per cent. commission for collection?

12. Allowing 5 per cent. for collection, what sum must be assessed to raise $16475 net?

13. Allowing 3½ per cent. for collection, what sum must be assessed to raise $32860 net?

FORMATION OF TAX BILLS.

459. In making out a tax bill for a Town, District, &c., having found the tax on $1, it is advisable to make a table, showing the amount of tax on any number of dollars from 1 to $10; then from $10 to $100; and from $100 to $1000.

14. A township composed of 16 citizens, levies a tax of $5700; the town contains 30 polls, which are assessed 50 cents each, and

QUEST.—458. How find what sum must be assessed to raise a tax of a given amount?

its taxable property is inventoried at $199500. What amount of tax must be raised to pay the debt and 5 per cent. commission for collection; and what is the tax on a dollar?

Solution.—The sum to be raised is $6000; (Art. 458;) and the tax is 3 cents on a dollar. (Art. 456.) Now, since the tax on $1 is $.03, it is obvious that multiplying $.03 by 2 will be the tax on $2; multiplying it by 3, will be the tax on $3, &c., as seen in the following

TABLE.

$1	pays	$.03	$10	pay	$.30	$100	pay	$3.00
2	"	.06	20	"	.60	200	"	6.00
3	"	.09	30	"	.90	300	"	9.00
4	"	.12	40	"	1.20	400	"	12.00
5	"	.15	50	"	1.50	500	"	15.00
6	"	.18	60	"	1.80	600	"	18.00
7	"	.21	70	"	2.10	700	"	21.00
8	"	.24	80	"	2.40	800	"	24.00
9	"	.27	90	"	2.70	900	"	27.00
10	"	.30	100	"	3.00	1000	"	30.00

15. In the above assessment, what is A. B.'s tax, who is rated at $2256, and pays for 3 polls?

Operation.

$2000	pay	$60.00
200	"	6.00
50	"	1.50
6	"	.18
3 polls	"	1.50
Amount,		$69.18

$2256=2000+200+50+6 dollars. Now if we add together the tax paid on each of these sums, as found in the table above, the amount will be the tax on $2256.

A. B.'s tax therefore is $69.18.

16. What is G. A.'s tax, who is assessed for 2 polls, and $2400?
17. What is H. B.'s tax, who is assessed for 1 poll, and $3850?
18. What is W. C.'s tax, who is assessed for 3 polls, and $15000?
19. E. D. is assessed for $16024, and 1 poll: what is his tax?
20. J. F. is assessed for $10450, and 2 polls: what is his tax?
21. T. G. is assessed for $20680, and 3 polls: what is his tax?
22. W. H. is assessed for $17530, and 1 poll: what is his tax?

23. L. J. is assessed for \$8760, and 1 poll: what is his tax?
24. W. L. is assessed for \$21000, and 2 polls: what is his tax?
25. J. K. is assessed for \$6530, and 2 polls: what is his tax?
26. G. L. is assessed for \$13480, and 1 poll: what is his tax?
27. F. M. is assessed for \$12300, and 3 polls: what is his tax?
28. C. P. is assessed for \$15240, and 2 polls: what is his tax?
29. J. S. is assessed for \$16000, and 1 poll: what is his tax?
30. R. W. is assessed for \$18000, and 2 polls: what is his tax?

Note.—Rate Bills for schools are generally apportioned according to the number of days each scholar has attended. Hence,

460. To make out Rate Bills for schools.

First find the number of days attendance of all the scholars, and the whole amount of expenses, including teacher's salary, fuel, repairs, &c. From the amount of expenses deduct the public money, if any, then divide the remainder by the whole number of days attendance, and the quotient will be the rate per day. Finally, multiply the rate per day by the number of days attendance of each man's children, and the product will be his tax.

OBS. In New York and some other states, the *general* principle is to include only the *Teacher's Salary* in the Rate Bill. (*Revised Statutes. N. Y.*)

31. A certain district paid \$130 for teacher's salary, \$34 for board, \$19.42 for fuel, and \$2.58 for repairs; the district drew \$30 public money, and the whole number of days attendance was 2400: what was the rate per day; and how much was A's tax, who sent 115 days?

Solution.—Amount of expenses, \$186—\$30=\$156; and \$156 ÷2400=\$.065, the rate per day. Now \$.065×115=\$7.475, A's tax is therefore \$7.475.

32. If the expenses of a district are \$313.20, and the whole attendance 3915 days, what is B's tax, who sends 167 days?

33. A district paid their teacher \$115, and their fuel cost \$21.20; it drew \$38.50 public money, and the number of days attendance was 1954: what was C's tax, who sent 69 days?

34. The expenses of a district were \$215.20, and the number of days attendance 2152: what was D's tax, who sent 134 days?

SECTION XIII.

ANALYSIS.

ART. **461.** The term *Analysis*, in physical science, signifies the *resolving* of a compound body into its *elements*, or *component parts*.

ANALYSIS, in arithmetic, signifies the *resolving of numbers* into the *factors* of which they are composed, and the *tracing of the relations* which they bear to each other. (Art. 95. Obs. 2.)

OBS. In the preceding sections the student has become acquainted with the method of *analyzing particular* examples and combinations of numbers, and thence deducing *general principles and rules*. But *analysis* may be applied with advantage not only to the *development* of mathematical truths, but also to the *solution* of a great variety of problems, both in arithmetic and practical life. Indeed, it is the method by which business men generally solve practical questions. A little practice will give the student great facility in its application.

462. No specific directions can be given for solving examples by *analysis*. None in fact are requisite. The *judgment*, from the conditions of the question, will *suggest the process*. Hence, *Analysis* may, with propriety, be called the COMMON SENSE RULE.

OBS. In solving questions analytically, it may be remarked in general, that we reason from the *given number* to 1, then from 1 to the *number required*.

Ex. 1. If 60 yards of cloth cost $240, what will 85 yards cost?

Analytic solution.—Since 60 yds. cost $240, 1 yd. will cost $\frac{1}{60}$ of $240; and $\frac{1}{60}$ of $240 is $4. Now if 1 yd. costs $4, 85 yds. will cost 85 times as much; and $4×85=$340. *Ans.*

Or, we may reason thus: 85 yds. are $\frac{85}{60}$ of 60 yds.; therefore 85 yds. will cost $\frac{85}{60}$ of $240, (the cost of 60 yds.) and $\frac{85}{60}$ of $240 is $240×$\frac{85}{60}$=$340, the same as before. (Arts. 210, 212.)

OBS. 1. Other solutions of this example might be given; but our present object is to show how this and similar questions may be solved by *analysis*. The

QUEST.—461. What is meant by analysis in physical science? What in arithmetic? To what may analysis be advantageously applied? 462. Can any particular rules be prescribed for solving questions by analysis? How then will you know how to proceed? Obs. What is the operation of solving questions by analysis called?

former method is the simplest and most strictly analytic, though not so short as the latter. It contains two steps:

First, we separate the given price of 60 yds. ($240) into 60 equal parts, to find the value of one part, or the cost of 1 yd., which is $4.

Second, we multiply the price of 1 yd. ($4) by 85, the number of yds. whose cost is required, and the product is the answer sought.

2. This and similar questions are usually placed under the rule of *Simple Proportion*, or the *Rule of Three.*

3. The operation of solving a question by analysis, is called an *analytic solution.* In reciting the following examples, each one should be analyzed, and the reason for every step given in full.

2. A man bought a horse, and paid $45 down, which was $\frac{5}{7}$ of the price of it: what did he give for the horse?

Analysis.—Since $45 is $\frac{5}{7}$ of the price, the question resolves itself into this: $45 is $\frac{5}{7}$ of what sum? If $45 is $\frac{5}{7}$ of a certain sum, $\frac{1}{7}$ is $\frac{1}{5}$ of $45; and $\frac{1}{5}$ of $45 is $9. Now if $9 is 1 seventh, 7 sevenths are 7 times as much; and $9×7=$63. *Ans.* $63.

PROOF.—$\frac{1}{7}$ of $63=$9, and 5 sevenths are 5 times as much, which is $45, the sum he paid down for the horse.

Note.—In solving examples of this kind, the learner is often perplexed in finding the value of $\frac{1}{7}$, &c. This difficulty arises from supposing that if $\frac{5}{7}$ of a certain number is 45, $\frac{1}{7}$ of it must be $\frac{1}{7}$ of 45. This mistake will be easily avoided by substituting in his mind the word *parts* for the given *denominator.* Thus, if 5 parts cost $45, 1 part will cost $\frac{1}{5}$ of $45, which is $9. But this part is a *seventh.* Now if 1 seventh cost $9, then 7 sevenths will cost 7 times as much.

3. If 40 cords of wood cost $120, how much will 100 cords cost?

4. Bought 35 tons of hay for $700: how much will 16 tons cost?

5. What cost 37 gallons of molasses, at $21 a hogshead?

6. What cost 1500 pounds of hay, at $14 per ton?

7. What cost 18 quarts of chestnuts, at $3 a bushel?

8. If 55 tons of hemp cost $660, what will 220 tons cost at the same rate?

9. If 165 bushels of apples cost $132, how much will 31 bushels cost?

10. If 72 bushels of peanuts cost $\$253.44$, what will a pint cost at the same rate?

11. If 150 acres of land cost $\$7000$, what will a square rod cost?

12. If 2 pipes of wine cost $\$315$, what is that per gill?

13. A farmer bought a yoke of oxen, and paid $\$40$ in work, which was $\frac{5}{8}$ of the cost: what did they cost?

14. Bought a house, and paid $\$630$ in goods, which was $\frac{7}{12}$ of the price of it: what was the cost of the house?

15. A young man lost $\$256$ by gambling, which was $\frac{8}{15}$ of all he was worth: how much was he worth?

16. A man having $\$1500$, paid $\frac{3}{5}$ of it for $112\frac{1}{2}$ acres of land: how much did his land cost per acre?

17. If a stack of hay will keep 350 sheep 90 days, how long will it keep 525 sheep?

18. If 440 bbls. of flour will last 15 men 55 months, how long will the same quantity last 28 men?

19. If 136 men can build a block of stores in 120 days, how long will it take 15 men to build it?

20. If $\frac{5}{8}$ of a pound of tea cost 40 cents, what will $\frac{7}{8}$ of a pound cost?

21. If $\frac{2}{3}$ of a yard of broadcloth cost $\$2.50$, how much will $\frac{4}{3}$ of a yard cost?

22. Bought $\frac{3}{11}$ of a ton of hay for $\$3.42$: how much will $\frac{7}{11}$ of a ton cost?

23. Bought $\frac{19}{20}$ of a hogshead of molasses for $\$38.19$: how much will $\frac{3}{20}$ of a hogshead cost?

24. If $\frac{3}{5}$ of an acre of land cost $\$108$, how much will $\frac{8}{9}$ of an acre cost?

25. If $\frac{3}{4}$ of a barrel of beef cost $\$6.48$, how much will $\frac{3}{7}$ of a barrel cost?

26. Paid $\$4200$ for $\frac{5}{7}$ of a sloop: how much can I afford to sell $\frac{7}{12}$ of the sloop for?

27. Sold $18\frac{1}{2}$ baskets of peaches for $\$34$: how much would $65\frac{1}{4}$ baskets come to?

28. If I pay $\$60.50$ for building $20\frac{1}{2}$ rods of wall, how much must I pay for $215\frac{2}{3}$ rods?

29. A man can hoe a field of corn in 6 days, and a boy can hoe it in 9 days: how long will it take them both together to hoe it?

Analysis.—Since the man can hoe the field in 6 days, in 1 day he can hoe $\frac{1}{6}$ of it; and since the boy can hoe it in 9 days, in 1 day he can hoe $\frac{1}{9}$ of it; consequently in 1 day they can both hoe $\frac{1}{6}+\frac{1}{9}=\frac{5}{18}$ of the field. (Art. 202.) Now if $\frac{5}{18}$ of the field requires them both 1 day, $\frac{1}{18}$ of it will require them $\frac{1}{5}$ of a day, and $\frac{18}{18}$ will require them 18 times as long, or $\frac{18}{5}$ of a day, which is equal to $3\frac{3}{5}$ days. *Ans.*

30. If A can chop a cord of wood in 4 hours, and B in 6 hours, how long will it take them both to chop a cord?

31. A can dig a cellar in 6 days, B in 9 days, and C in 12 days: how long will it take all of them together to dig it?

32. A man bought 25 pounds of tea at 6s. a pound, and paid for it in corn at 4s. a bushel: how many bushels did it take?

Analysis.—If 1 lb. of tea costs 6s., 25 lbs. will cost 25 times as much, which is 150s. Again, if 4s. will buy 1 bushel of corn, 150s. will buy as many bushels as 4s. is contained times in 150s.; and 150s.$\div$4$=37\frac{1}{2}$. *Ans.* $37\frac{1}{2}$ bushels.

463. The last and similar examples are frequently arranged under the rule of *Barter*.

Barter signifies an exchange of articles of commerce at prices agreed upon by the parties.

OBS. Such examples are so easily solved by *Analysis* that a *specific rule* for them is *unnecessary*.

33. A farmer bought 110 lbs. of sugar at 18 cents a pound, and paid for it in lard at $5\frac{1}{2}$ cents a pound: how much lard did it take?

34. How much butter, at $12\frac{1}{2}$ cents a pound, must be given for 250 lbs. of tea, at 75 cents a pound?

35. How many cords of wood, at $\$2\frac{1}{2}$ per cord, must be given for 56 yds. of cloth, at $\$4\frac{1}{4}$ per yard?

36. How many pair of boots, at $4.50 a pair, must be given for 50 tons of coal at $9 per ton?

37. A, B, and C, united in business; A put in \$250; B, \$270 and C, \$340; they gained \$258: what was each man's share of the gain?

Analysis.—The whole sum invested is \$250+\$270+\$340= \$860. Now since \$860 gain \$258, it is plain \$1 will gain $\frac{1}{860}$ of \$258, which is 30 cents. And

If \$1 gains 30 cts. \$250 will gain \$250×.30=\$75, A's share,
" \$1 " " \$270 " \$270×.30= 81, B's share,
" \$1 " " \$340 " \$340×.30=102, C's share.

Or, we may reason thus: Since the sum invested is \$860,
A's part of the investment is equal to $\frac{250}{860}$, or $\frac{25}{86}$;
B's " " " $\frac{270}{860}$, or $\frac{27}{86}$;
C's " " " $\frac{340}{860}$, or $\frac{34}{86}$. Consequently,
A must receive $\frac{25}{86}$ of the whole gain \$258=\$75;
B " " $\frac{27}{86}$ " " 258= 81;
C " " $\frac{34}{86}$ " " 258=102;
PROOF.—The whole gain is \$258. (Ax. 11.)

464. When two or more individuals associate themselves together for the purpose of carrying on a joint business, the union is called a *partnership* or *copartnership*.

OBS. The process by which examples like the last one are solved, is often called *Fellowship*.

38. A and B join in a speculation; A advances \$1500 and B \$2500; they gain \$1200: what was each one's share of the gain?

39. A, B, and C, entered into partnership; A furnished \$3000, B \$4000, and C \$5000; they lost \$1800: what was each one's share of the loss?

40. A's stock is \$4200; B's \$3600; and C's \$5400; the whole gain is \$2400: what is the gain of each?

41. A's stock is \$7560; B's \$8240; C's \$9300; and D's \$6200; the whole gain is \$625: what is the share of each?

42. A bankrupt owes one of his creditors \$400; another \$500; and a third \$600; his property amounts to \$1000: how much can he pay on a dollar; and how much will each of his creditors receive?

OBS. The solution of this example is the same in principle as that of Ex. 37

465. Examples like the preceding are commonly arranged under the rule of *Bankruptcy*.

Note.—A *bankrupt* is a person who is insolvent, or unable to pay his just debts.

43. A bankrupt owes $5000, and his property is worth $3500: how much can he pay on a dollar?

44. A man died owing $16400, and his effects were sold for $4100: what per cent. did his estate pay?

45. If a man owes A $6240, B $8760, and C $9000, and has but $11500, how much will each creditor receive?

46. If I owe $48000, and have property to the amount of $32000, what per cent. can I pay?

47. What per cent. can a man pay, whose liabilities are $120000, and whose assets are $45000?

48. What per cent. can a man pay, whose liabilities are $1500000, and whose assets are $150000?

466. It often happens in storms and other casualties at sea, that masters of vessels are obliged to throw portions of their cargo overboard, or sacrifice the ship and their crew. In such cases, the law requires that the loss shall be divided among the owners of the vessel and cargo, in proportion to the amount of each one's property at stake.

The process of finding each man's loss, in such instances, is called *General Average*.

OBS. The operation is the same as that in solving questions in bankruptcy and partnership.

49. A, B and C, freighted a ship from New York to Liverpool; A had on board 100 tons of iron, B 200 tons, and C 300 tons, in a storm 240 tons were thrown overboard: what was the loss of each?

50. A packet worth $36000 was loaded with a cargo valued at $65000. In a tempest the master threw overboard $25250 worth of goods: what per cent. was the general average?

51. A steam ship being in distress, the master threw $\frac{1}{4}$ of the cargo overboard; finding she still labored, he afterwards threw overboard $\frac{1}{3}$ of what remained. The steamer was worth

$120000, and the cargo $240000 : what per cent. was the general average, and what would be a man's loss who owned $\frac{1}{4}$ of the ship and cargo ?

52. A man mixed 25 bushels of peas worth 6s. a bushel, with 15 bushels of corn worth 4s. a bushel, and 20 bushels of oats worth 3s. a bushel : what was the mixture worth per bushel ?

Analysis.—25 bu. peas at 6s.=150s., value of the peas;
15 bu. corn at 4s.= 60s., " " corn;
and 20 bu. oats at 3s.= 60s., " " oats.
The mixture=60 bu. and 270s., value of whole mixture.

Now if 60 bu. mixture are worth 270s., 1 bu. mixture is worth $\frac{1}{60}$ of 270s.; and 270s.÷60=4½s. *Ans.*

PROOF.—60 bu. at 4½s.=270s., the value of the whole mixture.

467. The process of finding the value of a compound or mixture of articles of different values, or of forming a compound which shall have a given value, is called *Alligation.* Alligation is usually divided into two kinds, *Medial* and *Alternate.*

OBS. 1. When the prices of the several articles and the number or quantity of each are given, the process of finding the *value* of the mixture, as in the last example, is called *Alligation Medial.*

2. When the *price* of the mixture is given, together with the price of each article, the process of finding how much of the several articles must be taken to form the required mixture, is called *Alligation Alternate.* Alligation Alternate embraces *three* varieties of examples, which are pointed out in the following notes.

53. If you mix 40 gallons of sperm oil worth 8s. per gallon, with 60 gallons of whale oil worth 3s. per gallon, what will the mixture be worth per gallon ?

54. At what price per pound can a grocer afford to sell a mixture of 30 lbs. of tea worth 4s. a pound, and 40 lbs. worth 7s. a pound ?

55. If 120 lbs. of butter at 10 cts. a pound are mixed with 24 lbs. at 8 cts. and 24 lbs. at 5 cts. a pound, what is the mixture worth ?

56. A tobacconist had three kinds of tobacco, worth 15, 18, and 25 cents a pound : what is a mixture of 100 lbs. of each worth per pound ?

57. A liquor dealer mixed 200 gallons of alcohol worth 50 cts. a gallon, with 100 gallons of brandy worth $1.75 a gallon: what was the value of the mixture per gallon?

58. A grocer sells imperial tea at 10s. a pound, and hyson at 4s.: what part of each must he take to form a mixture which he can afford to sell at 6s. a pound?

Note.—1. It will be observed in this example that the price of the *mixture* and also the price of the several *articles* or *ingredients* are given, to find *what part* of each the mixture must contain.

Analysis.—Since the imperial is worth 10s. and the required mixture 6s., it is plain he would lose 4s. on every pound of imperial which he puts in. And since the hyson is worth 4s. a pound and the mixture 6s., he would gain 2s. on every pound of hyson he puts in. The question then is this: How much hyson must he put in to make up for the loss on 1 lb. of imperial? If 2s. profit require 1 lb. of hyson, 4s. profit will require twice as much, or 2 lbs. He must therefore put in 2 lbs. of hyson to 1 lb. of imperial.

PROOF—2 lbs. of hyson, at 4s. a pound, are worth 8s., and 1 lb. of imperial is worth 10s. Now 8s.+10s.=18s. And if 3 lbs. mixture are worth 18s., 1 lb. is worth $\frac{1}{3}$ of 18s., which is 6s., the price of the mixture required.

59. A farmer has oats which are worth 20 cts. a bushel, rye 55 cts., and barley 60 cts., of which he wishes to make a mixture worth 50 cts. per bushel: what part of each must the mixture contain?

Analysis.—The prices of the rye and barley must each be compared with the price of the oats. If 1 bu. oats gains 30 cts. in the mixture, it will take as many bu. of rye to balance it, as 5 cts. (the loss per bu.) are contained times in 30 cts., viz: 6 bu. Again, since 1 bu. oats gains 30 cts., it will take as many bushels of barley to balance it, as 10 cts. (the loss per bu.) are contained times in 30 cts., viz: 3 bu. Hence, the mixture must contain 2 parts of oats, 6 parts rye, and 3 parts barley.

60. If a man have four kinds of sugar worth, 8, 9, 11, and 12

cents a pound respectively, how much of each kind must he take to form a mixture worth 10 cents a pound?

Note.—2. In examples like the preceding, we compare two kinds together, one of a higher and the other of a lower price than the required mixture; then compare the other two kinds in the same manner. In selecting the pairs to be compared together, it is necessary that the price of one article shall be above, and the other below the price of the mixture. Hence, when there are several articles to be mixed, some cheaper and others dearer than the mixture, a variety of answers may be obtained. Thus, if we compare the highest and lowest, then the other two, the mixture will contain 1 part at 8 cts.; 1 part at 9 cts.; 1 part at 11 cts.; and 1 part at 12 cts. Again, by comparing those at 8 and 11 cts., and those at 9 and 12 cts. together, we obtain for the mixture 1 part at 8 cts.; 2 parts at 11 cts.; 2 parts at 9 cts.; and 1 part at 12 cts.

Other answers may be found by comparing the first with the third and fourth; and the second with the fourth, &c.

61. A goldsmith having gold 16, 18, 23, and 24 carats fine, wished to make a mixture 21 carats fine: what part of each must the mixture contain?

62. A farmer had 30 bu. of corn worth 6s. a bu., which he wished to mix with oats worth 3s. a bu., so that the mixture may be worth 4s. per bu.: how many bushels of oats must he use?

Note.—3. In this example, it will be perceived, that the price of the mixture, with the prices of the several articles and the *quantity* of one of them are given, to find *how much* of the other article the mixture *must contain.*

Analysis.—Reasoning as above, we find that the mixture (without regard to the specified quantity of corn) in order to be worth 4s. per bu., must contain 2 bu. of oats to 1 bu. of corn. Hence, if 1 bu. of corn requires 2 bu. of oats to make a mixture of the required value, 30 bu. of corn will require 30 times as much; and 2 bu.$\times$30$=$60 bu., the quantity of oats required.

63. A merchant wished to mix 100 gallons of oil worth 80 cts. per gallon, with two other kinds worth 30 cts. and 40 cts. per gallon, so that the mixture may be worth 60 cts. per gallon: how many gallons of each must it contain?

64. A merchant has Havana coffee at 12 cts. and Java at 18 cts. per pound, of which he wishes to make a mixture of 150 lbs., which he can sell at 16 cts. a pound: how much of each must he use?

Note.—4. In this example, the whole quantity to be mixed, the price of the mixture, and the prices of the several articles are given, to find *how much* of each must be taken.

Analysis.—On 1 lb. of the Havana it is obvious he will gain 4 cts., and on 1 lb. of the Java he will lose 2 cts.; therefore to balance the 4 cts. gain he must put in 2 lbs. of Java; that is, the mixture must contain 1 part of Havana to 2 parts of Java. Now if 3 lbs. mixture require 1 lb. Havana, 150 lbs. mixture, (the quantity required,) will require as many pounds of Havana as 3 is contained times in 150, viz: 50 lbs. But the mixture contains twice as much Java as Havana, and 50 lbs$\times 2=100$ lbs.

Ans. 50 lbs. Havana, and 100 lbs. Java.

65. It is required to mix 240 lbs. of different kinds of raisins, worth 8d., 12d., 18d., and 22d. a pound, so that the mixture may be worth 10d. a pound: how much of each must be taken?

66. If 10 horses consume 720 quarts of oats in 6 days, how long will it take 30 horses to consume 1728 quarts?

Analysis.—Since 10 horses will consume 720 qts. in 6 days, 1 horse will consume $\frac{1}{10}$ of 720 qts. in the same time; and $\frac{1}{10}$ of 720 qts. is 72 qts. And if 1 horse will consume 72 qts. in 6 days, in 1 day he will consume $\frac{1}{6}$ of 72 qts., which is 12 qts. Again, if 12 qts. last 1 horse 1 day, 1728 qts. will last him as many days as 12 qts. are contained times in 1728 qts., viz: 144 days. Now if 1 horse will consume 1728 qts. in 144 days, 30 horses will consume them in $\frac{1}{30}$ of the time; and 144 d.$\div 30=4\frac{4}{5}$.

Ans. 30 horses will consume 1728 qts. in $4\frac{4}{5}$ days.

468. This and similar examples are usually placed under the rule of *Compound Proportion*, or *Double Rule of Three.*

67. If 15 horses consume 40 tons of hay in 30 weeks, how many horses will it require to consume 56 tons in 70 weeks?

68. If 8 men can make 9 rods of wall in 12 days, how long will it take 10 men to make 36 rods?

69. If 35 bbls. of water will last 950 men 7 months, how many men will 1464 bbls. of water last 1 month?

70. If 13908 men consume 732 bbls. of flour in 2 months, in how long time will 425 men consume 175 bbls.?

71. If the interest of \$30 for 12 months is \$2.10, how much is the interest of \$1560 for 6 months?

72. If the interest of \$750 for 8 months is \$28, how much is the interest of \$16425 for 6 months?

73. A man being asked how much money he had, replied that $\frac{2}{3}$, $\frac{3}{4}$, and $\frac{5}{8}$ of it made \$980: what amount did he have?

Analysis.—It is plain that $\frac{2}{3}+\frac{3}{4}+\frac{5}{8}=\frac{49}{24}$. (Art. 202.) The question then resolves itself into this: \$980 are $\frac{49}{24}$ of what sum? Now if \$980 are $\frac{49}{24}$ of a certain sum, $\frac{1}{24}$ is $\frac{1}{49}$ of \$980; and \$980 $\div 49=$\$20, and $\frac{24}{24}$ is \$20$\times 24=$\$480. *Ans.*

PROOF.—$\frac{2}{3}$ of \$480$=$\$320; $\frac{3}{4}$ of \$480$=$\$360; and $\frac{5}{8}$ of \$480 $=$\$300. Now \$320$+$\$360$+$\$300$=$\$980, according to the conditions of the question.

469. This and similar examples are placed under the rule of *Position*. The shortest and easiest method of solving them is by *Analysis*.

74. A sailor having spent $\frac{1}{4}$ of his money for his outfit, deposited $\frac{2}{3}$ of it in a savings bank, and had \$50 left: how much had he at first?

75. A man laid out $\frac{1}{3}$ of his money for a house, $\frac{1}{4}$ for furniture, and had \$1500 left; how much had he at first?

76. A man lost $\frac{1}{6}$ of his money in gambling, $\frac{1}{6}$ in betting, and spent $\frac{2}{9}$ in drinking; he had \$259 left: how much had he at first?

77. What number is that $\frac{2}{3}$ and $\frac{3}{4}$ of which is 102?

78. What number is that $\frac{1}{2}$, $\frac{1}{3}$, $\frac{1}{4}$, and $\frac{1}{6}$ of which is 450?

79. What number is that $\frac{1}{5}$ and $\frac{1}{6}$ of which being added to itself, the sum will be 164?

80. What number is that $\frac{7}{5}$ of which exceeds $\frac{4}{5}$ of it by 18?

81. A post stands 40 feet above water, $\frac{1}{4}$ in the water, and $\frac{1}{6}$ in the ground: what is the length of the post?

82. What will 376 yds. of muslin cost, at 2s. and 6d. per yd.?

Analysis.—2s. 6d.$=$£$\frac{1}{8}$. Now if 1 yd. costs £$\frac{1}{8}$, 376 yds. will cost 376 times as much; and £$\frac{1}{8}\times 376=$£47. *Ans.*

83. If 1 yard of silk costs 50 cents, what will 256 yards cost?

Analysis.—50 cts.=$\$\frac{1}{2}$. Now if 1 yd. costs $\$\frac{1}{2}$, 256 yds. will cost 256 times as much; and $\$\frac{1}{2}\times256=\$128$. *Ans.*

470. Examples like the preceding, in which the price of a single article is an aliquot part of a dollar, &c., are usually classed under the rule of *Practice.*

Practice is defined by a late English author to be "an abridged method of performing operations in the rule of proportion by means of *aliquot parts;* and it is chiefly employed in computing the prices of commodities."

OBS. After giving several tables of aliquot parts in money, weight, and measure, the same author proceeds to divide his subject into *twelve* subdivisions or cases, and gives a *specific rule* for each case, to be committed to memory by the pupil. It is believed, however, that so *many specific rules* are *worse* than *useless.* They have a tendency to prevent the exercise of thought and reason, while they tax the time and memory of the student with a multiplicity of particular directions for the solution of a class of examples, which his common sense, if permitted to be exercised, will solve more expeditiously by *Analysis.*

TABLE OF ALIQUOT PARTS OF $1, £1, AND 1s.

Parts of a Dollar.	Parts of a Pound sterling.	Parts of a Shilling sterling.
50 cts.=$\$\frac{1}{2}$	10s. =£$\frac{1}{2}$	6 pence = $\frac{1}{2}$ shil.
$33\frac{1}{3}$ cts.=$\$\frac{1}{3}$	6s. 8d.=£$\frac{1}{3}$	4 pence = $\frac{1}{3}$ shil.
25 cts.=$\$\frac{1}{4}$	5s. =£$\frac{1}{4}$	3 pence = $\frac{1}{4}$ shil.
20 cts.=$\$\frac{1}{5}$	4s. =£$\frac{1}{5}$	2 pence = $\frac{1}{6}$ shil.
$16\frac{2}{3}$ cts.=$\$\frac{1}{6}$	3s. 4d.=£$\frac{1}{6}$	$1\frac{1}{2}$ pence = $\frac{1}{8}$ shil.
$12\frac{1}{2}$ cts.=$\$\frac{1}{8}$	2s. 6d.=£$\frac{1}{8}$	1 penny=$\frac{1}{12}$ shil.
10 cts.=$\$\frac{1}{10}$	2s. =£$\frac{1}{10}$	$\frac{1}{2}$ penny=$\frac{1}{24}$ shil.
$8\frac{1}{3}$ cts.=$\$\frac{1}{12}$	1s. 8d.=£$\frac{1}{12}$	7 pence = $\frac{1}{2}$s.+$\frac{1}{12}$s.
$6\frac{1}{4}$ cts =$\$\frac{1}{16}$	1s. =£$\frac{1}{20}$	8 pence = $\frac{1}{2}$s.+$\frac{1}{6}$s.
5 cts.=$\$\frac{1}{20}$	11s.=£$\frac{1}{2}$+£$\frac{1}{20}$	9 pence = $\frac{1}{2}$s.+$\frac{1}{4}$s.

Note.—If the price itself is not an aliquot part of $1, or £1, &c., it may sometimes be divided into such parts as will be aliquot parts of $1, £1, &c., or which will be aliquot parts of each other. Thus, $87\frac{1}{2}$ cts. is not an aliquot part of $1, but $87\frac{1}{2}$ cts.=50+25+$12\frac{1}{2}$ cts. Now 50 cts.=$\$\frac{1}{2}$; 25 cts.=$\$\frac{1}{4}$; and $12\frac{1}{2}$ cts.=$\$\frac{1}{8}$. Or thus: 50 cts.=$\$\frac{1}{2}$, 25 cts.=$\frac{1}{2}$ of 50 cts., and $12\frac{1}{2}$ cts.=$\frac{1}{2}$ of 25 cts.

84. What will 680 bu. of wheat cost, at $87\frac{1}{2}$ cts. per bushel?

Analysis.—It is plain, if the price were $1 per bu., the cost of 680 bu. would be $680. Hence,

Were the price 50 cts. the cost would be $\frac{1}{2}$ of $680, which is $340
" " 25 cts. " " $\frac{1}{4}$ of $680, which is $170
" " $12\frac{1}{2}$ cts. " " $\frac{1}{8}$ of $680, which is $ 85
But since the price is $50+25+12\frac{1}{2}$ cents, the cost must be $595

Or, thus: $1×680=$680, the cost at $1 per bushel.

At 50 cts., or $\$\frac{1}{2}$, it will be $\frac{1}{2}$ of $680, or $340
" 25 cts., $\frac{1}{2}$ of 50 cts., " " $\frac{1}{2}$ of $340, or $170
" $12\frac{1}{2}$ cts., $\frac{1}{2}$ of 25 cts., " " $\frac{1}{2}$ of $170, or $ 85
Therefore the whole cost is $595. *Ans.*

85. What cost 478 yards of cashmere, at 50 cts. per yard?
86. What cost 1560 lbs. of tea, at 75 cts. per pound?
87. What cost 2400 gals. of molasses, at $37\frac{1}{2}$ cts. per gal.?
88. What cost 1800 yds. of satinet, at $62\frac{1}{2}$ cts. per yard?
89. At 25 cts. per bushel, what cost 1470 bu. of oats?
90. At $33\frac{1}{3}$ cts. a pound, what cost 1326 lbs. of ginger?
91. At $6\frac{1}{4}$ cts. per roll, what cost 3216 rolls of tape?
92. At $8\frac{1}{3}$ cts. per pound, what cost 4200 lbs. of lard?
93. At $12\frac{1}{2}$ cts. per dozen, what cost 1920 doz. of eggs?
94. At $16\frac{2}{3}$ cts. a pound, what cost 4524 lbs. of figs?
95. At $66\frac{2}{3}$ cts. per yard, what cost 1620 yds. of sarcenet?
96. What cost 840 bu. of rye, at $\$\frac{3}{4}$ per bushel?
97. What cost 690 yds. of cloth, at 6s. 8d. per yard?

Analysis.—At £1 per yard the cost would be £690. But 6s. 8d. is £$\frac{1}{3}$; therefore the cost must be $\frac{1}{3}$ of £690, which is £230. *Ans.*

98. What cost 360 gals. of wine, at 16s. per gallon?

Analysis.—16s.=10s.+5s.+1s. Now 10s.=£$\frac{1}{2}$; 5s.=£$\frac{1}{4}$; 1s.=$\frac{1}{5}$ of 5s.

If the price were £1 per gal., the cost of 360 gals. would be **£360**

At 10s., £$\frac{1}{2}$, it will be $\frac{1}{2}$ of £360, or £180
" 5s., $\frac{1}{2}$ of 10s., " $\frac{1}{2}$ of £180, or £ 90
" 1s., $\frac{1}{5}$ of 5s., " $\frac{1}{5}$ of £ 90, or £ 18
Therefore the whole cost is £288. *Ans.*

99. What cost 1240 yds. of flannel, at 3s. 4d. per yard?
100. What cost 2128 lbs. of spice, at 2s. 6d. per pound?
101. What cost 5250 yds. of lace, at 6d. per yard?
102. What cost 56480 yds. of tape, at $1\frac{1}{2}$d. per yard?

471. Notwithstanding the law requires accounts to be kept in Federal Money, goods are frequently sold at prices stated in the denominations of the old state currencies.

When the *price* per yard, pound, &c., stated in those currencies, is an *aliquot part* of a *dollar*, the answer may be easily obtained in Federal Money.

TABLE OF ALIQUOT PARTS IN DIFFERENT STATE CURRENCIES.

Parts of a Dollar, New York Currency.	Parts of a Dollar, New England Currency.	Parts of a Shilling, N. E. and N. Y. Currency.
4 shil. $=\$\frac{1}{2}$	3 shil. $=\$\frac{1}{2}$	6 pence $=\frac{1}{2}$ shil.
2s. 8d. $=\$\frac{1}{3}$	2 shil. $=\$\frac{1}{3}$	4 pence $=\frac{1}{3}$ shil.
2 shil. $=\$\frac{1}{4}$	1s. 6d. $=\$\frac{1}{4}$	3 pence $=\frac{1}{4}$ shil.
1s. 4d. $=\$\frac{1}{6}$	1 shil. $=\$\frac{1}{6}$	2 pence $=\frac{1}{6}$ shil.
1 shil. $=\$\frac{1}{8}$	4s. $=\$\frac{1}{2}+\$\frac{1}{6}$	$1\frac{1}{2}$ pence $=\frac{1}{8}$ shil.
5s. $=\$\frac{1}{2}+\$\frac{1}{8}$	5s. $=\$\frac{1}{2}+\$\frac{1}{3}$	1 penny $=\frac{1}{12}$ shil.

Note.—1. In N. Y. currency 8s. make \$1; in N. E. currency 6s. make \$1. From example 103 to 119 inclusive, the prices are given in N. Y. currency; from example 120 to 133 inclusive, they are given in N. E. currency. For the mode of reducing the different State currencies to each other and to Federal Money, see Section XVII.

103. At 1s. 4d. per yard, what cost 726 yds. of cambric?

Analysis.—If the price were \$1 per yard, the cost would be $\$1\times726=\$726$. But 1s. 4d. $=\$\frac{1}{6}$; therefore the cost must be $\frac{1}{6}$ of \$726, which is \$121. *Ans.*

104. What cost 896 bu. of wheat at 6s. per bushel?

Analysis.—6s. = 4s. + 2s. Now 4s. $=\$\frac{1}{2}$; and 2s. $=\frac{1}{2}$ of 4s.

At \$1 a bushel the cost would be \$896.

At 4s., $\$\frac{1}{2}$, it will be $\frac{1}{2}$ of \$896, or \$448
" 2s., $\frac{1}{2}$ of 4s., " " $\frac{1}{2}$ of \$448, or \$224

Therefore the whole cost is \$672. *Ans.*

Or, thus: 6s $=\$\frac{3}{4}$; therefore the number of bu. minus $\frac{1}{4}$ of itself, will be the cost, and 896—224 ($\frac{1}{4}$ of 896) = 672. *Ans.* \$672.

105. What cost 752 yds. of balzorine, at 2s. 8d. per yard?
106. What cost 1232 yds. of calico, at 1s. 6d. per yard?
107. What cost 763 lbs. of pepper, at 1s. 3d. a pound?
108. What cost 1116 bu. of apples, at 1s. 4d. per bushel?
109. What cost 1920 yds. of shirting, at 1s. 2d. per yard?
110. At 6s. a basket, what will 1560 baskets of peaches cost?
111. At 5s. 4d. a pound, what will 1200 lbs. of tea come to?

Note.—2. Since 5s. 4d. is $\frac{1}{3}$ less than $1, it is plain 1200—400=$800. *Ans.*

112. At 7s. per yard, what will 432 yds. of crape cost?
113. At 6s. 8d. a pound, what cost 972 lbs. of nutmegs?
114. At 2s. 8d. a pair, what cost 864 pair of cotton hose?
115. At 1$\frac{1}{2}$d. a yard, how much will 2800 yds. of tape come to?
116. What cost 1628 yds. of flannel, at 4s. per yard?
117. What cost 2560 bu. of oats, at 2s. per bushel?
118. What cost 9600 lbs. of wool, at 2s. 6d. a pound?
119. What cost 3200 lbs. of sugar, at 6d. per pound?
120. What cost 600 yds. of damask, at 5s., N. E. cur., per yard?

Note.—3. 5s. N. E. cur. is $\frac{1}{6}$ less than $1; hence, 600—100=$500. *Ans.*

121. What cost 2500 bu. of potatoes, at 1s. 6d. per bushel?
122. What cost 1440 yds. of gingham, at 2s. per yard?
123. How much will 4848 chickens cost, at 1s. apiece?
124. How much will 1680 slates cost, at 1s. 6d. apiece?
125. How much will 920 turkeys cost, at 4s. 6d. apiece?
126. What cost 4860 lbs. of butter, at 1s. 1d. per pound?
127. What cost 1260 melons, at 8d. apiece?
128. What cost 2340 lbs. of tea, at 4s. a pound?
129. What cost 240 bu. of peas, at 4s. 6d. per bushel?
130. What cost 720 pair of gloves, at 5s. 3d. a pair?
131. What cost 360 bushels of corn, at 3s. per bushel?
132. What cost 7686 lbs. of butter, at 1s. per pound?
133. What cost 960 yds. of silk, at 5s. per yard?
134. What will 75 lbs. of butter cost, at $16.80 per cwt.?
135. What will 125 lbs. of wool cost, at $36 per hundred?
136. What will 15 cwt. of hemp cost at $60 per ton?
137. What will 2500 lbs. of iron cost, at $72 per ton?
138. What cost 1$\frac{1}{4}$ acre of land, at $120 per acre?

SECTION XIV.

RATIO AND PROPORTION.

ART. **472.** In comparing numbers or quantities with each other, we may inquire, either *how much* greater one of the numbers or quantities is than the other; or *how many times* one of them contains the other. In finding the answer to either of these inquiries, we discover what is called the *relation* between the two numbers or quantities.

473. The *relation* between the two quantities thus compared, is of *two* kinds:

First, that which is expressed by their *difference.*

Second, that which is expressed by the *quotient* of the one divided by the other.

474. RATIO is that relation between two numbers or quantities, which is expressed by the *quotient* of the one divided by the other. Thus, the ratio of 6 to 2 is $6 \div 2$, or 3; for 3 is the quotient of 6 divided by 2.

OBS. The relation between two numbers or quantities denoted by their difference, is sometimes called *arithmetical ratio;* while that denoted by the quotient of the one divided by the other, is called *geometrical ratio.* Thus 4 is the arithmetical ratio of 8 to 4; and 2 is the geometrical ratio of 8 to 4.

But as the term *arithmetical ratio* is merely a substitute for the word *difference,* the term difference, in the succeeding pages, is used in its stead; and when the word *ratio* simply is used, it signifies that which is denoted by the *quotient* of the one divided by the other, as in the article above.

475. The two given numbers thus compared, when spoken of together, are called a *couplet;* when spoken of separately, they are called the *terms* of the ratio.

The *first* term is the *antecedent;* and the *last,* the *consequent.*

QUEST.—472. In how many ways are numbers or quantities compared? 474. What is ratio? 475. What are the two given numbers called when spoken of together? When spoken of separately?

476. Ratio is expressed in two ways:

First, in the form of a fraction, making the *antecedent* the *numerator,* and the *consequent* the *denominator.* Thus, the ratio of 8 to 4 is written $\frac{8}{4}$; the ratio of 12 to 3, $\frac{12}{3}$, &c.

Second, by placing two points or a colon (:) between the numbers compared. Thus, the ratio of 8 to 4 is written 8 : 4; the ratio of 12 to 3, is 12 : 3, &c. The expressions $\frac{8}{4}$, and 8 : 4, are of the same import, and one may be exchanged for the other, at pleasure.

OBS. 1. The sign (:) used to denote *ratio,* is derived from the sign of division, (÷) the horizontal line being omitted. The English mathematicians put the antecedent for the numerator, and the consequent for the denominator as above; but the French put the consequent for the numerator and the antecedent for the denominator. The English method appears to be equally simple, while it is confessedly the most in accordance with reason.

2. In order that *concrete* numbers may have a *ratio* to each other, they must necessarily express objects so far of the same nature, that one can be properly said to be *equal* to, or *greater,* or *less* than the other. (Art. 294.) Thus a foot has a ratio to a yard; for one is *three times* as long as the other; but a foot has not properly a ratio to an hour, for one cannot be said to be *longer* or *shorter* than the other.

477. A *direct* ratio is that which arises from dividing the antecedent by the consequent; as 6÷2. (Art. 474.)

478. An *inverse,* or *reciprocal* ratio, is the ratio of the *reciprocals* of two numbers. (Art. 160. Def. 10.) Thus, the direct ratio of 9 to 3, is 9 : 3, or $\frac{9}{3}$; the reciprocal ratio is $\frac{1}{9} : \frac{1}{3}$, or $\frac{1}{9} \div \frac{1}{3} = \frac{3}{9}$; (Art. 229;) that is, the consequent 3, is divided by the antecedent 9.

Note.—The term *inverse,* signifies *inverted.* Hence,

An inverse, or reciprocal ratio is expressed by inverting the fraction which expresses the direct ratio; or when the notation is by points, by inverting the order of the terms. Thus, 8 is to 4, inversely, as 4 to 8.

QUEST.—476. In how many ways is ratio expressed? The first? The second? *Obs.* Which of the terms do the English mathematicians put for the numerator? Which do the French? In order that concrete numbers may have a ratio to each other, what kind of objects must they express? 477. What is a direct ratio? 478. What is an inverse or reciprocal ratio? How is a reciprocal ratio expressed by a fraction? How by points?

479. A *simple* ratio is a ratio which has but *one antecedent* and *one consequent*, and may be either direct or inverse; as 9 : 3, or $\frac{1}{9}:\frac{1}{3}$.

480. A *compound* ratio is the ratio of the *products* of the corresponding terms of two or more simple ratios. Thus,

The simple ratio of	9 : 3 is 3;
And " " of	8 : 4 is 2;
The ratio compounded of these is	72 : 12=6.

OBS. 1. A compound ratio is of the *same nature* as any other ratio. The term is used to denote the *origin* of the ratio in particular cases.

2. The compound ratio is equal to the *product* of the simple ratios.

Ex. 1. What is the ratio of 27 to 9? *Ans.* 3.

2. What is the ratio of 8 to 32? *Ans.* $\frac{1}{4}$.

Required the ratio of the following numbers:

3. 14 to 7.	13. 324 to 81.	23. 63 lbs. to 9 oz.
4. 36 to 9.	14. 802 to 99.	24. 68 yds. to 17 yds.
5. 54 to 6.	15. 9 to 45.	25. 40 yds. to 20 ft.
6. 108 to 18.	16. 17 to 68.	26. 60 miles to 4 fur.
7. 144 to 24.	17. 13 to 52.	27. 45 bu. to 3 pks.
8. 136 to 17.	18. 27 to 135.	28. 6 gals. to 1 hhd.
9. 261 to 29.	19. 53 to 212.	29. 3 qts. to 20 gal.
10. 567 to 63.	20. 47 to 329.	30. £1 to 15s.
11. 405 to 45.	21. 18 lbs. to 6 lbs.	31. 15s. to £3.
12. 576 to 64.	22. 28 lbs. to 4 lbs.	32. £10 to 10d.

481. From the definition of ratio and the mode of expressing it in the form of a fraction, it is obvious that the *ratio* of two numbers is the same as the *value* of a fraction whose numerator and denominator are respectively equal to the antecedent and consequent of the given couplet; for, *each* is the *quotient* of the numerator divided by the denominator. (Arts. 474, 185.)

OBS. From the principles of fractions already established, we may, therefore, deduce the following truths respecting ratios.

QUEST.—479. What is a simple ratio? 480. What is a compound ratio? *Obs.* Does it differ in its nature from other ratios? What is the term used to denote?

482. *To multiply the antecedent of a couplet by any number multiplies the ratio by that number; and to divide the antecedent, divides the ratio:* for, multiplying the numerator, multiplies the value of the fraction by that number, and dividing the numerator, divides the value. (Arts. 186, 187.)

Thus, the ratio of 16 : 4 is 4;
The ratio of 16×2 : 4 is 8, which equals 4×2;
And " 16÷2 : 4 is 2, which equals 4÷2.

OBS. With a given consequent the greater the *antecedent*, the greater the *ratio;* and on the other hand, the greater the ratio, the greater the antecedent. (Art. 187. Obs.)

483. *To multiply the consequent of a couplet by any number, divides the ratio by that number; and to divide the consequent, multiplies the ratio:* for, multiplying the denominator, divides the value of the fraction by that number, and dividing the denominator, multiplies the value. (Arts. 188, 189.)

Thus, the ratio of 16 : 4 is 4;
The " 16 : 4×2 is 2, which equals 4÷2;
And " 16 : 4÷2 is 8, which equals 4×2.

OBS. With a given antecedent, the greater the *consequent*, the less the *ratio*; and the greater the ratio, the less the consequent. (Art. 189. Obs.)

484. *To multiply or divide both the antecedent and consequent of a couplet by the same number, does not alter the ratio:* for, multiplying or dividing both the numerator and denominator by the same number, does not alter the value of the fraction. (Art. 191.)

Thus, the ratio of 12 : 4 is 3;
The " 12×2 : 4×2 is 3;
And " 12÷2 : 4÷2 is 3.

485. If the two numbers compared are *equal*, the *ratio* is a *unit* or 1, and is called a ratio of *equality*. Thus, the ratio of 6×2 : 12 is 1; for the value of $\frac{12}{12}=1$. (Art. 196.)

QUEST.—482. What is the effect of multiplying the antecedent of a couplet by any number? Of dividing the antecedent? 483. What is the effect of multiplying the consequent by any number? Of dividing the consequent? Why? 484. What is the effect of multiplying or dividing both the antecedent and consequent by the same number? Why? 485. When the two numbers compared are equal, what is the ratio? What is it called?

486. If the antecedent of a couplet is *greater* than the consequent, the ratio is *greater* than a unit, and is called a ratio of *greater inequality*. Thus, the ratio of 12 : 4 is 3; for the value of $\frac{12}{4}=3$. (Art. 196.)

487. If the antecedent is *less* than the consequent, the ratio is *less* than a unit, and is called a ratio of *less inequality*. Thus, the ratio of 3 : 6 is $\frac{3}{6}$, or $\frac{1}{2}$; for $\frac{3}{6}=\frac{1}{2}$. (Art. 195.)

OBS. 1. The *direct* ratio of two fractions which have a common *numerator*, is the same as the reciprocal ratio of their denominators. Thus, the ratio of $\frac{2}{4}:\frac{2}{8}$ is the same as $\frac{1}{4}:\frac{1}{8}$, or 8 : 4.

2. The ratio of two *fractions* which have a common denominator, is the same as the ratio of their *numerators*. Thus, the ratio of $\frac{8}{2}:\frac{4}{2}$ is the same as that of 8 : 4, viz: 2. Hence,

488. The ratio of any two fractions may be expressed in whole numbers, by reducing them to a common denominator, and then using the numerators for the terms of the ratio. (Art. 484.) Thus, the ratio of $\frac{1}{2}$ to $\frac{1}{6}$ is the same as $\frac{6}{12}:\frac{2}{12}$, or 6 : 2.

33. What is the direct ratio of 4 to 12, expressed in the lowest terms? *Ans.* $\frac{1}{3}$.

34. What is the inverse ratio of 4 to 12? *Ans.* $\frac{1}{4}\div\frac{1}{12}=3$

35. What is the direct ratio of 64 to 8? Of 9 to 63?

36. What is the direct ratio of 84 to 21? Of 256 to 32?

37. What is the inverse ratio of 4 to 16? Of 28 to 7?

38. What is the inverse ratio of 42 to 6? Of 8 to 72?

39. Which is the greater, the ratio of 63 to 9, or that of 72 to 8?

40. Which is the greater, the ratio of 86 to 240, or that of 45 to 72?

41. Which is the greater, the ratio of 120 to 85, or that of 240 to 170?

42. Which is the greater, the ratio of 624 to 416, or that of 936 to 560?

43. Is the ratio of 5×6 to 24, a ratio of greater, or less inequality?

QUEST.—486. When the antecedent is greater than the consequent, what is the ratio called? 487. If the antecedent is less than the consequent, what is the ratio called? 488. How may the ratio of two fractions be expressed in whole numbers?

44. Is the ratio of 6×9 to 7×8, a ratio of greater, or less in equality?

45. Is the ratio of $2\times4\times16$ to 4×32 a ratio of greater, or less inequality?

46. What is the ratio compounded of the ratios of 5 to 3, and 12 to 4?

47. What is the ratio compounded of 8 : 10, and 20 : 16?

48. What is the ratio compounded of 3 : 8, and 10 : 5?

49. What is the ratio compounded of 18 : 20, and 30 : 40?

50. What is the ratio compounded of 35 : 40, and 60 : 75, and 21 to 19?

51. What is the ratio compounded of 60 : 40, and 12 : 24, and 25 : 30?

489. In a series of ratios, if the consequent of each preceding couplet is the antecedent of the following one, the ratio of the *first antecedent* to the *last consequent*, is equal to that compounded of all the *intervening ratios.*

Thus, in the series of ratios 3 : 4
4 : 7
7 : 16

the ratio of 3 to 16, is equal to that which is compounded of the ratios of 3 : 4, of 4 : 7, and 7 : 16; for, the compound ratio is $\frac{3\times4\times7}{4\times7\times16}=\frac{3}{16}$, or 3 : 16.

490. *If to or from the terms of any couplet, two other numbers having the same ratio be added or subtracted, the sums or remainders will also have the same ratio.* (Thomson's Legendre, B. III., Prop. 1, 2.) Thus, the ratio of 12 : 3 is the same as that of 20 : 5. And the *ratio* of the *sum* of the *antecedents* $12+20$ to the *sum* of the *consequents* $3+5$, is the same as the ratio of either couplet. That is,

$12+20:3+5::12:3=20:5$, or $\frac{12+20}{3+5}=\frac{12}{3}=\frac{20}{5}=4$.

So also the *ratio* of the *difference* of the *antecedents*, to the *difference* of the *consequents*, is the same. That is,

$20-12:5-3::12:3=20:5$, or $\frac{20-12}{5-3}=\frac{12}{3}=\frac{20}{5}=4$.

491. If in several couplets the ratios are equal, *the sum of all the antecedents has the same ratio to the sum of all the consequents, which any one of the antecedents has to its consequent.*

$$\text{Thus, the ratio of} \begin{cases} 12:4=3 \\ 15:5=3 \\ 18:6=3 \end{cases}$$

Therefore the ratio of $(12+15+18):(4+5+6)=3$.

OBS. 1. A ratio of *greater inequality* is *diminished* by adding the *same number* to both terms. Thus, the ratio of $8:2$, is 4; and the ratio of $8+4:2+4$ is 2.

2. A ratio of *less inequality* is *increased* by adding the *same number* to both the terms. Thus, the ratio of $2:8$ is $\frac{1}{4}$, and the ratio of $2+16:8+16$ is $\frac{3}{4}$.

PROPORTION.

492. PROPORTION is an equality of ratios. Thus, the two ratios $6:3$ and $4:2$ form a proportion; for $\frac{6}{3}=\frac{4}{2}$, the ratio of each being 2.

OBS. The terms of the two couplets, that is, the numbers of which the proportion is composed, are called *proportionals.*

493. *Proportion* may be expressed in two ways.

First, by the sign of equality (=) placed between the two ratios.

Second, by four points (: :) placed between the two ratios.

Thus, each of the expressions, $12:6=4:2$, and $12:6::4:2$, is a proportion, one being equivalent to the other. The latter expression is read, "the ratio of 12 to 6 equals the ratio of 4 to 2," or simply, "12 is to 6 as 4 is to 2."

OBS. The sign (: :) is said to be derived from the sign of equality, the *four points* being merely the *extremities* of the lines.

494. The number of *terms* in a proportion must at least be *four,* for the equality is between the ratios of *two couplets,* and each couplet must have an antecedent and a consequent. (Art. 476.)

There may, however, be a proportion formed from *three numbers,* for one of the numbers may be repeated so as to form *two*

QUEST.—492. What is Proportion? 493. How many ways is proportion expressed? What is the first? The second? 494. How many terms must there be in a proportion? Why? Can a proportion be formed of three numbers? How?

terms. Thus, the numbers 8, 4, and 2, are proportional; for the ratio of 8 : 4=4 : 2. It will be seen that 4 is the consequent in the first couplet, and the antecedent in the last. It is therefore a *mean proportional* between 8 and 2.

OBS. 1. In this case, the number repeated is called the *middle term* or *mean proportional* between the other two numbers.

The *last* term is called a *third* proportional to the other two numbers. Thus 2 is a third proportional to 8 and 4.

2. Care must be taken not to confound *proportion* with *ratio*. (Arts. 474, 492.) In a simple ratio there are but *two* terms, an antecedent and a consequent; whereas in a proportion there must at least be *four* terms, or *two couplets.*

Again, one *ratio* may be *greater* or *less* than another; the ratio of 9 to 3 is greater than the ratio of 8 to 4, and less than that of 18 to 2. *One proportion,* on the other hand, cannot be *greater* or *less* than another; for *equality* does not admit of degrees.

495. The *first* and *last* terms of a proportion are called the *extremes;* the other two, the *means.*

OBS. *Homologous* terms are either the two antecedents, or the two consequents. *Analogous* terms are the antecedent and consequent of the same couplet.

496. *Direct* proportion is an equality between two *direct* ratios. Thus, 12 : 4 : : 9 : 3 is a direct proportion.

OBS. In a direct proportion, the first term has the same ratio to the second, as the third has to the fourth.

497. *Inverse* or *reciprocal* proportion is an equality between a *direct* and a *reciprocal* ratio. Thus, 8 : 4 : : $\frac{1}{3}$: $\frac{1}{6}$; or 8 is to 4, reciprocally, as 3 is to 6.

OBS. In a reciprocal or inverse proportion, the first term has the same ratio to the second, as the fourth has to the third.

498. *If four numbers are proportional, the product of the extremes is equal to the product of the means.* Thus, 8 : 4 : : 6 : 3 is a proportion; for $\frac{8}{4}=\frac{6}{3}$, (Art. 492,) and $8\times3=4\times6$.

QUEST.—*Obs.* What is the number repeated called? What is the last term called in such a case? What is the difference between proportion and ratio? 495. Which terms are the extremes? Which the means? *Obs.* What are homologous terms? Analogous terms? 496. What is direct proportion? *Obs.* In direct proportion what ratio has the first term to the second? 497. What is inverse proportion? *Obs.* What ratio has the first term to the second in this case? 498. If four numbers are proportional, what is the product of the extremes equal to?

Again, $12 : 6 :: \frac{1}{3} : \frac{1}{6}$ is a proportion. (Art. 496.)
And $12 \times \frac{1}{6} = 6 \times \frac{1}{3}$.

OBS. 1. The truth of this proposition may also be illustrated in the following manner:

The numbers $2 : 3 :: 6 : 9$ are obviously proportional. (Art. 492.)
For, $\frac{2}{3} = \frac{6}{9}$. (Art. 195.) Now,
Multiplying each ratio by 27, (the product of the denominators,)
The proportion becomes $\frac{2 \times 27}{3} = \frac{6 \times 27}{9}$. (Art. 21. Ax. 6.)

Dividing both the numerator and the denominator of the first couplet by 3, (Art. 191,) or canceling the denominator 3 and the same factor in 27, (Art. 221,) also canceling the 9, and the same factor in 27, we have $2 \times 9 = 6 \times 3$. But 2 and 9 are the extremes of the given proportion, and 3 and 6 are the means; hence, the product of the extremes is equal to the product of the means.

2. Conversely, if the product of the extremes is equal to the product of the means, the four numbers are proportional; and if the products are not equal, the numbers are not proportional.

499. *Proportion,* in arithmetic, is usually divided into *Simple* and *Compound.*

SIMPLE PROPORTION.

500. SIMPLE PROPORTION is an equality between two *simple* ratios. It may be either *direct* or *inverse.* (Arts. 479, 496, 497.)

The most important application of simple proportion is the *solution* of that class of examples in which *three terms are given to find a fourth.*

501. We have seen that, if four numbers are in proportion, the product of the extremes is equal to the product of the means. (Art. 498.) Hence,

If the product of the means is divided by one of the extremes, the quotient will be the other extreme; and if the product of the extremes is divided by one of the means, the quotient will be the

---

QUEST.—*Obs.* If the product of the extremes is equal to the product of the means, what is true of the four numbers? If the products are not equal, what is true of them? 499. How is proportion usually divided? 500. What is simple proportion? What is the most important application of it? 501. If the product of the means is divided by one of the extremes, what will the quotient be? If the product of the extremes is divided by one of the means, what will the quotient be?

other mean. For, if the product of two factors is divided by one of them, the quotient will be the other factor. (Art. 156.)

Take the proportion 8 : 4 : : 6 : 3.

Now the product $8 \times 3 \div 4 = 6$, one of the means;
So the product $8 \times 3 \div 6 = 4$, the other mean.
Again, the product $4 \times 6 \div 8 = 3$, one of the extremes;
And the product $4 \times 6 \div 3 = 8$, the other extreme.

502. *If, therefore, any three terms of a proportion are given, the fourth may be found by dividing the product of two of them by the other term.*

OBS. Simple Proportion is often called the *Rule of Three*, from the circumstance that three terms are given to find a fourth. In the older arithmetics, it is also called the *Golden Rule*. But the fact that these names convey no idea of the nature or object of the rule, seems to be a strong objection to their use, not to say a sufficient reason for discarding them.

Ex. 1. If the product of the means is 84, and one of the extremes is 7, what is the other extreme, or term of the proportion?

2. If the product of the means is 54, and one of the extremes is 18, what is the other extreme?

3. If the product of the means is 720, and one of the extremes is 45, what is the other extreme?

4. If the product of the means is 639, and one of the extremes is 213, what is the other extreme?

5. If the first three terms of a proportion are 8, 12, and 16, what is the fourth term?

Solution.—$12 \times 16 = 192$, and $192 \div 8 = 24$, the fourth term, or number required; that is, 8 : 12 : : 16 : 24.

6. It is required to find the fourth term of the proportion, the first three terms of which are 36, 30, and 24.

7. Required the fourth term of the proportion, the first three terms of which are 15, 27, and 31.

8. Required the fourth term of the proportion whose first three terms are 45, 60, and 90.

QUEST.—*Obs.* What is simple proportion often called? Do these terms convey an idea of the nature or object of the rule?

9. If 8 yds. of broadcloth cost $96, how much will 20 yds. cost at the same rate?

Solution.—It is plain that 8 yds. has the same ratio to 20 yds. as the cost of 8 yds., viz: $96, has to the cost of 20 yds. That is,

8 yds. : 20 yds. : : $96 : to the cost of 20 yds.

Now $96×20=$1920; and $1920÷8=$240. *Ans.*

10. If 35 men will consume a certain quantity of flour in 20 days, how long will it take 50 men to consume it?

Note.—Since the answer is days, we put the given days for the third term. Then, as the flour will not last 50 men so long as it will 35 men, we put the smaller number of men for the second term, and the larger for the first.

Operation.

```
Men.  Men.   Days.
50 : 35 : : 20 : to the number of days required.
     20
50 ) 700
     14 days. Ans.
```

Multiply the second and third terms together, and divide the product by the first term, as in the last example.

PROOF.—50×14=35×20. (Art. 498.)

503. From the preceding illustrations and principles, we deduce the following general

RULE FOR SIMPLE PROPORTION.

I. *Place that number for the third term, which is of the same kind as the answer or number required.*

II. *Then, if by the nature of the question the answer must be greater than the third term, place the greater of the other two numbers for the second term; but if it is to be less, place the less of the other two numbers for the second term, and the other for the first.*

III. *Finally, multiply the second and third terms together, divide the product by the first, and the quotient will be the answer in the same denomination as the third term.*

PROOF.—*Multiply the first term and the answer together, and if the product is equal to the product of the second and third terms, the work is right.* (Art. 498.)

QUEST.—503. In arranging the terms in simple proportion, which number is put for the third term? How arrange the other two numbers? Having stated the question how is the answer found? Of what denomination is the answer? How is simple proportion proved

Demonstration.—If four numbers are proportional, we have seen that the product of the *means* is equal to the product of the *extremes*; (Art. 498;) therefore the product of the *second* and *third* terms must be equal to that of the first and fourth. But if the product of two factors is divided by one of them, the quotient will be the other; (Art. 156;) consequently, when the first three terms of a proportion are given, the product of the *second* and *third* terms divided by the *first*, must give the *fourth term* or *answer*.

The object of placing that number, which is of the same kind as the answer, for the *third* term, instead of the *second*, as is sometimes done, is twofold: 1st, it avoids the necessity of the *Rule of Three Inverse*; 2d, the third term, in many cases, has no *ratio* to the first; consequently it is inconsistent with the principles of proportion to put it for the second term. Thus, in the ninth example, if we put $96 for the second term, it would read, 8 yds. : $96 : : 20 yds. : $240, the answer. But a *yard* can have no ratio to a *dollar*; for one cannot be said to be *greater* nor *less* than the other. (Art. 476. Obs. 2.)

OBS. 1. If the first and second terms are compound numbers, reduce them to the lowest denomination mentioned in either, before the multiplication or division is performed.

When the third term contains different denominations, it must also be reduced to the lowest denomination mentioned in it.

2. The process of arranging the terms of a question for solution, or putting it into the form of a proportion, is called *stating the question*.

3. Questions in Simple Proportion, we have seen, may be solved by *Analysis*. After solving the following examples by proportion, it will be an excellent exercise for the student to solve them by analysis. (Art. 462. Obs. 2.)

11. If 16 barrels of flour cost $112, what will 129 barrels cost?

12. If 40 acres of land cost $540, what will 97 acres cost?

13. If 641 sheep cost $1923, what will 75 sheep cost?

14. At the rate of 155 miles in 12 days, how far can a man travel in 60 days?

15. How much hay, at $17.50 per ton, can you buy for $350?

16. If $45 buy 63 lbs. of tea, how much will $1540 buy?

17. If 90 lbs. of pepper are worth 72 lbs. of ginger, how many lbs. of ginger are 64 lbs. of pepper worth?

18. A bankrupt compromised with his creditors, at 64 cts. on a dollar; how much will be received on a debt of $2563.50?

19. An emigrant has a draft for £1460 sterling: how much is it worth, allowing $4.84 to a pound?

QUEST.—*Obs.* If the first and second terms contain different denominations, how proceed? When the third term contains different denominations, what is to be done? What is meant by stating a question?

SIMPLE PROPORTION BY CANCELATION.

20. If 72 tons of coal cost \$648, how much will 9 tons cost?

Operation.

Tons. Tons. Dolls.
72 : 9 :: 648 : Ans.
8 : 1

Now \$648÷8=\$81. *Ans.*

Having stated the question as before, we perceive the factor 9 is common to the first two terms, and therefore may be canceled. (Art. 151.)

Or thus, $\frac{9\times 648}{72}$=the answer. (Art. 503.)

But $\frac{9\times 648}{72}=\frac{9\times 648}{72,8}$=\$81, the same as before. Hence,

504. When the first term has factors common to either of the other two terms.

Cancel the factors which are common, then proceed according to the rule above. (Arts. 151, 221.)

PROOF.—*Place the answer in the denominator, or on the left of the perpendicular line, as the case may be, and if the factors of the divisor exactly cancel those of the dividend, the work is right.*

OBS. 1. The question should be *stated*, before attempting to *cancel* the common factors. When the terms are of different denominations, the reduction of them may sometimes be shortened by *Cancelation*.

2. Instead of points, it may sometimes be more convenient to place a perpendicular line between the first and second terms, as in division of fractions. (Art. 231.) In this case the third term should be placed under the second, with the sign of proportion (::) before it to denote its origin, and its relation to the fourth term or the answer.

3. It will be perceived that cancelation is applicable in Simple Proportion to all those examples, whose *first* term has one or more factors *common* to it and either of the *other* terms.

21. If 24 yds. of cloth cost \$63, what will 320 yds. cost?

Operation.

24 yds.	320 yds., 40
3	:: \$63, 21
Ans.	\$21×40=\$840.

When arranged in this way, the question is read, 24 yds. is to 320 yds., as \$63 is to the answer required.

22. If 20 bu. of oats cost £1, how much will 2 quarts cost?

23. If 12 bbls. of flour cost \$88, what will 108 barrels cost?

24. If 30 cows cost \$480, what will 173 cows cost?

25. If a man can travel 240 miles in 16 days, how far can he travel in 29 days?

26. If 48 men can build a ship in 84 days, how long would it take 16 men to build it?

27. If $\frac{1}{4}$ of a ton of hay costs £$\frac{4}{5}$, what will $\frac{5}{8}$ of a ton cost?

Solution.—$\overset{\text{ton.}}{\frac{1}{4}} : \overset{\text{ton.}}{\frac{5}{8}} :: \overset{\text{£}}{\frac{4}{5}}$: Ans. Now, $\frac{4}{1}\times\frac{5}{8}\times\frac{4}{5}=$ £2. *Ans.* Hence,

505. If the terms in a proportion are *fractional*, the question is stated, and the answer obtained in the same manner as if they were whole numbers.

OBS. When the first and second terms are fractions, we may reduce them to a common denominator, and then employ the numerators only; for the ratio of two fractions which have a common denominator, is the *same* as the ratio of their numerators. (Art. 487. Obs. 2.)

28. If $\frac{3}{7}$ of a cord of wood cost \$1.35, what will $\frac{6}{7}$ of a cord cost?

29. If $\frac{7}{9}$ of a yard of berége cost 6 shillings, what will $\frac{3}{7}$ of a yard cost?

30. If $\frac{6}{7}$ of a yard of sarcenet cost $\frac{3}{4}$ of a dollar, what will $3\frac{3}{5}$ yds. cost?

31. If $\frac{3}{5}$ of a pound of chocolate cost $\frac{1}{5}$ of a dollar, what will $25\frac{2}{3}$ pounds cost?

32. What will 165 melons cost, at $\frac{4}{5}$ of a dollar for 5 melons?

33. A man had 420 acres of land which he wished to divide among his three sons A, B, and C, in proportion to the numbers 7, 5, and 3: how much land would each receive?

Solution.—Since the several parts are to be proportional to the numbers 7, 5, and 3, the sum of which is 15, it is evident that the sum of all the given numbers is to any one of them, as the whole quantity to be divided to the part corresponding to the number used as the second term.

That is 15 : 7 :: 420A. to A's share, which is 196 acres;
Also 15 : 5 :: 420A. to B's " " " 140 acres;
And 15 : 3 :: 420A. to C's " " " 84 acres.

PROOF.—196+140+84=420A. the given number. (Ax. 11.)

506. Hence, to divide a given number or quantity into parts which shall be *proportional* to any given numbers.

Place the whole number or quantity to be divided for the third term, the sum of the given numbers for the first term, and each of the given numbers respectively for the second; then multiply and divide as before. (Art. 503.)

34. A farmer wishes to mix 100 bushels of provender of oats and corn in the ratio of 3 to 7: how many bushels of each must he put in?

35. Bell metal is composed of 3 parts of copper, and 1 of tin: how much of each ingredient will be used in making a bell which weighs 2567 pounds?

36. Gunpowder is composed of 76 parts of nitre, 14 of charcoal, and 10 of sulphur: how much of each of these ingredients will it take to make a ton of powder?

37. If 40.12 lbs. of sugar are worth $5.13, how much can be bought for $125.375?

38. The Vice-President's salary is $5000 a year: if his daily expenses are $10, how much can he lay up?

39. If $\frac{5}{7}$ lb. of snuff cost £$\frac{3}{14}$, what will 150 lbs. cost?

40. If $\frac{2}{3}$ of $\frac{3}{4}$ of $\frac{8}{9}$ of a sloop cost $1500, what will the whole cost?

41. If $\frac{1}{2}$ of $\frac{4}{5}$ of an acre of land on Broadway is worth $8200, how much is $\frac{1}{4}$ of $\frac{5}{8}$ of an acre worth?

42. A man bought $\frac{5}{8}$ of a vessel and sold $\frac{4}{5}$ of what he bought for $8240, which was just the cost of it: what was the whole vessel worth?

43. How many times will the fore wheel of a carriage which is 7 ft. 6 in. in circumference turn round in going 100 miles?

44. How many times will the hind wheel of a carriage 9 ft. 2 in. in circumference, turn round in going the same distance?

45. There are two numbers which are to each other as 15 to 34, the smaller of which is 75: what is the larger?

46. What two numbers are those which are to each other as 5 to 6, the greater of which is 240?

47. If two numbers are as 8 to 12, and the less is 320, what is the greater?

48. There are two flocks of sheep which are to each other as 15 to 20, and the greater contains 500 : how many does the less contain?

49. An express traveling 60 miles a day had been dispatched 5 days, when a second was sent after him traveling 75 miles a day: how long will it take the latter to overtake the former?

50. A fox has 150 rods the start of a hound, but the hound runs 8 rods while the fox runs 5 rods: how far must the hound run before he catches the fox?

51. A stack of hay will keep a cow 20 weeks, and a horse 15 weeks: how long will it keep them both?

52. A traveler divided 80s. among 4 beggars in such a manner, that as often as the first received 10s., the second received 9s., the third 8s., and the fourth 7s.: what did each receive?

53. Pure water is composed of oxygen and hydrogen in the ratio of 8 to 1 by weight: what is the weight of each in a cubic foot of water, or 1000 ounces avoirdupois?

COMPOUND PROPORTION.

507. COMPOUND PROPORTION is an equality between a *compound* ratio and a *simple* one. (Arts. 479, 480.)

Thus, $\left.\begin{matrix}6 : 3\\ 4 : 2\end{matrix}\right\} :: 12 : 3$, is a compound proportion.
Into

That is, $6\times4 : 3\times2 :: 12 : 3$; for, $6\times4\times3=3\times2\times12$.

OBS. Compound proportion is chiefly applied to the solution of examples which would require *two or more statements* in simple proportion. It is sometimes called *Double Rule of Three.*

Ex. 1. If 8 men can reap 32 acres in 6 days, how many acres can 12 men reap in 15 days?

Suggestion.—When stated in the form of a *compound proportion,* the question will stand thus:

$$\left.\begin{matrix}8m. : 12m.\\ 6d. : 15d.\end{matrix}\right\} :: 32 A. : \text{to the answer.}$$

That is, the product of the antecedents 8×6, has the same

QUEST.—507. What is compound proportion? *Obs.* To what is it chiefly applied? What is it sometimes called?

ratio to the product of the consequents 12×15, as 32 has to the answer; or simply, 8 into 6 : 12 into 15 :: 32 : to the answer.

Operation.	
32×12×15=5760,	The product of the numbers standing in the 2d and 3d places divided by the product of those standing in the first place, will give the answer.
And 8× 6=48.	
Now 5760÷48=120.	
Ans. 120 acres.	

Note.—The learner will observe that it is not the ratio of 8 to 12 alone, nor that of 6 to 15, which is equal to the ratio of 32 to the answer, as it is sometimes stated; but it is the ratio *compounded* of 8 to 12, and 6 to 15, which is equal to the ratio of 32 to the answer. Thus, 8×6 : 12×15 :: 32 : 120, the answer. A compound proportion when stated as above, is read, "the ratio of 8×6 is to 12×15 as 32 is to the answer."

2. If 6 men can earn £42 in 60 days working 8 hours per day, how much can 10 men earn in 84 days working 12 hours a day?

Operation.

6m. : 10m.
60d. : 84d. } :: £42 : to Ans.
8hrs. : 12hrs.

State the question, then multiply and divide, as before.

10×84×12×42=423360; and 6×60×8=2880.
Now 423360÷2880=147. *Ans.* £147.

508. From the foregoing illustrations we derive the following general

RULE FOR COMPOUND PROPORTION.

I. *Place that number which is of the same kind as the answer required for the third term.*

II. *Then take the other numbers in pairs, or two of a kind, and arrange them as in simple proportion.* (Art. 503.)

III. *Finally, multiply together all the second and third terms, divide the result by the product of the first terms, and the quotient will be the fourth term or answer required.*

QUEST.—508. In stating a question in compound proportion, which number do you put for the third term? How arrange the other numbers? Having stated the question, how is the answer found?

PROOF.—*Multiply the answer into all of the first terms or antecedents of the first couplets, and if the product is equal to the continued product of all the second and third terms, the work is right.* (Art. 498.)

OBS. 1. Among the given numbers there is but one which is of the same kind as the answer. This is sometimes called the *odd* term, and is always to be placed for the *third* term.

2. If the *antecedent* and *consequent* of any couplet are compound numbers, they must be reduced to the lowest denomination mentioned in either, before the multiplication is performed. When the *third* term contains different denominations, it must also be reduced to the lowest mentioned in it.

3. Questions in Compound Proportion may be solved by *Analysis;* also by *Simple Proportion*, by making *two* or *more* separate statements.

3. If 12 horses can plough 11 acres in 5 days, how many horses can plough 33 acres in 18 days?

4. If a man walking 12 hours a day, can travel 250 miles in 10 days, how long will it take him to travel 400 miles, if he walks but 10 hours a day?

5. If 40 gallons of water will last 20 persons 5 days, how many gallons will 9 persons drink in a year?

6. If 16 laborers can earn £15, 12s. in 18 days, how many laborers will it take to earn £35, 2s. in 24 days?

COMPOUND PROPORTION BY CANCELATION.

7. If a person can make 60 rods of wall in 45 days, working 12 hours a day, how many rods can he make in 72 days, working 8 hours a day?

Statement.

$$\left.\begin{matrix}45\text{d.} & : & 72\text{d.} \\ 12\text{hrs.} & : & 8\text{hrs.}\end{matrix}\right\} :: \overset{\text{Rods.}}{60} : \text{to the answer.} \quad \text{That is,}$$

$$\frac{72\times8\times60}{45\times12}=\frac{\overset{\not{6},\,2}{\not{7}\not{2}}\times8\times\overset{4}{\not{6}\not{0}}}{\underset{\not{3}}{\not{4}\not{5}}\times\not{1}\not{2}}=64 \text{ rods. } \textit{Ans.} \quad \text{Hence,}$$

QUEST.—How are questions in compound proportion proved? *Obs.* Among the given numbers, how many are of the same kind as the answer? Can questions in compound proportion be solved in any other way?

509. When the first terms have factors common to the second or third terms.

Cancel the factors which are common, then divide the product of those remaining in the second and third terms by the product of those remaining in the first, and the quotient will be the answer.

PROOF.—*Place the answer in the denominator, or on the left of the perpendicular line, and if the factors of the divisor and dividend exactly cancel each other, the work is right.*

OBS. 1. Instead of placing points between the antecedents and consequents of the left hand couplets of the proportion, it is sometimes more convenient to put a perpendicular line between them, as in division of fractions. (Art. 232.) This will bring all the terms whose product is to be divided on the right of the line, and those whose product is to form the divisor, on the left. In this case the third term should be placed below the second terms, with the sign of proportion (: :) before it, to show its *origin*, and its *relation* to the answer.

2. It will be observed that *Cancelation* can be applied in Compound Proportion to all those examples whose *first* terms have factors common to the *second* terms, or to the *third* term.

8. If 24 men can saw 90 cords of wood in 6 days, when the days are 9 hours long, how many cords can 8 men saw in 36 days, when they are 12 hours long?

Operation.

3, 24m. (cancelled)	8m. (cancelled)
6d. (cancelled)	36d., 2 (36 cancelled)
9hrs. (cancelled)	12hrs.
	: : 90c., 10 (90 cancelled)
Ans.	2×12×10=240 cords.

9. If 6 men can make 120 pair of boots in 20 days, working 8 hours a day, how long will it take 12 men to make 360 pair, working 10 hours a day?

10. If 12 men can build a wall 30 ft. long, 6 ft. high, and 4 ft. thick, in 18 days, how long will it take 36 men to build one 360 ft. long, 8 ft. high, and 6 ft. thick.

11. If a horse can travel 120 miles in 4 days when the days are 8 hours long, how far can he travel in 30 days when the days are 10 hours long?

QUEST.—509. When the first terms have factors common to the second or third terms how proceed?

12. If $250 gain $30 in 2 years, what will be the interest of $750 for 5 years?

13. What will be the interest of $500 for 4 years, if $600 will gain $42 in 1 year?

14. If $360 gain $14.40 in 8 months, what will $4800 gain in 32 months?

15. If a family of 8 persons spend $200 in 9 months, how much will 18 persons spend in 12 months?

16. If 15 men, working 12 hours a day, can hoe 60 acres in 20 days, how long will it take 30 boys, working 10 hours a day, to hoe 96 acres, 6 men being equal to 10 boys?

CONJOINED PROPORTION.

510. When each *antecedent* of a compound ratio is *equal* in value to its *consequent*, the proportion is called *Conjoined Proportion.*

OBS. Conjoined Proportion is often called the *chain rule.* It is chiefly used in comparing the coins, weights and measures of two countries, through the medium of those of other countries, and in the higher operations of exchange. The *odd term* is sometimes called the *demand.*

17. If 20 lbs. United States make 12 lbs. in Spain; and 15 lbs. Spain 20 lbs. in Denmark; and 40 lbs. Denmark 60 lbs. in Russia: how many pounds in Russia are equal to 100 lbs. U. S.?

Operation.

20 lbs. U. S. = 12 lbs. Spain
15 lbs. Spain = 20 lbs. Den.
40 lbs. Den. = 60 lbs. Rus.
How many lbs. R. = 100 lbs. U. S.

Arrange the given terms in pairs, making the first term the antecedent, and its equal the consequent; then since it is required to find how many of the last kind are equal to a given number (100 lbs.) of the first, place the odd term or demand under the consequents.

Then, 20×15×40 : 12×20×60 : : 100 : Ans.

That is	
20	12
15	20
10, 40	60, 4
	: : 100, 10
Ans.	12×10=120 lbs.

Cancel the factors common to both sides, and the product of those remaining on the right divided by the product of those on the left, is the answer.

511. From these illustrations we derive the following

RULE FOR CONJOINED PROPORTION.

I. *Taking the terms in pairs, place the first term on the left of the sign of equality or a perpendicular line for the antecedent, and its equal on the right for the consequent, and so on. Then, if the answer is to be of the same kind as the first term, place the odd term under the antecedents; but if not, place it under the consequents.*

II. *Cancel the factors common to both sides, and if the odd term falls under the consequents, divide the product of the factors remaining on the right by the product of those on the left, and the quotient will be the answer; but if the odd term falls under the antecedents, divide the product of the factors remaining on the left by the product of those on the right, and the quotient will be the answer.*

PROOF.—*Reverse the operation, taking the consequents for the antecedents, and the answer for the odd term, and if the result thus obtained is the same as the odd term in the given question, the work is right.*

OBS. In arranging the terms, it should be observed that the *first antecedent* and the *last consequent* will always be of the *same kind.*

18. If 100 lbs. United States, make 95 lbs. Italian; and 19 lbs. Italian, 25 lbs. in Persia; how many pounds in the U. S. are equal to 50 lbs. in Persia? *Ans.* 40 lbs.

19. If 10 yds. at New York make 9 yds. at Athens; and 90 yds. at Athens, 112 yds. at Canton; how many yds. at Canton are equal to 50 yds. at New York?

20. If 50 yds. of cloth in Boston are worth 45 bbls. of flour in Philadelphia; and 90 bbls. of flour in Philadelphia 127 bales of cotton in New Orleans; how many bales of cotton at New Orleans are worth 100 yds. of cloth in Boston?

21. If $18 U. S. are worth 8 ducats at Frankfort; 12 ducats at Frankfort 9 pistoles at Geneva; and 50 pistoles at Geneva, 24 rupees at Bombay: how many rupees at Bombay are equal to $100 United States?

SECTION XV.

DUODECIMALS.

ART. **512.** DUODECIMALS are a species of compound numbers, the *denominations* of which increase and decrease uniformly in a *twelvefold ratio.* The denominations are *feet, inches* or *primes, seconds, thirds, fourths, fifths,* &c.

Note.—The term *duodecimal* is derived from the Latin numeral *duodecim,* which signifies *twelve.*

TABLE.

12 fourths ('''')	make	1 third,	marked	'''
12 thirds	"	1 second,	"	''
12 seconds	"	1 inch or prime,	"	*in.* or '
12 inches or primes	"	1 foot,	"	*ft.*

Hence $1' = \frac{1}{12}$ of 1 foot.

$1'' = \frac{1}{12}$ of 1 in., or $\frac{1}{12}$ of $\frac{1}{12}$ of 1 ft. $= \frac{1}{144}$ of 1 ft.

$1''' = \frac{1}{12}$ of $1''$, or $\frac{1}{12}$ of $\frac{1}{12}$ of $\frac{1}{12}$ of 1 ft. $= \frac{1}{1728}$ of 1 ft.

OBS. The accents used to distinguish the different denominations below feet, are called *Indices.*

513. Duodecimals may be added and subtracted in the same manner as the other compound numbers. (Arts. 300, 302.)

MULTIPLICATION OF DUODECIMALS.

514. Duodecimals are principally applied to the measurement of *surfaces* and *solids.* (Arts. 285, 286.)

Ex. 1. How many square feet are there in a board 12 ft. 7 in. long, and 4 ft. 3 in. wide?

QUEST.—512. What are duodecimals? What are the denominations? *Note.* What the meaning of the term duodecimal? Repeat the Table. *Obs.* What are the accents called, which are used to distinguish the different denominations? 513. How are duodecimals added and subtracted? 514. To what are duodecimals chiefly applied?

Operation.		
12 ft.	7′	
4 ft.	3′	
50 ft.	4′	
3 ft.	1′	9″
53 ft.	5′	9″

We first multiply each denomination of the multiplicand by the feet in the multiplier, beginning at the right hand. Thus, 4 times 7′ are 28′, equal to 2 ft. and 4′. Set the 4 under inches, and carry the 2 feet to the next product. 4 times 12 ft. are 48 ft. and 2 to carry make 50 ft. Again, since $3'=\frac{3}{12}$ of a ft. and $7'=\frac{7}{12}$ of a ft., 3′ into 7′ is $\frac{21}{144}$ of a ft.$=21''$, or 1′ and 9″. Write the 9″ one place to the right of inches, and carry the 1′ to the next product. Then 3′ or $\frac{3}{12}$ of a ft. multiplied into 12 ft.$=$ $\frac{36}{12}$ of a ft., or 36′, and 1′ to carry make 37′; but 37′=3 ft. and 1′. Now adding the partial products, the sum is 53 ft. 5′ 9″.

OBS. It will be seen from this operation, that feet multiplied into feet, produce feet; feet into inches, produce inches; inches into inches, produce seconds, &c. That is, the product of any two factors has as many accents as the factors themselves have. Hence,

515. To find the denomination of the product of any two factors in duodecimals.

Add the indices of the two factors together, and the sum will be the index of their product.

Thus, feet into feet, produce *feet;* feet into inches, produce *inches;* feet into seconds, produce *seconds;* feet into thirds, produce *thirds;* &c.

Inches into inches, produce *seconds;* inches into seconds, produce *thirds;* inches into fourths, produce *fifths,* &c.

Seconds into seconds, produce *fourths;* seconds into thirds, produce *fifths;* seconds into sixths, produce *eighths,* &c.

Thirds into thirds, produce *sixths;* thirds into fifths, produce *eighths;* thirds into sevenths, produce *tenths,* &c.

Fourths into fourths, produce *eighths;* fourths into eighths, produce *twelfths,* &c.

Note.—The foot is considered the *unit* and has no *index.*

QUEST.—515. How find the denomination of the product in duodecimals? What do feet into feet produce? Feet into inches? Feet into seconds? What do inches into inches produce? Inches into thirds? Inches into fourths? Seconds into seconds? Seconds into thirds? Seconds into eighths? Thirds into thirds? Thirds into sixths?

516. From these illustrations we derive the following

RULE FOR MULTIPLICATION OF DUODECIMALS.

I. *Place the several terms of the multiplier under the corresponding terms of the multiplicand.*

II. *Multiply each term of the multiplicand by each term of the multiplier separately, beginning with the lowest denomination in the multiplicand, and the highest in the multiplier, and write the first figure of each partial product one place to the right of that of the preceding product, under its corresponding denomination.* (Art. 515.)

III. *Finally, add the several partial products together, carrying 1 for every 12 both in multiplying and adding, and the sum will be the answer required.*

OBS. 1. It is sometimes asked whether the inches in duodecimals, are *linear*, *square*, or *cubic*. The answer is, they are *neither*. An *inch* is 1 *twelfth* of a foot. Hence, in measuring surfaces an inch is $\frac{1}{12}$ of a *square* foot; that is, a surface 1 foot *wide* and 1 inch *long*. In measuring solids, an inch denotes $\frac{1}{12}$ of a *cubic* foot. In measuring lumber, these inches are commonly called *carpenter's inches*.

2. Mechanics, also surveyors of wood and lumber, in taking dimensions of their work, lumber, &c., often call the inches a *fractional* part of a foot, and then find the contents in feet and a *fraction* of a foot. Sometimes inches are regarded as *decimals* of a foot.

3. We have seen that one of the factors in multiplication, is always to be considered an *abstract* number. (Art. 82. Obs. 2.) How then, can feet be multiplied by feet, inches by inches, &c.

It should be observed, that when one geometrical quantity is multiplied by another, some particular *extent* is to be considered the *unit*. It is immaterial what this extent is, provided it remains the same in the different parts of the same calculation. Thus, if one of the factors is *one foot* and the other *half* a foot, the former being 12 in., and the latter 6 in., the product is 72 in. Though it would be nonsense to say that a given length is repeated *as often as another is long*, yet there is no impropriety in saying that one is *repeated* as *many times* as there are feet or inches in another.

4. On the principles of duodecimals, it has been supposed that pounds, shillings, pence, and farthings can be multiplied by pounds, shillings, pence, and farthings. But it may be asked, *what denomination* shillings multiplied by pence, or pence by farthings, will produce? It is absurd to say that 2s. and 6d. *is repeated* 2s. and 6d. *times*.

QUEST.—516. What is the rule for multiplication of duodecimals? *Obs.* What kind of inches are those spoken of in measuring surfaces by duodecimals? In measuring solids

Ex. 2. How many square feet are there in a piece of marble 9 ft. 7 in. 2″ long, and 3 ft. 4 in. 7″ wide?

Note.—It is not absolutely necessary to begin to multiply by the highest denomination of the multiplier, or to place the lower denomination to the right of the multiplicand. The result will be the same if we begin with the lowest denomination of the multiplier, and place the first figure of each partial product under the figure by which we multiply.

Common Method.

	9 ft.	7′	2″		
	3 ft.	4′	7″		
	28 ft.	9′	6″		
	3 ft.	2′	4″	8‴	
		5′	7″	2‴	2⁗
Ans.	32 ft.	5′	5″	10‴	2⁗.

Second Method.

			9 ft.	7′	2″
			3 ft.	4′	7″
		5′	7″	2‴	2⁗
	3 ft.	2′	4″	8‴	
	28 ft.	9′	6″		
Ans.	32 ft.	5′	5″	10‴	2⁗.

3. How many square feet are there in a board 15 ft. 7 in. long, and 1 ft. 10 in. wide?

4. How many cubic feet in a stick of timber 15 ft. 3 in. long 2 ft. 4 in. wide, and 1 ft. 8 in. thick?

5. How many cubic feet in a block of granite 18 ft. 5 in. long, 4 ft. 2. in. wide, and 3 ft. 6 in. thick?

6. How many square feet in a stock of 10 boards, 15 ft. 8 in. long, and 1 ft. 6 in. wide?

7. How many square feet in a stock of 15 boards, 20 ft. 3 in. long, and 2 ft. 5 in. wide?

8. Multiply 16 ft. 3′ 4″ by 6 ft. 5′ 8″ 10‴.

9. Multiply 20 ft. 4′ 8″ 5‴ by 7 ft. 6′ 9″ 4″.

10. Multiply 18 ft. 0′ 5″ 10‴ by 4 ft. 8′ 7″ 9‴.

11. Multiply 50 ft. 6′ 0″ 2‴ 6⁗ by 3 ft. 10′ 5″.

12. How many cords in a pile of wood 50 ft. 6 in. long, 8 ft. 3 in. wide, and 7 ft. 4 in. high?

13. If a cistern is 30 ft. 10 in. long, 12 ft. 3 in. wide, and 10 ft. 2 in. deep, how many cubic feet will it contain?

14. What will it cost to plaster a room 20 ft. 6 in. long, 18 ft. wide, and 10 ft. high, at $12\frac{1}{2}$ cts. per square yard?

15. How many bricks 8 in. long, 4 in. wide, and 2 in. thick, will make a wall 50 ft. long, 10 ft. high, and 2 ft. 6 in. thick?

SECTION XVI.

EQUATION OF PAYMENTS.

ART. **517.** EQUATION OF PAYMENTS is the process of finding the *equalized* or *average time* when two or more payments due at different times, may be made *at once*, without loss to either party

OBS. The *equalized* or *average* time for the payment of several debts, due at different times, is often called the *mean* time.

518. From principles already explained, it is manifest, when the *rate* is fixed, the *interest* depends both upon the *principal* and the *time*. (Art. 404.) Thus, if a given principal produces a certain interest in a given time,

Double that principal will produce *twice* that interest;
Half that principal will produce *half* that interest; &c.
In *double* that time the same principal will produce *twice* that interest;
In *half* that time, *half* that interest; &c.

519. Hence, it is evident that any given principal will produce the same interest in any given time, as

One half that principal will produce in *double* that time;
One third that principal will " " *thrice* that time;
Twice that principal will " " *half* that time;
Thrice that principal will " " *a third* of that time; &c.

For example, at any given per cent.

The int. of $2 for 1 year, is the same as the int. of $1 for 2 yrs.;
The int. of $3 for 1 year, " " " $1 for 3 yrs.; &c.
The int. of $4 for 1 mo. " " " $1 for 4 mos.;
The int. of $5 for 1 mo. " " " $1 for 5 mos.; &c.

QUEST.—517. What is equation of payments? *Obs.* What is the average time for the payment of several debts sometimes called? 518. When the rate is fixed, upon what does the interest depend?

520. *The interest, therefore, of any given principal for* 1 *year or* 1 *month, &c., is the same, as the interest of* 1 *dollar for as many years, or months, as there are dollars in the given principal.*

Ex. 1. Suppose you owe a man \$15, and are to pay him \$5 in 10 months, and \$10 in 4 months, at what time may both payments be made without loss to either party ?

Analysis.—Since the interest of \$5 for 1 month is the same as the interest of \$1 for 5 months, (Art. 519,) the interest of \$5 for 10 months must be equal to the interest of \$1 for 10 times 5 months. And 5 mo.$\times$10$=$50 mo. In like manner the interest of \$10 for 4 months is equal to the interest of \$1 for 4 times 10 months; and 10 mo.$\times$4$=$40 months. Now 50 months added to 40 months make 90 months; that is, you are entitled to the use of \$1 for 90 months. But \$1 is $\frac{1}{15}$ of \$15, consequently you are entitled to the use of \$15, $\frac{1}{15}$ of 90 months, and 90$\div$15$=$6.

Ans. 6 months.

Proof.

The interest of \$5 at 6 per cent. for 10 mo. is	\$5 $\times$.05$=$	\$.25
The interest of \$10 " " " 4 mo. is	\$10$\times$.02$=$	.20
	Sum of both	\$.45

The interest of \$15 at 6 per cent. for 6 mo. is 15$\times$.03$=$\$.45.

521. From these principles we derive the following general

RULE FOR EQUATION OF PAYMENTS.

First multiply each debt by the time before it becomes due; then divide the sum of the products thus obtained by the sum of the debts, and the quotient will be the average time required.

OBS. 1. If one of the debts is *paid down*, its product will be *nothing*; but in finding the *sum* of the debts, this payment must be added with the others.

2. When there are months and days, the months must be reduced to days, or the days to the fractional part of a month.

3. This rule is based upon the supposition that *discount* and *interest paid in advance* are *equal*. But this is not exactly true; consequently, the rule, though in general use, is not strictly accurate. (Art. 432. Obs. 1.)

QUEST.—521. What is the rule for equation of payments ?

2. If you owe a man \$60, payable in 4 months, \$120 payable in 6 months, and \$180 payable in 3 months, at what time may you justly pay the whole at once?

Operation.

\$ 60×4 = \$240, the same as \$1 for 240 months. (Art. 520.)
\$120×6 = \$720, " " " \$1 for 720 "
\$180×3 = \$540, " " " \$1 for 540 "

\$360 debts. \$1500, sum of products.

Now 1500÷360=$4\frac{1}{6}$ months. *Ans.*

3. A merchant bought one lot of goods for \$1000 on 5 months; another for \$1000 on 4 months; another for \$1500 on 8 months: what is the average time of all the payments?

4. If a man has one debt of \$150, due in 3 months; another of \$200, due in $4\frac{1}{2}$ months; another of \$500, due in $7\frac{1}{2}$ months; what is the average time of the whole?

5. A man bought a house for \$3500, and agreed to pay \$500 down, and the balance in 6 equal annual instalments: at what time may he pay the whole?

6. If you owe one bill of \$175, due in 30 days; another of \$180 due in 60 days; another of \$120, due in 65 days, and another of \$200, due in 90 days: when may you pay the whole at once?

PARTNERSHIP.

522. PARTNERSHIP *is the associating of two or more individuals together for the transaction of business.* (Art. 464.) The persons thus associated are called *partners;* and the association itself, a *company* or *firm.*

The money employed is called the *capital* or *stock;* and the profit or loss to be shared among the partners, the *dividend.*

CASE I.—*When stock is employed an equal length of time.*

Ex. 1. A and B formed a partnership; A furnished \$600 capital, and B \$900; they gained \$300: what was each partner's share of the gain?

QUEST.—522. What is partnership? What are the persons thus associated called? What is the association itself called? What is the money employed called? What the profit or loss?

Analysis.—Since the whole stock is \$600+\$900=\$1500, A's part of it was $\frac{600}{1500}=\frac{2}{5}$, and B's part was $\frac{900}{1500}=\frac{3}{5}$. Now since A put in $\frac{2}{5}$ of the stock, he must have $\frac{2}{5}$ of the gain; and \$300 $\times\frac{2}{5}$=\$120. For the same reason B must have $\frac{3}{5}$ of the gain; and \$300$\times\frac{3}{5}$=\$180.

Or, we may reason thus: As the whole stock is to the whole gain or loss, so is each man's particular stock to his share of the gain or loss.

That is, \$1500 : \$300 : : \$600 : A's gain; or \$120.
And \$1500 : \$300 : : \$900 : B's gain; or \$180.

PROOF.—\$120+\$180=\$300, the whole gain. (Art. 21. Ax. 11.)

523. Hence, to find each partner's share of the gain or loss, when the *stock* of each is employed for the *same time.*

Multiply each man's stock by the whole gain or loss; divide the product by the whole stock, and the quotient will be his share of the gain or loss.

Or, make each man's stock the numerator, and the whole stock the denominator of a common fraction; multiply the gain or loss by the fraction which expresses each man's share of the stock, and the product will be his share of the gain or loss.

PROOF.—*Add the several shares of the gain or loss together, and if the sum is equal to the whole gain or loss, the work is right.* (Art. 21. Ax. 11.)

OBS. 1. The preceding case is often called *Single* Fellowship. But since a *partnership* is necessarily composed of *two* or *more* individuals, it is somewhat difficult to see the propriety of calling it *single.*

2. This rule is applicable to questions in Bankruptcy, and all other operations in which there is to be a division of property in *specified proportions.* (Arts. 465, 466)

2. A, B, and C formed a partnership; A put in \$1200 of the capital, B \$1600, and C \$2000; they gained \$960: what was each man's share of the gain?

QUEST.—523. How is each man's share of the gain or loss found, when the stock of each is employed for the same time? How is the operation proved? *Obs.* What is it sometimes called? To what is this rule applicable?

3. A, B, and C entered into partnership; A furnished $2350, B $3200, and C $1820; they lost $860: what was each man's share of the loss?

4. A bankrupt owes A $2400, B $4600, C $6800, and D $9000; his whole effects are worth $11200: how much will each creditor receive?

5. A, B, C, and D, engaged in an adventure; A put in $170, B $160, C $140, and D $130; they made $3000: what was each man's share?

CASE II.—*When the stocks are employed unequal lengths of time.*

6. A and B formed a partnership; A put in $900 for 4 months, and B put in $400 for 12 months; they gained $763: what was each man's share of the gain?

Note.—It is obvious that the gain of each depends both upon the *capital* he furnished, and the *time* it was employed. (Art. 518.)

Analysis.—Since A's capital $900, was employed 4 months, his share of the gain is the same as if he had put in $3600 for 1 month; (Art. 519;) for $900×4=$3600. Also B's capital $400, being employed 12 months, his share of the gain is the same as if he had put in $4800 for 1 month; for $400×12= $4800. The sum of $3600 and $4800 is $8400. Therefore,

A's share of the gain must be $\frac{3600}{8400}=\frac{3}{7}$.
B's " " " " $\frac{4800}{8400}=\frac{4}{7}$.

Now $763×$\frac{3}{7}$=$327, A's share.
And $763×$\frac{4}{7}$=$436, B's share. Hence,

524. To find each partner's share of the gain or loss, when the *stock* of each is employed *unequal lengths of time.*

Multiply each partner's stock by the time it is employed; make each man's product the numerator, and the sum of the products the denominator of a common fraction; then multiply the whole gain or loss by each man's fractional share of the stock, and the product will be his share of the gain or loss.

OBS. This case is often called *Compound* or *Double* Fellowship.

QUEST.—524. When the stock of each partner is employed unequal lengths of time how is each man's share found? *Obs.* What is this case sometimes called?

7. The firm of X, Y, and Z lost $4500; X had $3200 employed for 6 months, Y $2400 for 7 months, and Z $1800 for 9 months: what was each partner's loss?

8. A, B, and C hired a pasture for $60; A put in 15 oxen for 20 days, B 17 oxen for $16\frac{1}{2}$ days, and C 22 oxen for 10 days: what rent ought each man tó pay?

9. In a certain adventure A put in $12000 for 4 months, then adding $8000 he continued the whole 2 months longer; B put in $25000, and after 3 months took out $10000, and continued the rest for 3 months longer; C put in $35000 for 2 months, then withdrawing $\frac{2}{7}$ of his stock, continued the remainder 4 months longer; they gained $15000: what was the share of each?

GENERAL AVERAGE.

525. The term *General Average*, in commerce, signifies the *apportionment* of certain losses among the different interests concerned, when a *part* of the cargo, furniture, &c., of a ship has been voluntarily sacrificed to preserve the rest. (Art. 466.)

The property thus sacrificed is called the *jettison.*

526. Losses thus incurred are charged to the *ship*, the *cargo*, and the *freight, pro rata;* or according to the *value* of each.

The contributory interests are to be freed from all charges upon them before the average is made.

OBS. 1. In estimating the freight, in New York, *one-half*, but in most ports *one-third* is deducted from the gross amount, for *seamen's* wages, pilotage, and other small charges.

2. In the valuation of masts, spars, cables, rigging, &c., of the ship, it is customary to deduct a *third* from the cost of replacing them; thus calling the old, two-thirds the value of the new, in making the average.

3. The cargo is valued at the price it would bring at its destined port, after the storage and other necessary charges are deducted. The property sacrificed must be taken into the account as well as that which is saved.

527. General Average may be calculated both by *Analysis* and *Partnership.* (Arts. 466, 522.)

OBS. 1. Losses arising from the ordinary wear and tear, or from a sacrifice made for the safety of the ship only, or a particular part of the cargo, must be borne by the individuals who own the property lost, and not by *general average.*

2. General average is not allowed, unless the *peril* was *imminent*, and the sacrifice indispensable for the safety of the ship and crew.

10. The ship Minerva from London to New York, had on board a cargo valued at \$75000, of which A owned \$30000; B \$27000; and C \$18000; the gross amount of freight and passage money was \$11040. The ship was worth \$40000, and the owner paid \$520 for insurance on her. Being overtaken by a severe tempest, the master threw \$18000 worth of A's goods overboard, and cut away her mainmast and anchors; finally, he brought her into port, where it cost \$2796.75 to repair the injury: what was the loss of each owner of the ship and cargo?

Operation.

Ship valued at	\$40000.00	
Less premium for insurance .	520.00	\$39480.00
Cargo worth		75000.00
Freight and passage money .	\$11040.00	
Less one-half for wages of crew	5520.00	5520.00
Amount of contributory interests		\$120000.00
Goods thrown overboard valued at		\$18000.00
Cost new masts, spars, &c. .	\$2796.75	
Less one third for wear of old .	932.25	1864.50
Commissions on repairs . .		15.13
Port duties and incidentals .		120.37
Amount of loss		\$20000.00

Now \$20000×30000÷\$120000=\$5000, loss of A.
\$20000×27000÷\$120000=\$4500 " B.
\$20000×18000÷\$120000=\$3000 " C.
\$20000×39480÷\$120000=\$6580 " Ship.
\$20000× 5520÷\$120000=\$ 920 " Freight.

PROOF.—Whole loss (Ax. 11.) \$20000, the same as above.

Note.—We may also find what per cent. the loss is; then multiply each contributory interest by the per cent. Thus, since \$120000 lose \$20000, \$1 will lose $\frac{1}{120000}$ of \$20000; and 20000÷\$120000=.16$\frac{2}{3}$; that is, the loss is 16$\frac{2}{3}$ per cent. Now \$30000×.16$\frac{2}{3}$=\$5000, A's share of the loss. The loss of the others may be found in a similar manner.

EXCHANGE OF CURRENCIES.

528. The term *currency,* signifies *money,* or the *circulating medium* of trade.

529. The *intrinsic value* of the coins of different nations, depends upon their *weight* and the *purity* of the metal of which they are made. (Art. 245. Obs.)

OBS. For the present standard weight and purity of the coins of the United States, see Arts. 245, 246. For that of British coin, see Art. 248. Obs.

530. The *relative value* of foreign coins is determined by the laws of the country and commercial usage.

OBS. The *legal* value of a pound sterling in this country has been different at different times. By act of Congress, 1799, it was fixed at $4.44\frac{4}{9}$. In 1832 its value was raised by the same authority to $4.80; and in 1842, to $4.84.

531. The process of changing money from the denominations of one country to its equivalent value in the denominations of another country, is called *Exchange of Currencies.*

CASE I.—*Reduction of Sterling to Federal Money.*

Ex. 1. Change £60 sterling to Federal money.

Solution.—Since £1 is worth $4.84, £60 are worth 60 times as much, and $4.84×60=$290.40. *Ans.*

2. Change £8, 7s. 6d. to Federal money.

Operation.	
$4.84	We first reduce the 7s. 6d. to the decimal of a pound; (Art. 346;) then multiply $4.84, and £8.375 together, and point off the product as in multiplication of decimals. Hence,
8.375	
$40.535 *Ans.*	

532. To reduce Sterling to Federal Money.

Multiply the legal value of one pound, $4.84, by the given number of pounds, point off the product as in multiplication of decimals, and it will be the answer required. (Art. 324.)

If the example contains shillings, pence, and farthings, they must be reduced to the decimal of a pound.

QUEST.—528. What is meant by currency? 529. On what does the intrinsic value of the coins of different countries depend? 530. How is the relative value of foreign coins determined? *Obs.* What is the value of a pound sterling? 531. What is meant by exchange of currencies? 532. How is Sterling money reduced to Federal?

OBS. 1. The *reason* of this rule is obvious from the principle that **£5 are** worth 5 times as much as £1, &c.

2. The rule usually given for reducing Sterling to Federal Money, is to reduce the shillings, pence, and farthings to the decimal of a pound, and placing it on the right of the given pounds, divide the whole sum by $\frac{9}{40}$. This rule is based on the law of 1799, which fixed the value of a pound at \$4.44$\frac{4}{9}$, and that of a dollar at 4s. 6d. But \$4.44$\frac{4}{9}$ is 9 per cent. of itself, or 40 cents less than \$4.84, which is the present *legal* value of a pound; consequently, the result or answer obtained by it, must be 9 per cent. too *small*. A dollar is now equal to 49.6d. very nearly, instead of 54d. as formerly.

533. From the preceding rule it is plain that *Guineas, Francs, Doubloons,* and all *foreign coins,* may be reduced to *Federal money,* by multiplying *the legal value of one* by the given number.

Change the following sums of Sterling to Federal money:

3. £850, 10s.	8. £1000, 4s. 6d.	13. £50173, 12s. 6$\frac{3}{4}$d.
4. £175, 15s.	9. £1600, 8s. 7$\frac{1}{4}$d.	14. £53262, 13s. 8$\frac{1}{2}$d.
5. £85, 13s. 6d.	10. £12531, 10s. 4$\frac{1}{2}$d.	15. £76387, 15s. 7$\frac{3}{8}$d.
6. £200, 7s. 6d.	11. £43116, 9s. 10d.	16. £58762, 18s. 9$\frac{1}{4}$d.
7. £421, 16s. 4d.	12. £68318, 10s. 3$\frac{1}{4}$d.	17. £1000000.

CASE II.—*Reduction of Federal to Sterling Money.*

18. Change \$40.535 to sterling money.

Solution.—Since \$4.84 are worth £1, \$40.535 are worth as many pounds as \$4.84 are contained times in \$40.535; and \$40.535÷\$4.84=8.375; that is £8.375. Now reducing the decimal .375 to shillings and pence, (Art. 348,) we have £8, 7s. 6d. for the answer. Hence,

534. To reduce Federal to Sterling money.

Divide the given sum by \$4.84, (*the value of* £1,) *and point off the quotient as in division of decimals. The figures on the left hand of the decimal point will be pounds; those on the right, decimals of a pound, which must be reduced to shillings, pence, and farthings.* (Art. 348.)

OBS. Federal money may be reduced to Guineas, Francs, or any foreign coin, by *dividing* the *given sum* by the value of *one guinea, one franc,* &c.

QUEST.—533. How may foreign coins be reduced to Federal money? 534. How is Federal money reduced to Sterling?

Change the following sums of Federal to Sterling money:

19. $396.88.	23. $2160.50.	27. $25265.
20. 435.60.	24. 975.66.	28. 41470.
21. 876.25.	25. 4275.10.	29. 50263.
22. 1265.33.	26. 5300.75.	30. 100000.

535. Previous to the adoption of Federal money in 1786, accounts in the United States were kept in pounds, shillings, pence, and farthings.

OBS. At the time Federal money was adopted, the *colonial currency* or *bills of credit* issued by the colonies, had more or less *depreciated* in value: that is, a colonial pound was worth less than a pound Sterling; a colonial shilling, than a shilling Sterling, &c. This depreciation being greater in some colonies than in others, gave rise to the *different values* of the *State currencies*.

In N. E. cur., Va., Ky., Tenn., Ia., Ill., Miss., Missou.,	6s. or £$\frac{3}{10}$=$1.
In N. Y. cur., N. C., Ohio, and Mich., - -	8s. or £$\frac{2}{5}$=$1.
In Penn. cur., New Jer., Del., and Md.,	7s. 6d. (7½s.) or £$\frac{3}{8}$=$1.
In Georgia cur., and South Carolina,	4s. 8d. (4⅔s.) or £$\frac{7}{30}$=$1.
In Canada cur., and Nova Scotia, - - -	5s. or £$\frac{1}{4}$=$1

Ala., La., Ark., and Florida use Federal Money exclusively.

CASE III.—*Reduction of Federal Money to State currencies.*

31. Reduce $63.25 to New England currency.

Solution.—Since $1 contains 6s. N. E. cur., $63.25 contains 63.25 times as many; and 6s.×63.25=379.50s. Now 379÷20 =£18, 19s., and .5s.×12=6d. (Art. 348.) *Ans.* £18. 19s. 6d.

536. Hence, to reduce Federal money to State currencies.

Multiply the given sum by the number of shillings which, in the required currency, make $1, *and the product will be the answer in shillings, and decimals of a shilling. The shillings should be reduced to pounds, and the decimals to pence and farthings.* (Art. 348.)

32. Reduce $450 to New England currency.
33. Reduce $567.50 to New York currency.
34. Reduce $840.10 to Pennsylvania currency.
35. Reduce $1500 to Canada currency.

QUEST.—535. Previous to the adoption of Federal money, in what were accounts kept? 536. How is Federal money reduced to the State currencies?

CASE IV.—*Reduction of State currencies to Federal Money.*

36. Reduce £23, 12s. 6d. N. E. currency, to Federal money.

Solution.—£23, 12s. 6d.=472.5s. (Art. 348.) Now since 6s. N. E. cur. make $1, 472.5s. will make as many dollars as 6s. is contained times in 472.5s.; and 472.5s.÷6s.=78.75. *Ans.* $78.75.

537. Hence, to reduce State currencies to Federal money.

Reduce the pounds to shillings, and the given pence and farthings to the decimal of a shilling; then divide this sum by the number of shillings which, in the given currency, make $1, and the quotient will be the answer in dollars and cents.

OBS. One state currency may be reduced to another by first reducing the given currency to Federal money, then to the currency required.

37. Reduce £160, 5s. N. E. currency, to Federal money.
38. Reduce £245, 13s. 6d. N. Y. currency, to Federal money.
39. Reduce £369, 15s, 7½d. Penn. currency, to Federal money.
40. Reduce £1800, Georgia currency, to Federal money.
41. Reduce £5000, Canada currency, to Federal money.

FOREIGN COINS AND MONEYS OF ACCOUNT.

538. The *denominations of money,* in which the laws of a country require accounts to be kept, are called *Moneys of account.* They are generally represented by a *coin* of the same name; sometimes, however, they are merely nominal, like *mills* in Federal money. (Art. 245.)

539. *Foreign Moneys of Account, with the par value of the unit established by commercial usage, expressed in Federal Money.**

Austria.—60 kreutzers=1 florin; 1 florin, (silver) is equal to	$0 485
Belgium.—100 cents=1 guilder or florin; 1 guilder, (silver)	.40
The coinage of Belgium in 1832, was made similar to that of France.	
Bencoolen.—8 satellers=1 soocoo; 4 soocoos=1 dollar or rial, -	1.10
Brazil.—1000 rees=1 milree=$.828. The silver coin, 1200 rees	.994
Bremen.—5 schwares=1 grote; 72 grotes=1 rix dollar, (silver)	.787
British India.—12 pice=1 anna; 16 annas=1 Co. rupee, (silver)	.445
The current (silver) rupee of Bengal, Bombay and Madras, is worth	.444

QUEST.—537. How are the several State currencies reduced to Federal Money?

* M'Culloch's Commercial Dictionary; Kelly's Universal Cambist

Buenos Ayres.—8 rials=1 dollar currency, (fluctuating) - -	$0.93
Canton.—10 cash=1 candarine; 10 can.=1 mace; 10 mace=1 tael	1.48
The cash, which is made of copper and lead, is said to be the only money coined in China.	
Cape of Good Hope.—6 stivers=1 schilling; 8 schillings=1 rix dollar	.313
Ceylon.—4 pice=1 fanam; 12 fanams=1 rix dollar - -	.40
Cuba.—8 rials plate=1 dollar; 1 dollar - - - - -	1.00
*Colombia.**—8 rials=1 dollar; 1 dollar, (variable) mean value -	1.00
Chili.—8 rials=1 dollar; 1 dollar, (silver) - - -	1.00
Denmark.—12 pfenings=1 skilling; 16 skillings=1 marc; 6 marcs=1 rigsbank or rix dollar, (silver) - - - - - -	.52
Egypt.—3 aspers=1 para; 40 paras=1 piastre, (silver) - -	.048
France and Great Britain.—See Tables. (Arts. 247, 272.)	
Greece.—100 lepta=1 drachmè; 1 drachmè, (silver) - -	.166
Holland.—100 cents=1 florin or guilder; 1 florin, (silver) -	.40
Hamburg.—12 pfenings=1 schilling or sol; 16 schillings=1 marc Lubs; 3 marcs=1 rix dollar. The current marc, (silver)=$.28; marc banco - - - - - - - - -	.35
The term Lubs, signifies money of Lubec. The *marc currency* is the common coin; the *marc banco* is based upon certificates of deposit of bullion and jewelry in the bank of Hamburg.	
Invoices and accounts are sometimes made out in *pounds, schillings*, and *pence*, Flemish, whose subdivisions are like sterling money; the pound Flemish=7½ marcs banco.	
Japan.—10 candarines=1 mace; 10 mace=1 tael - -	.75
Java.—100 cents=1 florin; 1 florin, as in Netherlands - -	.40
Also 5 doits=1 stiver; 2 stivers=1 dubbel; 3 dub.=1 schilling; 4 schillings=1 florin - - - - - - -	.40
Malta.—20 grani†=1 taro; 12 tari=1 scudo; 2½ scudi=1 pezza	1.00
Mauritius.—In public accounts 100 cents=1 dollar - -	.968
In mercantile accounts 20 sols=1 livre; 10 livres=1 dollar.	
Manilla.—34 maravedis=1 rial; 8 rials=1 dollar, (Spanish) -	1.00
Milan.—12 denari=1 soldo; 20 soldi=1 lira† - - -	.20
Mexico.—8 rials=1 dollar; 1 dollar - - - -	1.00
Monte Video.—100 centesimos=1 rial; 8 rials=1 dollar - -	.833
Naples.—10 grani=1 carlino; 10 carlini=1 ducat, (silver) -	.80
Netherlands.—Accounts are kept throughout the kingdom in florins or guilders, and cents, as adopted in 1815. See Holland.	
New South Wales.—Accounts are kept in sterling money.	
Norway.‡—120 skillings=1 rix dollar specie, (silver) - -	1.06
Papal States.—10 bajocchi=1 paolo; 10 paoli=1 scudo or crown	1.06
Peru.—8 rials=1 dollar, (silver) - - - - - -	1.00

* Venezuela, New Grenada, and Ecuador.

† Grani is the plural of grano, tari of taro, scudi of scudo, lire of lira, pezze of pezza.

‡ Norway has no national gold coin.

Portugal.—400 rees=1 cruzado; 1000 rees=1 milree or crown - $1.12
Prussia.—12 pfenings=1 grosch, (silver) 30 groschen=1 thaler or dol. .69
*Russia.**—100 copecks=1 rouble, (silver) - - - - - .78
Sardinia.—100 centesimi=1 lira; 1 lira=1 franc, French - .186
Sweden.—12 rundstycks=1 skilling; 48 skillings=1 rix dol., specie 1.06
Sicily.—20 grani=1 taro; 30 tari=1 oncia, (gold) - - 2.40
Spain.—2 maravedis=1 quinto; 16 quintos=1 rial of old plate - .10
20 rials vellon=1 Spanish dollar - - - - - 1.00
The rial of old plate is not a coin, but it is the denomination in which invoices and exchanges are generally computed.
St. Domingo.—100 centimes=1 dollar; 1 dollar - - - - .33⅓
Tuscany.—12 denari di pezza=1 soldo di pezza; 2 soldi di pezza=1 pezza of 8 rials; 1 pezza, (silver) - - - - - .90
Turkey.—3 aspers=1 para; 40 paras=1 piastre, (fluctuating) - .05
Venice.—100 centesimi=1 lira; 1 lira=1 franc, French - - .186
Formerly accounts were kept in ducats, lire, &c. 12 denari=1 soldo; 20 soldi=1 lira piccola; $6\frac{1}{5}$ lire piccole=1 ducat current; 8 lire pic.=1 ducat effective. The value of the lira piccola is .096
West Indies, British.—Accounts are kept in pounds, shillings, pence and farthings, of the same relative value as in England. The value of the pound varies very much in the different islands, and is in all cases less than the pound sterling.

540. *The following coins and moneys of account have been made current in the United States, by act of Congress, at the rates annexed.*†

Pound sterling of Gt. Britain,	$4.84	Rix Dollar of Bremen,	$0.78¾
Pound of Canada, Nova Scotia,	4.00	Specie Dollar of Denmark,	1.05
Do. New Brunswick and Newfoundland,	4.00	Do. Sweden and Norway,	1.06
Franc of France and Belgium,	.186	Rouble, silver, of Russia,	.75
Livre Tournois of France,	.185	Florin of Austria,	.485
Florin of Netherlands,	.40	Lira of Lombardo, Venetian kingdom,	.16
Do. Southern States Germany,	.40	Lira of Tuscany,	.16
Guilder of Netherlands,	.40	Do. of Sardinia,	.186
Real Vellon of Spain,	.05	Ducat of Naples,	.80
Do. Plate of Spain,	.10	Ounce of Sicily,	2.40
Milree of Portugal,	1.12	Leghorn Livres,	.16
Do. Azores,	.83⅓	Tael of China,	1.48
Marc Banco of Hamburg,	.35	Rupee, Company,	.445
Thaler or Rix Dollar, Prussia, and North. States Germany,	.69	Do. of British India,	.445
		Pagoda of India,	1.84

* Previous to 1840, accounts were kept in paper roubles, 3½ of which made a silver rouble.

† Laws of United States

541. *Foreign gold and silver coins, at the rates established by the Custom Houses and commercial usage.**

Guinea, English,	(*gold*)	$5.00	Leghorn Dollar,	(*s.*)	$0.90
Crown, "	(*silver*)	1.12	Scudo of Malta,	(*s.*)	.40
Shilling piece, "	(*s.*)	.23	Doubloon, Mexico,	(*g.*)	15.60
Bank token, "	(*s.*)	.25	Livre of Neufchatel,	(*s.*)	.26½
Florin of Basle,	(*s.*)	.41	Half Joe, Portugal,	(*g.*)	8.53
Moidore, Brazil,	(*g.*)	4.80	Florin, Prussia,	(*s.*)	.22¾
Livre of Catalonia.	(*s.*)	.53⅓	Imperial, Russia,	(*g.*)	7.83
Florence Livre,	(*s.*)	.15	Rix Dollar, Rhenish,	(*s.*)	.60¾
Louis d'or, French,	(*g.*)	4.56	Rix Dollar of Saxony,	(*s.*)	.69
Crown, "	(*s.*)	1.06	Pistole, Spanish,	(*g.*)	3.97
40 Francs, "	(*g.*)	7.66	Rial "	(*s.*)	.12½
5 Francs, "	(*s.*)	.93	Cross Pistareen,	(*s.*)	.16
Geneva Livre,	(*s.*)	.21	Other Pistareens,	(*s.*)	.18
10 Thalers, German	(*g.*)	7.80	Swiss Livre,	(*s.*)	.27
10 Pauls, Italy,	(*s.*)	.97	Crown of Tuscany,	(*s.*)	1.05
Jamaica Pound, *nominal*,		3.00	Turkish Piastre,	(*s.*)	.05

Note.—The true method of estimating the value of foreign coins, is by their *weight* and *purity*.

EXCHANGE.

542. EXCHANGE, in commerce, signifies the receiving or paying of money in one place, for an equal sum in another, *by draft or bill of Exchange.*

OBS. 1. A *Bill of Exchange* is a written order, addressed to a person, directing him to pay at a specified time, a certain sum of money to another person, or to his order.

2. The person who *signs* the bill is called the *drawer* or *maker;* the person in whose favor it is drawn, the *buyer* or *remitter;* the person on whom it is drawn, the *drawee*, and after he has accepted it, the *accepter;* the person to whom the money is directed to be paid, the *payee;* and the person who has legal possession of it, the *holder.*

3. On the reception of a bill of exchange, it should be immediately presented to the *drawee* for his *acceptance.*

543. The *acceptance* of a bill or draft is a promise to pay it at *maturity* or the *specified time.* The common method of accept-

QUEST.—542. What is meant by exchange? *Obs.* What is a bill of exchange? Who is the drawer of a bill? The drawee? The payee? The holder? 543. What is meant by the acceptance of a bill? What is the common method of accepting a bill?

* See Manual of Gold and Silver Coins by Eckfeldt & Du Bois; Ogden on the Tariff of 1846; Taylor's Gold and Silver Coin Examiner.

ing a bill, is for the drawee to write his *name* under the word *accepted,* across the bill, either on its face or back. The drawee is not responsible for its payment, until he has accepted it.

OBS. 1. If the *payee* wishes to *sell* or *transfer* a bill of exchange, it is necessary for him to *endorse* it, or write his name on the back of it.

2. If the endorser directs the bill to be paid to a particular person, it is called a *special endorsement,* and the person named, is called the *endorsee.* If the endorser simply writes his name upon the back of the bill, the endorsement is said to be *blank.* When the endorsement is *blank,* or when a bill is drawn payable to the *bearer,* it may be transferred from one to another at pleasure, and the drawee is bound to pay it to the holder at maturity. If the drawee or accepter of a bill fail to pay it, the endorsers are responsible for it.

544. When *acceptance* or *payment* of a bill is refused, the holder should duly notify the endorsers and drawer of the fact by a legal *protest,* otherwise they will not be responsible for its payment.

OBS. 1. A *protest* is a formal declaration in writing, made by a civil officer termed a *notary public,* at the request of the holder of a bill, for its *non-acceptance,* or *non-payment.*

2. When a bill is returned *protested* for non-acceptance, the drawer must pay it immediately, though the specified time has not arrived, otherwise he is liable to prosecution.

3. The *time specified* for the payment of a bill is a matter of agreement between the parties at the time it is negotiated. Some are payable at *sight;* others in a certain *number* of *days* or *months* after sight, or after date. When payable after sight or date, the day on which they are presented is not reckoned. When the time is expressed in months, they are always understood to mean *calendar* months. Hence, if a bill payable in one month is dated the 25th of January, it will be due on the 25th of February. And if it is dated the 28th, 29th, 30th, or 31st of January, it will be due on the last day of February. It is customary to allow *three days grace* on bills of exchange.

545. Bills of exchange are usually divided into *inland* and *foreign bills.* When the *drawer* and *drawee* both reside in the same country, they are termed *inland bills or drafts;* when they reside in different countries, *foreign bills.*

OBS. In negotiating foreign bills, it is customary to draw *three* of the *same date* and *amount,* which are called the *First, Second* and *Third of Exchange;* and collectively, a *Set of Exchange.* These are sent by different ships or

QUEST.—544. When the acceptance or payment of a bill is refused, what should be done? *Obs.* What is a protest? 545. How are bills of exchange divided? *Obs.* What is meant by a set of exchange?

conveyances, and when the *first* that arrives, is accepted or paid, the *others* become *void*. The object of this arrangement is to avoid delays, which might arise from accidents, miscarriage, &c.

FORM OF A FOREIGN BILL OF EXCHANGE.

Exchange £1000. BOSTON, Oct. 3d, 1847.

At ninety days sight of this first of Exchange, (the second and third of the same date and tenor unpaid,) pay George Lewis, Esq., or order, One Thousand Pounds sterling, with or without farther advice.

JOHN W. ADAMS.

To Messrs. ROTHSCHILD & Co.
Brokers, London.

FORM OF AN INLAND BILL OR DRAFT.

$2500. NEW YORK, Sept. 27th, 1847.

Thirty days after sight, pay to the order of Messrs. Newman & Co., Twenty-five Hundred Dollars, value received, and charge the same to

MACY & WOODBURY.

To Messrs. D. BAKER & Co.
Merchants, New Orleans.

546. The term *par of exchange*, denotes the *standard* by which the comparative worth of the money of different countries is estimated. It is either *intrinsic* or *commercial*.

The *intrinsic par* is the real value of the money of different countries, determined by the *weight* and *purity* of their coin.

The *commercial par* is a nominal value, fixed by law or commercial usage, by which the worth of the money of different countries is estimated.

OBS. 1. The intrinsic par remains the same, so long as the *standard coins* of each country are of the same *metals*, and of the same *weight* and *purity*; but in case the standard coins are of *different* metals, the intrinsic par must vary, as the comparative values of the metals vary.

2. The commercial par is conventional, and may at any time be changed by law or custom.

547. By the term *course of exchange* is meant the *current price* which is paid in one place for bills of a given amount drawn on another place.

OBS. 1. The course of exchange is seldom *stationary* or at *par*. It varies

QUEST.—546. What is meant by par of exchange? Intrinsic par? Commercial par?

according to the circumstances of trade. When the balance of trade is against a country, that is when the exports are *less* than the imports, bills on the foreign country will be *above par*, for the reason that there will be a greater demand for them to pay the balance due abroad. On the other hand, when the balance of trade is in favor of a country, foreign bills will be *below par*, for the reason that fewer will be required.

2. It should be remarked that the course of exchange can never exceed very much the *intrinsic par value;* for it is plain that *coin* or *bullion* instead of bills will be remitted, whenever the course of exchange is such that the expense of insuring and transporting it from the debtor to the creditor country, is less than the *premium* for bills, and the exchange will soon sink to par.

548. *Rates of exchange on Great Britain* are commonly reckoned at a certain per cent., on the *old commercial par,* instead of the *new par.*

OBS. 1. According to the *old par*, the value of a pound sterling is 4.44\frac{4}{9}$, as fixed by act of Congress, in 1799. According to the *new par* it is $4.84. The *intrinsic* value of a £ ster. or sovereign, according to assays at the U. S. mint, is $4.861. The *new par* is the value fixed by the government in 1842, and is used in calculating *duties*, when the invoice is in sterling money.

2. The old par is *nine* per cent. *less* than the *new par* or legal value; consequently the rate of exchange must reach the nominal premium of 9 per cent. before it is *at par* according to the *new standard.*

Table of Exchange showing the value of £1 Sterling from 1 to to 12½ per cent. premium on the old par of 4.44\frac{4}{9}$.

Old Par	$4.444	5½ per ct.	$4.689	8 per ct.	$4.800	9¾ per ct.	$4.878
1 per ct.	4.489	6 "	4.711	8¼ "	4.811	10 "	4.889
2 "	4.533	6½ "	4.733	8½ "	4.822	10½ "	4.911
3 "	4.578	7 "	4.756	8¾ "	4.833	11 "	4.933
4 "	4.622	7¼ "	4.767	9 New Par	4.844	11½ "	4.956
4½ "	4.644	7½ "	4.778	9¼ per ct.	4.856	12 "	4.978
5 "	4.667	7¾ "	4.789	9½ "	4.867	12½ "	5.000

Note.—1. When exchange is 10 per cent. advance or over, on the old par, it will cause a shipment of specie to England; for the freight, interest and insurance will not amount to so much as the premium. When the premium is less than 9 per cent. English funds are, in reality, below their intrinsic par.

2. The practice of quoting rates of exchange at the *old par*, is calculated to lead persons unacquainted with the subject into serious *mercantile mistakes*, and to *degrade* our *national currency* by making it *appear* to foreign nations to be so much *below par*.

Ex. 1. A merchant negotiated a bill of exchange on London for £500, 10s., at 8 per cent. premium on the old par: how much did he pay for the bill?

Solution.—£500, 10s.=£500.5. (Art. 346.)

Now 4.44\frac{4}{9}$×500.5=$2224.44$\frac{4}{9}$ at the old par value.

Then 2224.44\frac{4}{9}$× .08= 177.95$\frac{5}{9}$ the premium.

The sum paid $2402.40 *Ans.*

Or, the val. of £1 by table, $4.80×500.5=$2402.40. *Ans.*

2. A merchant negotiated a bill on Liverpool for £1000, at 1 per cent. discount from the new par: what did he pay for it?

3. What will a bill cost, on England, for £5265, 13s. 6d., at 8$\frac{1}{2}$ per cent. advance on the old par?

4. How much is a bill worth on France for 1500 francs, at 2 per cent. above par, which is $.186 per franc?

5. What will a bill cost on Paris for 56245 francs, exchange being 5 francs and 54 centimes to the dollar?

6. What cost a bill of exchange on Hamburg for 2000 marcs banco, at 1 per cent. above par, which is 35 cts. per marc?

7. What cost a bill of exchange on St. Petersburg for 2560 roubles, at 2 per cent. discount, the par being 75 cts. per rouble?

8. What cost an inland bill of exchange at Boston, on New Orleans, for $15265.85, at 1 per cent. advance?

9. What cost a draft at Albany, on Mobile, for $20260, at 2 per cent. premium?

10. What cost a draft at St. Louis, on New York, for $35678, at 2$\frac{1}{4}$ per cent. premium?

ARBITRATION OF EXCHANGE.

549. *Arbitration of Exchange* is the method of finding the *exchange* between two countries through the medium of that of other countries.

OBS. 1. When there is but *one* intervening country, the operation is termed *simple arbitration*, when *more than one*, it is termed *compound arbitration.*

2. Problems in Arbitration of exchange are usually solved by *conjoined proportion.* (Art. 511.) Care must be taken to reduce all the quantities which are of the same kind, to the same denomination.

Ex. 1. If the exchange of New York on London is 8 per cent. advance on old par, or $4.80 for £1 sterling, and that of Amsterdam on London is 12 florins for £1, what is the arbitrated exchange of New York on Amsterdam; that is, how many florins are equal to $1 U. S.? *Ans.* $1 = 2½ florins.

2. A merchant in Baltimore wishes to remit 1200 marcs banco to Hamburg, and the exchange of Baltimore on Hamburg is 35 cents for 1 marc. He finds the exchange of Baltimore on Paris is 18 cents for 1 franc; that of Paris on London, is 25 francs for £1 sterling; that of London on Lisbon, is 180 pence for 3 milrees; that of Lisbon on Hamburg, is 5 milrees for 18 marcs banco. How much will he gain by the circuitous exchange?

Ans. Direct Ex. $420; circuitous Ex. $375: Gain $45.

3. A man in England owes a man in Portugal £420; the direct exchange from London to Lisbon is 70d. for 1 milree; but the exchange between London and Amsterdam is 48 florins for £1 sterling; between Amsterdam and Paris it is 16 florins for 3 francs; and between Paris and Lisbon it is 6 francs for 2 milrees. Is it better for the man in Portugal to have a direct remittance from London to Lisbon, or a circuitous one through Amsterdam and Paris?

ALLIGATION.

550. *Alligation* is the method of finding the value of a compound or mixture of articles of different values, or of forming a compound which shall have a given value. (Art. 467.)

OBS. 1. The term *alligation* is derived from the Latin *alligo*, which signifies to *bind* or *tie* together. It had its origin in the manner of connecting the numbers together by a curve line in the solution of a certain class of examples.

2. Alligation is usually divided into *Medial* and *Alternate*. (Art. 467.)

NOTE.—For a *new method* of *Alligation Alternate*, see KEY, p. 72.

MEDIAL ALLIGATION.

551. *Medial Alligation* is the process of finding the *mean price* of a mixture of two or more ingredients or articles of *different values.*

Note.—The term *medial* is derived from the Latin *medius*, signifying a *mean* or *average.*

552. To find the mean value of a mixture, when the quantity and the price of each of the ingredients are given.

Divide the whole cost of the ingredients by the whole quantity mixed, and the quotient will be the mean price of the mixture.

PROOF.—*Multiply the whole mixture by the mean price, and if the product is equal to the whole cost, the work is right.*

Ex. 1. A grocer mixed 10 lbs. of tea worth 5s. a pound, with 18 lbs. worth 3s. a pound, and 20 lbs. worth 2s. a pound: what is the mixture worth per pound?

Solution.— 10 lbs. at 5s.=50s.
18 lbs. at 3s.=54s.
20 lbs. at 2s.=40s.
Whole quantity 48 lbs. and 144s. whole cost of mixture.
Now 144s.÷48=3. *Ans.* 3s. a pound.

2. A drover bought 870 lambs at 75 cts. apiece, and 290 sheep at $1.25 apiece: what is the mean price of the lot per head?

3. A grocer mixed 12 gals. of wine at 4s. 10d. per gal., with 21 gals. at 5s. 3d., and 29¼ gals. at 5s. 8d.: what is a gallon of the mixture worth?

ALTERNATE ALLIGATION.

553. *Alternate Alligation* is the process of finding what quantity of any number of ingredients, whose prices are given, will form a mixture of a *given mean price*.

Note—The term *alternate* is derived from the Latin *alternatus*, signifying *by turns*, and in its present application, refers to the connection of the prices which are *less* than the *mean price*, with those which are *greater*. Alternate alligation embraces three varieties of examples.

CASE I.

554. To find the quantity of each ingredient, when its price and that of the required mixture are given.

I. *Write the prices of the ingredients under each other, beginning with the least; then connect, with a curve line, each price which is less than that of the mixture with one or more of those that are greater; and each greater price with one or more of those that are less.*

II. *Write the difference between the price of the mixture and that of each of the ingredients opposite the price with which they are connected. If only one difference stands against any price, it will denote the quantity to be taken of that price; but if there are more than one, their sum will be the quantity.*

OBS. It is immaterial in what manner we select the pairs of ingredients, provided the price of one of the ingredients is *less* and the other *greater* than the *mean* price of the mixture required.

PROOF.—*Find the value of all the ingredients at their given prices; if this is equal to the value of the whole mixture at the given price, the work is right.*

4. A man mixed four kinds of oil, worth 8s., 9s., 11s., and 12s. per gal.; the mixture was worth 10s. per gal.: required, the quantity of each.

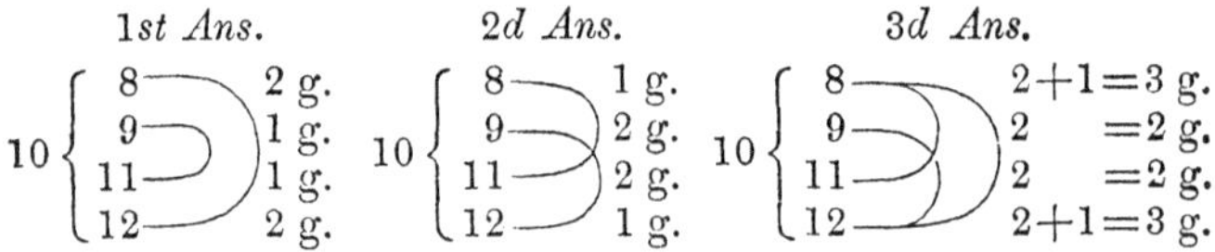

OBS. 1. It is manifest that other answers may be obtained by connecting the prices in a different manner.

2. It is also manifest, if we multiply or divide the answers already obtained by any number, the results will fulfil the conditions of the question; consequently the number of answers is unlimited.

5. A goldsmith has gold of 18, 20, 22, and 24 carats fine: how much may be taken of each to form a mixture 21 carats fine?

CASE II.

555. When the quantity of one of the ingredients and the mean price of the mixture are given.

Find the difference between the price of each ingredient and the mean price of the required mixture, as before; then by proportion,

As the difference of that ingredient whose quantity is given, is to each particular difference, so is the quantity given to the quantity required of each ingredient.

6. How many pounds of sugar at 10, and 15 cents a pound, must be mixed with 20 lbs. at 9 cents, so that the mixture may be worth 12 cents a pound?

Solution.—Connecting the prices as directed, the differences between them and the mean, are 3 cts., 3 cts. and 5 cts.

Then 3 cts. : 3 cts. : : 20 lbs. : to the lbs. at 10 cts.
Also 3 cts. : 5 cts. : : 20 lbs. : " " 15 cts.

Ans. 20 lbs. at 10 cts., and $33\frac{1}{3}$ lbs. at 15 cts.

7. How much gold of 16, 18, and 22 carats fine must be mixed with 10 oz. 24 carats fine, that the mixture may be 20 carats fine?

8. How much wool at 20, 30, and 54 cts. a pound must be mixed with 95 lbs. at 50 cts. to form a mixture worth 40 cts. a pound?

CASE III.

556. When the quantity to be mixed and the mean price of the required mixture are given.

Find the difference between the price of each ingredient and the mean price of the required mixture, as before; then by proportion,

As the sum of the differences is to each particular difference, so is the whole quantity to be mixed, to the quantity required of each ingredient.

9. A grocer has raisins worth 8, 10, and 16 cents a pound: how many of each kind may be taken to form a mixture of 112 lbs. worth 12 cents a pound?

Solution.—The sum of the differences between the prices of the ingredients, and the mean price, 6 cts.+4 cts.+4 cts.=14 cts.

Then 14 cts. : 6 cts. : : 112 lbs. : to the lbs. at 16 cts.
And 14 cts. : 4 cts. : : 112 lbs. : to the lbs. at 8 and 10 cts.

Ans. 48 lbs. at 16 cts., 32 lbs. at 10 cts., and 32 lbs. at 8 cts.

10. How much wine at 15, 17, 18, and 22 shillings per gallon, may be mixed to form a mixture of 320 gals. worth 20 shillings per gallon?

11. How much water must be mixed with wine worth 9s. per gal. to fill a pipe, so that the mixture may be worth 7s. per gal.?

SECTION XVII.

INVOLUTION.

ART. **557.** When any number or quantity is *multiplied* into *itself*, the *product* is called a *power*. Thus, 5×5=25; 3×3×3=27; 2×2×2×2=16; the products 25, 27, and 16 are powers.

The *original* number, that is, the number which being multiplied into itself, produces a power, is called the *root* of all the powers of that number, because they are derived from it.

558. *Powers* are divided into *different orders;* as the *first, second, third, fourth, fifth power*, &c. They take their name from the *number of times* the given number is used as a *factor*, in producing the *given power*.

OBS. 1. The *first* power of a number is said to be the number itself. Strictly speaking, it is not a *power*, but a *root*. (Art. 557.)

3 yards.

2. The *second* power of a number is also called the *square;* (Art. 257. Obs. 1;) for, if the side of a square is 3 yards, then the product of 3×3=9 yards, will be the *area* of the given *square*. (Art. 285.) But 3×3=9 is also the *second* power of 3; hence, it is called the *square*.

3 yards.

3×3=9 yards.

3. The *diagonal* of a square is a line connecting two of its *opposite* corners.

4. The *third* power of a number is also called the *cube;* (Art. 258. Obs. 1;) for, if the side of a cube is 3 feet, then the product of 3×3×3=27 feet, will be the solidity of the given cube. (Art. 286.) But 3×3×3=27 is also the *third* power of 3; hence it is called the *cube*. (Leg. VII. 11. Sch.)

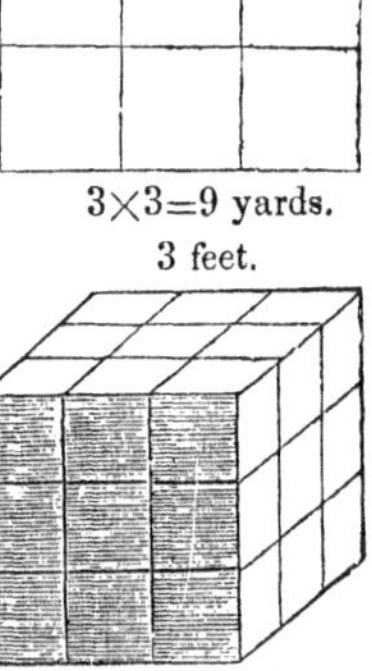

3×3×3=27 feet.

5. The *fourth* power of a number is called the *biquadrate*.

QUEST.—557. What is a power? 558. How are powers divided? From what do they take their name? *Obs.* What is said to be the first power? What is the second power called? The third? The fourth?

559. Powers are denoted by a *small figure* placed above the given number at the right hand.

This figure is called the *index* or *exponent.* It shows how many times the given number is employed as a factor to produce the required power. Thus,

The index of the *first* power is 1; but this is commonly omitted; for, $(2)^1=2$.

The index of the *second* power is 2;

The index of the *third* power is 3;

The index of the *fifth* power is 5; &c. That is,

$2^1=2$, the first power of 2;

$2^2=2\times2$, the square, or second power of 2;

$2^3=2\times2\times2$, the cube, or third power of 2;

$2^4=2\times2\times2\times2$, the biquadrate, or fourth power of 2;

$2^5=2\times2\times2\times2\times2$, the fifth power of 2;

$2^6=2\times2\times2\times2\times2\times2$, the sixth power of 2; &c.

Ex. 1. Express the square of 17, and the cube of 19.

Ans. 17^2, 19

Express the given powers of the following numbers:

2. The square of 54.
3. The cube of 43.
4. The square of 87.
5. The biquadrate of 91.
6. The 3d power of 416.
7. The 2d power of 299.
8. The 4th power of 785.
9. The 5th power of 228.
10. The 8th power of 693.
11. The 32d power of 999.

560. *The process of finding a power of a given number by multiplying it into itself, is called* INVOLUTION.

561. Hence, to *involve* a number to any required power.

Multiply the given number into itself, till it is taken as a factor, as many times as there are units in the index of the power to which the number is to be raised. (Art. 558.)

OBS. 1. The *number* of *multiplications* in raising a number to any given power, is *one* less than the index of the required power. Thus, $3^2=3\times3$; the 3 is taken twice as a factor, but there is but one multiplication.

QUEST.—559. How are powers denoted? What is this figure called? What does it show? What is the index of the first power? Of the second? The third? Fifth? 560. What is involution? 561. How is a number involved to any required power?

2. A *Fraction* is raised to a power by multiplying it into itself. Thus, the square of $\frac{2}{3}$ is $\frac{2}{3}\times\frac{2}{3}=\frac{4}{9}$.

Mixed numbers should be reduced to improper fractions, or the common fraction to a decimal. They may however be involved without reducing them. (Art. 220. Obs.)

3. The process of raising a number to a high power, may often be *contracted* by multiplying together powers already found. The *index* of the power thus found, is equal to the *sum* of the indices of the powers multiplied together. Thus, $2\times2=4$; and $4\times4=2\times2\times2\times2$, or 2^4. So $3^2\times3^3=3\times3\times3\times3\times3$, or 3^5; and $5^4\times5^3=5^7$.

12. What is the square of 23?

Common Operation.	*Analytic Operation.*
23	23=2 tens or 20+3 units.
23	23=2 tens or 20+3 units.
69	60+9
46	400+ 60
529 *Ans.*	And 400+120+9=529. *Ans.*

It will be seen from this operation that the square of 20+3 contains the square of the first part, viz: $20\times20=400$, added to twice the product of the two parts, viz: $20\times3+20\times3=120$, added to the square of the last part, viz: $3\times3=9$. Hence,

562. *The square of the sum of two numbers is equal to the square of the first part, added to twice the product of the two parts, and the square of the last part.*

OBS. 1. The product of any two factors cannot have more figures than both factors, nor but one less than both. For example, take 9, the greatest number which can be expressed by one figure. (Art. 34.) And $(9)^2$, or $9\times9=81$, has two figures, the same number which both factors have. 99 is the greatest number which can be expressed by two figures; (Art. 34;) and $(99)^2$, or $99\times99=9801$, has four figures, the same as both factors have.

Again, 1 is the smallest number expressed by one figure, and $(1)^2$, or $1\times1=1$, has but one figure less than both factors. 10 is the smallest number which can be expressed by two figures; and $(10)^2$, or $10\times10=100$, has one figure less than both factors. Hence,

QUEST.—*Obs.* How many multiplications are there in raising a number to a given power? How is a fraction involved? A mixed number? 562. What is the square of the sum of two numbers equal to? *Obs.* How many figures are there in the product of any two factors? How many figures will the square of a number contain? The cube?

2. *A square cannot have more figures than double the number of the root or first power, nor but one less.*

3. *A cube cannot have more figures than triple the number of the root or first power, nor but two less.*

4. All powers of 1 are the same, viz: 1; for, $1\times1\times1\times1$, &c.$=1$.

13. What is the square or second power of 123?

14. The cube of 135?

15. The square of 2880?

16. The 4th power of 10?

17. The 5th power of 5?

18. The 7th power of 6?

19. The 6th power of 7?

20. The 8th power of 4?

21. The 9th power of 9?

22. The square of 2.5?

23. The cube of .012?

24. The square of .00125?

25. The square of $\frac{2}{3}$?

26. The cube of $\frac{4}{5}$?

27. The square of $\frac{25}{34}$?

28. The cube of $\frac{30}{100}$?

29. The square of $4\frac{1}{2}$?

30. The square of $7\frac{3}{8}$?

31. The square of $38\frac{21}{44}$?

EVOLUTION.

563. If we resolve 25 into *two equal factors,* viz: 5 and 5, each of these equal factors is called a *root* of 25. So if we resolve 27 into *three equal factors,* viz: 3, 3, and 3, each factor is called a *root of* 27; if we resolve 16 into *four* equal factors, viz: 2, 2, 2, and 2, each factor is called a *root* of 16. And, universally, when a number is resolved into *any number* of *equal factors,* each of those factors is said to be a *root* of that number. Hence,

564. A *root* of a number is a *factor,* which, being *multiplied into itself* a certain number of times, will produce that number.

OBS. *Roots,* as well as powers, are divided into *different orders.* Thus, when a number is resolved into *two equal* factors, each of these factors is called the *second* or *square* root; when resolved into *three equal* factors, each of these factors is called the *third* or *cube* root, &c. Hence,

The name of the root expresses the number of equal factors into which the given number is to be resolved.

Roots.	1	2	3	4	5	6	7	8	9	10	11	12
Squares.	1	4	9	16	25	36	49	64	81	100	121	144
Cubes.	1	8	27	64	125	216	343	512	729	1000	1331	1728

QUEST.—*Obs.* What are all powers of 1? 564. What is a root of a number? *Obs.* What does the name of the root express?

565. *The process of resolving numbers into equal factors is called* EVOLUTION, *or the Extraction of Roots.*

OBS. 1. Evolution is the *opposite* of involution. (Art. 560.) One is finding a *power* of a number by multiplying it into itself; the other is finding a *root* by resolving a number into *equal factors. Powers* and *roots* are therefore *correlative* terms. If one number is a *power* of another, the latter is a *root* of the former. Thus, 27 is the cube of 3; and 3 is the cube root of 27.

2. The learner will be careful to observe, that
In *subtraction*, a number is resolved into *two parts;*
In *division*, a number is resolved into *two factors;*
In *evolution*, a number is resolved into *equal factors.*

566. Roots are expressed in *two* ways; one by the *radical sign* ($\sqrt{\ }$) placed before a number; the other by a *fractional index* placed above the number on the right hand. Thus, $\sqrt{4}$, or $4^{\frac{1}{2}}$ denotes the *square* or 2d root of 4; $\sqrt[3]{27}$, or $27^{\frac{1}{3}}$ denotes the *cube* or 3d root of 27; $\sqrt[4]{16}$, or $16^{\frac{1}{4}}$ denotes the 4th root of 16.

OBS. 1. The figure placed over the radical sign, denotes the *root*, or the number of *equal factors* into which the given number is to be resolved. The figure for the *square* root is usually omitted, and simply the radical sign $\sqrt{\ }$ is placed before the given number. Thus the square root of 25 is written $\sqrt{25}$.

2. When a root is expressed by a *fractional* index, the *denominator*, like the figure over the radical sign, denotes the *root* of the given number. Thus, $(25)^{\frac{1}{2}}$ denotes the square root of 25; $(27)^{\frac{1}{3}}$ denotes the cube root of 27.

3. A *fractional* index whose *numerator* is greater than 1, is sometimes used. In such cases the *denominator* denotes the *root*, and the *numerator* the *power* of the given number. Thus, $8^{\frac{2}{3}}$ denotes the *square* of the *cube* root of 8, or the *cube* root of the *square* of 8, each of which is 4.

4. The radical sign $\sqrt{\ }$, is derived from the letter *r*, the initial of the Latin *radix, a root.*

1. Express the cube root of 74. *Ans.* $\sqrt[3]{74}$, or $74^{\frac{1}{3}}$.
2. The square root of 119.
3. The 4th root of 231.
4. The 9th root of 685.
5. The square root of $\frac{4}{9}$.
6. The cube root of $\frac{3}{4}$.
7. The 4th root of $\frac{17}{26}$.
8. Express the 3d power of the 4th root of 6. *Ans.* $6^{\frac{3}{4}}$
9. Express the 2d power of the 3d root of 81.

QUEST.—565. What is evolution? *Obs.* Of what is it the opposite? Into what are numbers resolved in subtraction? In division? In evolution? 566. How many ways are roots expressed? What are they? *Obs.* What does the figure over the radical sign denote? What the denominator of the fractional index?

567. A number which can be resolved into *equal* factors, or whose root can be *exactly* extracted, is called a *perfect power*, and its root is called a *rational number*. Thus, 16, 25, 27, &c., are perfect powers, and their roots 4, 5, 3, are rational numbers.

568. A number, which *cannot* be resolved into *equal* factors, or whose root *cannot* be *exactly* extracted, is called an *imperfect power;* and its root is called a *Surd*, or *irrational number*. Thus, 15, 17, 45, &c., are imperfect powers, and their roots 3.8+; 4.1+; 6.7+, &c., are surds, for their roots cannot be exactly extracted.

OBS. A number may be a *perfect* power of one degree and an *imperfect* power of another degree. Thus, 16 is a perfect power of the second degree, but an imperfect power of the third degree; that is, it is a perfect *square* but not a perfect *cube*. Indeed numbers are seldom perfect powers of more than one degree. 16 is a perfect power of the 2d and 4th degrees; 64 is a perfect power of the 2d, 3d and 6th degrees.

569. Every *root*, as well as power of 1, is 1. (Art. 562. Obs. 4.) Thus, $(1)^2$, $(1)^3$, $(1)^6$, and $\sqrt{1}$, $\sqrt[3]{1}$, $\sqrt[6]{1}$, &c., are all equal.

PROPERTIES OF SQUARES AND CUBES.

570. The *properties* of numbers in general, have already been given. The following pertain to *square* and *cubic* numbers.

1. The product of any *two* or *more square* numbers, is a *square;* and the product of any *two* or *more* cubic numbers, is a *cube*. Thus $2^2 \times 3^2 = 36$, the square of 6; and $2^3 \times 3^3 = 216$, is the cube of 6.

2. If a *square* number is divided by a square, the *quotient* will be a square. Thus, $144 \div 9 = 16$, which is the square of 4.

3. If a square number is either *multiplied* or *divided* by a number that is not a square, neither the *product* nor *quotient* will be a square.

4. If you *double* the number of times a number is taken as a factor, it will not produce *double* the *product*, but the *square* of it. Thus, $3 \times 3 = 9$, and $3 \times 3 \times 3 \times 3 = 81$, and not 18.

5. The product of *two different prime* numbers cannot be a square.

6. The product of no two different numbers, which are *prime to each other*, will make a square, unless each of those numbers is a square.

7. The square and cube of an *even* number are *even;* and the square and cube of an *odd* number are *odd*. (Art. 161. Prop. 6, 10.) Hence,

---

QUEST.—567. What is a perfect power? What is a rational number? 568. An imperfect power? A surd? *Obs.* Are numbers ever perfect powers of one degree and imperfect powers of another degree? 569. What are all roots and powers of 1?

8. The *square* or *cube root* of an *even* number, is even; and the square or cube root of an *odd* number, is odd.

9. Every square number necessarily *ends* with one of these figures, 1, 4, 5 6, 9; or with an *even number* of *ciphers* preceded by one of these figures.

10. No number is a *square* that ends in 2, 3, 7, or 8.

11. A *cubic* number may end in any of the natural numbers, 1, 2, 3, 4, 5, 6, 7, 8, 9, or 0.

12. All the *powers* of any number, ending in 5, will also end in 5; and if a number ends in 6, all its powers will end in 6.

13. Every square number is *divisible* by 3, and also by 4, or becomes so when *diminished* by *unity*. Thus, 4, 9, 16, 25, &c., are all divisible by 3, and by 4, or become so when diminished by 1.

14. Every square number is *divisible* also by 5, or becomes so when *increased* or *diminished* by unity. Thus, $36-1$, and $49+1$, are divisible by 5.

15. Any *even* square number is divisible by 4.

16. An *odd* square number, divided by 4, leaves a remainder of 1.

17. Every *odd* square number, *decreased* by unity, is divisible by 8.

18. Every number is either a *square*, or is divisible into *two*, or *three*, or *four* squares. Thus 30 is equal to $25+4+1$; $33=16+16+1$; $63=49+9+4+1$.

19. The product of the *sum* and *difference* of two numbers, is equal to the *difference* of their *squares*. Thus, $(5+3)\times(5-3)=16$; also $5^2-3^2=16$.

20. If two numbers are such, that their *squares*, when added together, *form* a *square*, the product of these two numbers is divisible by 6. Thus, 3 and 4, the sum of whose squares, $9+16=25$, is a square number, and their product 12, is divisible by 6. Hence,

21. To find two numbers, the *sum* of whose squares shall be a *square* number.

Take any two numbers and multiply them together; the double of their product will be one of the numbers sought, and the difference of their squares will be the other. Thus, take any two numbers, as 2 and 3; the double of their product is 12, and the difference of their squares is 5; now $12^2+5^2=169$, the square of 13.

22. When two numbers are such, that the *difference* of their squares is a *square* number; the *sum* and *difference* of these numbers are themselves square numbers, or the double of square numbers. Thus, 8 and 10 give for the difference of their squares 36; and 18, the sum of these numbers, is the double of 9, which is a square number, and 2, their difference, is the double of 1, which is also a square number.

23. If two numbers, the *difference* of which is 2, be multiplied together, their product increased by unity, will be the square of the *intermediate* number.

24. The *sum* or *difference* of two numbers, will measure the *difference* of their squares.

25. The *sum* of two numbers, differing by unity, is equal to the *difference* of their squares.

26. The *sum* of two numbers will measure the *sum* of their *cubes*; and the *difference* of two numbers will measure the *difference* of their cubes.

27. If a square *measures* a square, or a cube a cube, the *root* will also *measure* the root.

28. If one number is *prime* to another, its square, cube, &c., will also be prime to it.

29. The difference between an *integral cube* and its *root*, is always divisible by 6.

30. If any series of numbers beginning from 1, be in continued geometrical proportion, the 3d, 5th, 7th, &c., will be squares; the 4th, 7th, 10th, &c., cubes; and the 7th will be both a square and a cube. Thus, in the series, 1, 2, 4, 8, 16, 32, 64, &c., the 3d, 5th, and 7th terms are squares; the 4th and 7th are cubes; and the 7th is both a square and a cube.

EXTRACTION OF THE SQUARE ROOT.

571. *To extract the* SQUARE ROOT, *is to resolve a given number into two equal factors; or, to find a number which being multiplied into itself, will produce the given number.* (Art. 564. Obs.)

Ex. 1. What is the square root of 36?

Solution.—Resolving the given number into two equal factors, we have $36=6\times6$. *Ans.* The square root of 36 is 6.

2. What is the length of one side of a square field which contains 529 square rods?

Operation.

```
   5̇2̇9(23
   4
43)129
   129
```

Since we may not see what the root of 529 is at once, we will separate it into two periods by placing a point over the 9 and another over the 5. Now the greatest square in 5, the left hand period, is 4, the root of which is 2. Placing the 2 on the right of the number, we subtract its square from the period 5, and to the right of the remainder bring down the next period. We then double the 2, the part of the root already found, and, placing it on the left of the dividend for a partial divisor, we perceive it is contained in the dividend, omitting its right hand figure, 3 times. Placing the 3 on the right of the root, also on the right of the partial divisor, we multiply the divisor thus completed by 3, and subtract the product from the dividend. The answer is 23 rods.

QUEST.—571. What is it to extract the square root of a number?

Note.—Since the root is to contain 2 figures, the 2 stands in tens place, hence the first part of the root found is properly 20; which being doubled, gives 40 for the divisor. For convenience we omit the cipher on the right; and to compensate for this, we omit the right hand figure of the dividend. This is the same as dividing both the divisor and dividend by 10, and therefore does not alter the quotient. (Art. 146.)

572. Hence, we derive the following general

RULE FOR EXTRACTING THE SQUARE ROOT.

I. *Separate the given number into periods of two figures each, by placing a point over the units figures, then over every second figure towards the left in whole numbers, and over every second figure towards the right in decimals.*

II. *Find the greatest square number in the first or left hand period, and place its root on the right of the number for the first figure in the root. Subtract the square of this figure of the root from the period under consideration; and to the right of the remainder bring down the next period for a dividend.*

III. *Double the root just found and place it on the left of the dividend for a partial divisor; find how many times it is contained in the dividend, omitting its right hand figure; place the quotient on the right of the root, also on the right of the partial divisor; multiply the divisor thus completed by the figure last placed in the root; subtract the product from the dividend, and to the remainder bring down the next period for a new dividend.*

IV. *Double the root already found for a new partial divisor, divide, &c., as before, and thus continue the operation till the root of all the periods is extracted.*

If there is a remainder after all the periods are brought down, the operation may be continued by annexing periods of ciphers.

PROOF.—*Multiply the root into itself; and if the product is equal to the given number, the work is right.* (Art. 564.)

573. *Demonstration.*—Take any number as that in the last example; then separating it into parts, $529=500+29$. Now the greatest square in 500 is 400, the root of which is 20, with a remainder of 100; consequently, the first part of

QUEST.—572. What is the first step in extracting the square root? The second? Third Fourth? When there is a remainder, how proceed? How is the square root proved?

the root must be 20, and the true remainder is 100+29, or 129. And since there are three figures in the given number, there must be two figures in the root; (Art. 562. Obs. 2;) but the square of the sum of two numbers, is equal to the *square* of the first part added to *twice* the product of the two parts and the *square* of the last part; it follows therefore that the remainder 129, must be twice the product of 20 into the part of the root still to be found, together with the square of that part. (Art. 562.) Now dividing 129 by 40 the double of 20, the quotient is 3, which being added to 40 makes 43; finally, multiplying 43 by 3, the product is 129, which is manifestly twice the product of 20 into 3, together with the square of 3. In the same manner the operation may be proved in every case. (For illustration of this rule by geometrical figures, see Practical Arithmetic, p. 318.)

1. The *reason* for separating the given numbers into *periods of two figures* each, is that a *square* number can not have *more figures* than *double* the number of figures in the root, nor but *one less*. It also shows *how many figures* the root will contain, and thus enables us to find *part* of it at a time. (Art. 562. Obs. 2.)

2. The reason for doubling that part of the root already found for a divisor, is because the remainder is *double* the product of the first part of the root into the second part, together with the *square* of the second part.

3. In dividing, the *right hand figure* of the dividend is *omitted*, because the cipher on the right of the divisor being omitted, the quotient would be 10 times *too large* for the next figure in the root. (Arts. 130, 146.)

4. The last figure of the root is placed on the right of the divisor simply for convenience in multiplying it into itself.

OBS. 1. The product of the divisor completed into the figure last placed in the root, cannot *exceed* the *dividend*. Hence, in finding the figure to be placed in the root, some allowance must be made for *carrying*, when the product of this figure into itself *exceeds* 9.

2. If the right hand period of decimals is deficient, it must be completed by annexing a cipher to it.

3. There will always be as many decimal figures in the root, as there are *periods* of decimals in the given number.

574. The square root of a common fraction is found by extracting the root of the numerator and denominator.

A mixed number should be reduced to an improper fraction. When either the numerator or denominator of a common fraction is not a *perfect square*, the fraction may be reduced to a decimal, and the *approximate* root be found as above.

QUEST.—573. *Dem.* Why do we separate the given number into periods of two figures each? Why double the root thus found for a divisor? Why omit the right hand figure of the dividend? Why place the last figure of the root on the right of the divisor? *Obs.* How many decimal figures will there be in the root? 574. How is the square root of a common fraction found? Of a mixed number?

Required the square root of the following numbers:

3. 2601.	10. 27889.	17. 566.44.	24. $\frac{25}{49}$.
4. 5329.	11. 961.	18. 7.3441.	25. $\frac{121}{256}$.
5. 784.	12. 97.	19. .81796.	26. $\frac{5}{8}$.
6. 87.	13. 7.	20. 1169.64.	27. $17\frac{3}{8}$.
7. 4761.	14. 190.	21. 627264.	28. $794\frac{1}{5}$.
8. 7056.	15. 43681.	22. 3.172181.	29. $207\frac{25}{36}$.
9. 9801.	16. 47089.	23. 10342656.	30. $34967\frac{8}{11}$.

31. What is the square root of 152399025?
32. What is the square root of 119550669121?
33. What is the square root of 964.5192360241?

575. When the root is to be extracted to many figures, the operation may be contracted in the following manner.

First find half, or one more than half the number of figures required in the root; then having found the next true divisor, cut off its right hand figure, and divide the remainder by it; place the quotient in the root, and continue the operation as in contraction of division of decimals. (Art. 333.)

34. Required the square root of 365 to eleven figures in the root. *Ans.* 19.104973174.

35. Required the square root of 2 to twelve figures.

36. Required the square root of 3 to seventeen figures.

APPLICATIONS OF THE SQUARE ROOT.

576. A *triangle* is a figure which has *three sides* and *three angles*. When one of the sides of a triangle is *perpendicular* to another side, the angle between them is called a *right-angle*.

577. A *right-angled triangle* is a triangle which has a right-angle.

The side opposite the right-angle is called the *hypothenuse*, and the other two sides, the *base* and *perpendicular*. The triangle ABC is right-angled at B, and the side BC is the hypothenuse.

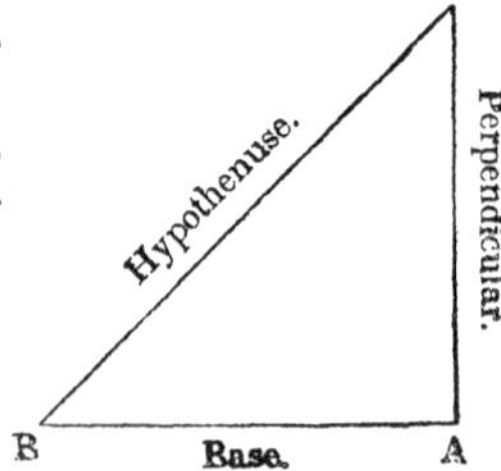

578. The square described on the *hypothenuse* of a right-angled triangle, is equal to the *sum* of the squares described on the *other two* sides. (Thomson's Legendre, B. IV. 11, Euc. I. 47.)

The truth of this principle may be seen from the following geometrical illustration. Thus,

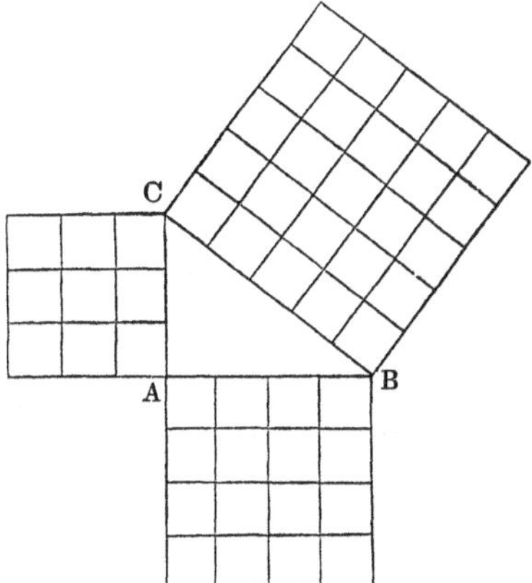

Let the base AB of the right-angled triangle ABC be 4 feet, the perpendicular AC, be 3 feet; then will the squares described on the base AB, and the perpendicular AC, contain as many square feet as the square described on the hypothenuse BC. Now $(4)^2+(3)^2=25$ sq. ft.; and the square described on BC also contains 25 sq. ft. Hence, *the square described on the hypothenuse of any right-angled triangle, is equal to the sum of the squares described on the other two sides.*

OBS. Since the square of the hypothenuse BC, is 25, it follows that the $\sqrt{25}$, or 5, must be the hypothenuse itself. Hence,

579. When the base and perpendicular are given, to find the *hypothenuse.*

Add the square of the base to the square of the perpendicular, and the square root of the sum will be the hypothenuse.

Thus, in the right-angled triangle ABC, if the base is 4 and the perpendicular 3, then $(4)^2+(3)^2=25$, and $\sqrt{25}=5$, the hypothenuse.

580. When the hypothenuse and base are given, to find the *perpendicular.*

From the square of the hypothenuse subtract the square of the base, and the square root of the remainder will be the perpendicular.

QUEST.—576. What is a triangle? What is a right-angle? 577. What is a right angled triangle? What is the side opposite the right-angle called? What are the other two sides called? 578. What is the square described on the hypothenuse equal to? 579. When the base and perpendicular are given, how is the hypothenuse found? 580. When the hypothenuse and base are given, how is the perpendicular found?

Thus, if the hypothenuse is 5 and the base 4, then $(5)^2-(4)^2=9$, and $\sqrt{9}=3$, the perpendicular.

581. When the hypothenuse and the perpendicular are given, to find the *base.*

From the square of the hypothenuse subtract the square of the perpendicular, and the square root of the remainder will be the base.

Thus, if the hypothenuse is 5 and the perpendicular 3, then $(5)^2-(3)^2=16$, and $\sqrt{16}=4$, the base.

OBS. 1. From the preceding principles it is manifest that the *area* of a square may be found by dividing the square of its *diagonal* by 2. (Arts. 285, 578.)

2. The areas of *all similar figures* are to each other as the *squares* of their *homologous sides*, or their *like dimensions*. (Leg. IV. 25, 27. V. 10.) Hence,

The sum of the areas of *equilateral* or other *similar triangles*, also of *similar polygons*, *circles* and *semicircles* described on the base and perpendicular of a right-angled triangle, is equal to the area of a *similar figure* described on the hypothenuse.

3. The *square* of a simple ratio is called a *duplicate* ratio; the cube of a simple ratio, a *triplicate* ratio.

The ratio of the *square roots* of two numbers is called a *sub-duplicate* ratio, that of the *cube* roots, a *sub-triplicate* ratio.

Ex. 1. If a street is 28 feet wide, and the height of a tower is 96 feet, how long must a rope be to reach from the top of the tower to the opposite side of the street?

Solution.—$(96)^2+(28)^2=10000$, and $\sqrt{10000}=100$ ft. *Ans.*

2. A ladder 40 feet long being placed at the opposite side of a street 24 feet wide, just reached the top of a house: how high was the house?

3. Two ships, one sailing 7 miles, the other 12 miles an hour, spoke each other at sea; one was going due east, the other due south: how far apart were they at the expiration of 12 hours?

4. What is the length of the side of a square farm which contains 360 acres; and how far apart are its opposite corners?

582. A *mean proportional* between two numbers is equal to the *square root* of their product. (Arts. 494, 498. Obs. 2.)

QUEST.—580. When the hypothenuse and perpendicular are given, how is the base found?

5. Find a mean proportional between 2 and 8.

Solution.—$8\times 2=16$; and $\sqrt{16}=4$. *Ans.*

Find a mean proportional between the following numbers:

6. 4 and 25.
7. 9 and 36.
8. 16 and 81.
9. 64 and 25.
10. 121 and 36.
11. 196 and 144.
12. 2.56 and 49.
13. 6.25 and 729.
14. $\frac{4}{9}$ and $\frac{16}{81}$.
15. $\frac{25}{36}$ and $\frac{64}{100}$.
16. $\frac{49}{81}$ and $\frac{144}{256}$.
17. $\frac{36}{49}$ and $\frac{169}{576}$.

583. To find the side of a square equal in area to any *given surface.*

Extract the square root of the given area, and it will be the side of the square sought.

OBS. When it is required to find the dimensions of a rectangular field, equal in *area* to a given surface, and whose length is double, triple, or quadruple, &c., of its breadth, the square root of $\frac{1}{2}$, $\frac{1}{3}$, $\frac{1}{4}$, of the given surface, will be the *width;* and this being *doubled, tripled,* or *quadrupled,* as the case may be, will be the *length.*

18. What is the side of a square equal in area to a rectangular field 81 rods long, and 49 rods wide?

19. What is the side of a square equal in area to a triangular field which contains 160 acres?

20. What is the side of a square equal in area to a circular field which contains 640 acres?

21. What are the length and breadth of a rectangular field which contains 480 acres, and whose length is triple its breadth?

22. A general arranged 10952 soldiers, so that the number in rank was double the file: how many were there in each?

584. When the *sum* of two numbers and the *difference* of their *squares* are given, to find the numbers.

Divide the difference of their squares by the sum of the numbers, and the quotient will be their difference; then proceed as in Art. 155.

23. The sum of two numbers is 42, and the difference of their squares is 756: what are the numbers? *Ans.* 12 and 30.

24. The sum of two numbers is 65, and the difference of their squares is 975: what are the numbers?

585. When the *difference* of two numbers and the *difference* of their *squares* are given, to find the numbers.

Divide the difference of the squares by the difference of the numbers, and the quotient will be their sum; then proceed as in Art. 155.

25. The difference of two numbers is 29, and the difference of their squares is 1885: what are the numbers?

EXTRACTION OF THE CUBE ROOT.

586. *To extract the cube root, is to resolve a given number into three equal factors; or, to find a number which being multiplied into itself twice, will produce the given number.* (Art. 564.)

Ex. 1. What is the cube root of 64?

Solution.—Resolving the given numbers into three equal factors, we have 64=4×4×4. *Ans.* 4.

2. What is the cube root of 12167?

Operation.

	Col. I.	Col. II.	12̇167̇(23
1st term	2	4×2=	8
2d "	4	1200 divisor,)	4167
3d "	63	1389×3=	4167

We first separate the given number into two periods, by placing a point over the units' figure, then over thousands. This shows us that the root must have two figures, (Art. 562. Obs. 3,) and thus enables us to find part of it at a time.

Beginning with the left hand period, we find the greatest cube in 12 is 8, the root of which is 2. Place the 2 on the right of the given number for the first part of the root, and also in Col. I. on the left of the number. Multiplying the 2 into itself, write the product 4 in Col. II.; and multiplying 4 by 2, subtract its product from the period, and to the right of the remainder bring down the next period for a dividend. Then adding 2, the first figure of the root, to the first term of Col. I., and multiplying the sum by 2, we add the product 8 to the 1st term of Col. II., and to this sum

QUEST.—586. What is it to extract the cube root?

annex two ciphers, for a divisor; also add 2, the first figure of the root, to the 2d term of Col. I. Finding the divisor is contained in the dividend 3 times, we place the 3 in the root, also on the right of the 3d term of Col. I. Then multiply the 3d term thus increased, by 3, the figure last placed in the root, and add the product to the divisor. Finally, we multiply this sum by 3 and subtract the product from the dividend. *Ans.* 23.

587. Hence, we derive the following general

RULE FOR EXTRACTING THE CUBE ROOT.

I. *Separate the given number into periods of three figures each, placing a point over units, then over every third figure towards the left in whole numbers, and over every third figure towards the right in decimals.*

II. *Find the greatest cube in the first period on the left hand; place its root on the right of the number for the first figure of the root, and also in Col. I. on the left of the number. Then multiplying this figure into itself, set the product for the first term in Col. II.; and multiplying this term by the same figure again, subtract this product from the period, and to the remainder bring down the next period for a dividend.*

III. *Adding the figure placed in the root to the first term in Col. I., multiply the sum by the same figure, add the product to the first term in Col. II., and to this sum annex two ciphers, for a divisor; also add the figure of the root to the second term of Col. I.*

IV. *Find how many times the divisor is contained in the dividend, and place the result in the root, and also on the right of the third term of Col. I. Next multiply the third term thus increased by the figure last placed in the root, and add the product to the divisor; then multiply this sum by the same figure, and subtract the product from the dividend. To the remainder bring down the next period for a new dividend.*

V. *Find a new divisor in the same manner that the last divisor was found, then divide, &c., as before; thus continue the operation till the root of all the periods is found.*

QUEST.—587. What is the first step in extracting the cube root? The second? Third? Fourth? Fifth? How is the cube root proved?

PROOF.—*Multiply the root into itself twice, and if the last product is equal to the given number, the work is right.*

OBS. 1. When there is a remainder, *periods of ciphers* may be added, as in square root.

2. If the right hand period of decimals is deficient, this deficiency must be supplied by ciphers. The root must contain as many decimals as there are *periods of decimals* in the given number.

588. *Demonstration.*—This rule depends upon the principle that the *cube* of the *sum* of two numbers is equal to the cube of the first part, added to 3 times the square of the first part into the last part, added to 3 times the first part into the square of the last, added to the cube of the last part. Take any number, as 23; we have $23=20+3$.

Then $(23)^3=(20)^3+(3\times20^2\times3)+(3\times20\times3^2)+3^3=12167$.

Or, $(23)^3=8000+3600+540+27=12167$.

After subtracting the greatest cube from the left hand period, it is plain the remainder must contain 3 times the square of the first part of the root into the last part, &c. Hence, if we divide the remainder by 3 times the square of the first part of the root, the quotient will be the last part. But it will be seen that the divisor is 3 times the square of the first part of the root, consequently the quotient must be the last part of the root required.

1. The *reason* for separating the given number into *periods of three figures*, is that the *cube* of a number can not have *more* figures than *triple* the number of figures in the root, nor but *two less*. It also shows how many figures the *root* will contain, and thus enables us to find part of it at a time. (Art. 562. Obs. 3.)

2. The reason for annexing 2 ciphers to the divisor is (manifestly) because the first figure of the root, of which the divisor is 3 times the square, stands in tens' place.

3. Placing the last figure of the root on the right of the 3d term in Col. I., then multiplying it by this figure, and adding the product to the divisor, and this sum being multiplied by the figure last placed in the root, the product will evidently be 3 times the square of the first part of the root into the last part, together with 3 times the first into the square of the last part, and the cube of the last part. In a similar manner the operation may be illustrated in all other cases.

Note.—The preceding method of extracting the cube root was discovered by the late Mr. Horner of Bath, England, and is often called Horner's Method. (For the common method, and its demonstration by cubical blocks, see Practical Arithmetic, p. 325.)

QUEST.—*Obs.* When there is a remainder, how proceed? When the right hand period of decimals is deficient, what must be done? How many decimals must the root contain? 588. Why separate the given number into periods of three figures? Why annex two ciphers to the right of the divisor?

589. The cube root of a *common fraction* is found by extracting the root of its numerator and denominator, or by first reducing it to a decimal.

A *mixed* number should be reduced to an improper fraction, or the fractional part to a decimal.

3. Required the cube root of 78314.6.

Operation.

	Col. I.	Col. II.	78314.600(42.78+
1st term	4	16×4 =	64
2d "	8	4800, 1st divisor)	14314
3d "	122	5044×2 =	10088
4th "	124	529200, 2d divisor)	4226600
5th "	1267	538069×7 =	3766483
6th "	1274	54698700, 3d divisor)	460117000
7th "	12818	54801244×8 =	438409952

590. When the root is required to many places of decimals, the operation may be contracted in the following manner.

First find one more than half the number of decimal figures required. For a new divisor, take as many figures plus one on the left of the last term in Col. II. as remain to be found in the root; and for a dividend retain one more figure on the left of the remainder than the divisor has; then proceed as in the contraction of division of decimals. (Art. 333.)

Required the cube root of the following numbers:

4. 91125.	8. 10218313.	12. 37.	16. $\frac{125}{343}$.
5. 140608.	9. 11543.176.	13. 6.	17. $\frac{2744}{6859}$.
6. 571787.	10. 20.570824.	14. 376.	18. $44\frac{3}{5}$.
7. 2515456.	11. .241804367.	15. 575.	19. $49\frac{8}{27}$

20. What is the cube root of 2 to eight decimals?

21. What is the cube root of $\frac{4}{15}$ to eleven decimals?

22. What is the side of a cubical mound which contains 314432 solid feet?

OBS. Similar solids are to each other as the cubes of their homologous sides, or like dimensions. (Leg. VII. 20, VIII. 11. Cor.) Hence,

QUEST.—589. How find the cube root of a common fraction? Of a mixed number?

591. To find the side of a cube whose solidity shall be double, triple, &c., that of a cube whose side is given.

Cube the given side, multiply it by the given proportion, and the cube root of the product will be the side of the cube required.

23. What is the side of a cubical bin which contains 8 times as many solid feet as one whose side is 4 feet? *Ans.* 8 ft.

24. What is the side of a cubical block which contains 4 times as many solid yards as one whose side is 6 feet?

25. If a ball 6 inches in diameter weighs 32 lbs., what is the weight of a ball whose diameter is 3 inches?

26. If a globe 4 ft. in diameter weighs 900 lbs., what is the weight of a globe 3 ft. in diameter?

592. To find two mean proportionals between two given numbers.

Divide the greater number by the less, and extract the cube root of the quotient. Multiply the root thus found by the least of the given numbers, and the product will be the least proportional sought; then multiply the least mean proportional by the same root, and this product will be the greater mean proportional required.

Find two mean proportionals between the following numbers:

27. 8 and 216. 29. 12 and 1500. 31. 71 and 15336.
28. 64 and 512. 30. 40 and 2560. 32. 83 and 60507.

EXTRACTION OF ROOTS OF HIGHER ORDERS.

593. When the index denoting the root to be extracted is a composite number.

First extract the root denoted by one of the prime factors of the given index; then of this root extract the root denoted by another prime factor, and so on. Thus,

For the 4th root, extract the *square* root *twice.*
For the 6th root, extract the *cube* root of the *square* root.
For the 8th root, extract the *square* root *three times.*
For the 27th root, extract the *cube* root *three times.*

1. What is the 4th root of 81? *Ans.* 3.

2. What is the 8th root of 256?
3. The 4th root of 65536?
4. The 4th root of 19987173376?
5. The 6th root of 46656?
6. The 6th root of 308915776?
7. The 8th root of 390625?
8. The 9th root of 40353607?
9. The 18th root of 387420489?
10. The 27th root of 134217728?

594. When the index denoting the root is not a **composite** number, we have the following general

RULE FOR EXTRACTING ALL ROOTS.

I. *Point off the number into periods of as many figures each, as there are units in the given index, commencing with the units figure.*

II. *Find the first figure of the root, and subtract its power from the left hand period; then to the right of the remainder bring down the first figure in the next period for a dividend.*

III. *Involve the root to the power next inferior to that of the index of the required root, and multiply it by the index itself, for a divisor.*

IV. *Find how many times the divisor is contained in the dividend, and the quotient will be the next figure of the root.*

V. *Involve the whole root to the power denoted by the index of the required root, and subtract it from the two left hand periods of the given number.*

VI. *Finally, bring down the first figure of the next period to the remainder, for a new dividend, and find a new divisor as before. Thus proceed till the whole root is extracted.*

OBS 1. The *reason* of this rule may be illustrated in the same manner as that for the extraction of the Square and Cube Roots.

2 The *proof* of all roots is by *involution.*

3. *Any root whatever* may be extracted by an *extension* of the principle applied to the extraction of the cube root. In this general application of the principle, the given number must be divided into periods, each consisting of as many figures as there are units in the index of the required root, and the number of columns employed will be one less than there are units in the given index. The operation then proceeds exactly as in the extraction of the cube root; and if there be a remainder, a like contraction is admissible.

11. Required the 5th root of 35184372088832.

Operation.

$$
\begin{array}{rl}
 & 35\dot{1}84372\dot{0}88832\dot{}(512 \ \textit{Ans.} \\
 & 3125 \\
5^4\times5=3125) & 3934 \\
51^5= & 345025251 \\
51^4\times5=33826005) & 68184698 \\
512^5= & 35184372088832
\end{array}
$$

12. Required the 5th root of $95\frac{2}{11}$.

13. Required the 7th root of 2103580000000000000.

Note.—The preceding method in most of the practical cases, gives perhaps as easy solutions, as the nature of the case will admit. But when roots of a very high order are required, the process may be shortened by the following *

APPROXIMATE RULE.

595. *Call the index of the given power* n; *and find by trial a number nearly equal to the required root, and call it the assumed root. Raise the assumed root to the power whose index is* n. *Then,*

As n+1 *times this power, added to* n—1 *times the given number, is to* n—1 *times the same power added to* n+1 *times the given number, so is the assumed root to the true root nearly.*

The number thus found may be employed as a new assumed root, and the operation repeated to find a result still nearer the true root.

14. Required the 365th root of 1.06.

Solution.—Take 1 for the assumed root, the 365th power of which is 1; and n being 365, we have $n+1=366$, and $n-1=364$. Then proceed in the following manner:

$$
\begin{array}{rcl}
1\times366=366 & & 1\times364=364 \\
1.06\times364=385.84 & & 1.06\times366=387.96 \\
\text{As } 751.84 & : & 751.96 :: 1 : \text{Ans.}
\end{array}
$$

Ans. 1.0001596

15. The 7th root of 2?

16. The 9th root of 2?

17. The 12th root of 1.05?

18. The 100th root of 100?

* Hutton's Mathematical Tracts; also Bonnycastle's Arithmetic.

SECTION XVIII

PROGRESSION.

ART. **596.** When there is a *series* of numbers such, that the *ratios* of the first to the second, of the second to the third, &c., are all equal, the numbers are said to be in *Continued Proportion*, or *Progression.* Progression is commonly divided into *arithmetical* and *geometrical.*

Note.—The terms *arithmetical* and *geometrical* are used simply to *distinguish* the *different kinds* of progression. They both belong equally to arithmetic and geometry.

ARITHMETICAL PROGRESSION.

597. Numbers which *increase* or *decrease* by a *common difference,* are in *arithmetical progression.* (Art. 474. Obs.)

OBS. 1. Arithmetical progression is sometimes called *progression by difference,* or *equidifferent series.*

2. When the numbers *increase,* the series is called *ascending;* as, 3, 5, 7, 9, 11, &c. When they *decrease,* the series is called *descending;* as, 11, 9, 7, 5, &c.

598. When *four* numbers are in arithmetical progression *the sum of the extremes is equal to the sum of the means.*

Thus, if 5—3=9—7, then will 5+7=3+9.

Again, if *three* numbers are in arithmetical progression, the sum of the extremes is double the mean.

Thus, if 9—6=6—3, then will 9+3=6+6.

599. In any arithmetical progression, the sum of the *two extremes* is equal to the sum of *any other two terms equally distant* from the extremes, or equal to *double the middle term,* when the number of terms is *odd.* Thus, in the series 1, 3, 5, 7, 9, it is obvious that 1+9=3+7=5+5.

600. In an *ascending* series, each succeeding term is found by adding the common difference to the preceding term. Thus, if the first term is 3, and the common difference 2, the series is 3, 5, 7, 9, 11, 13, 15, 17, 19, 21, &c.

In a *descending* series, each succeeding term is found by subtracting the common difference from the preceding term. Thus, if the first term is 15, and the common difference 2, the series is 15, 13, 11, 9, 7, &c.

601. In arithmetical progression there are five parts to be considered, viz: *the first term, the last term, the number of terms, the common difference, and the sum of all the terms.* These parts have such a relation to each other, that if any *three* of them are given, the other *two* may be easily found.

602. If the sum of the two extremes of an arithmetical progression is multiplied by the number of the terms, the product will be double the sum of all the terms in the series.

Take the series	2,	4,	6,	8,	10,	12.
The same inverted	12,	10,	8,	6,	4,	2.
The sums of the terms are	14,	14,	14,	14,	14,	14.

Thus, the sum of *all* the terms in the double series, is equal to the sum of the extremes *repeated* as many times as there are terms; that is, the sum of the double series is equal to 12+2 multiplied by 6. But this is *twice* the sum of the *single* series. Hence,

603. To find the *sum* of all the terms, when the *extremes* and the *number* of terms are given.

Multiply half the sum of the extremes by the number of terms, and the product will be the sum of the given series.

OBS. The *reason* of this process is manifest from the preceding illustration.

Ex. 1. The extremes of a series are 3 and 25, and the number of terms is 12: what is the sum of all the terms? *Ans.* 168.

2. What is the sum of the natural series of numbers, 1, 2, 3, 4, 5, &c., up to 100?

3. How many strokes does a common clock strike in 12 hours?

604. To find the *common difference*, when the *extremes* and the *number* of terms are given.

Divide the difference of the extremes by the number of terms less 1, and the quotient will be the common difference required.

OBS. The *truth* of this rule is manifest from Art. 602.

4. The extremes are 5 and 56, and the number of terms is 18: what is the common difference? *Ans.* 3.

5. If the extremes are 3 and 300, and the number of terms 10, what is the common difference?

605. To find the *number* of terms, when the *extremes* and *common difference* are given.

Divide the difference of the extremes by the common difference, and the quotient increased by 1 *will be the number of terms.*

OBS. The *truth* of this principle is manifest from the manner in which the successive terms of a series are formed. (Art. 600.)

6. If the extremes are 6 and 470, and the common difference is 8, what is the number of terms? *Ans.* 59.

7. If the extremes are 500 and 70, and the common difference is 10, what is the number of terms?

606. When the *sum* of the series, the *number* of terms, and *one* of the extremes are given, to find the *other extreme.*

Divide twice the sum of the series by the number of terms, and from the quotient take the given extreme.

OBS. The *reason* of this rule is manifest from Art. 602.

8. If the sum of a series is 576, the number of terms 24, and the first term 1, what is the last term? *Ans.* 47.

9. If the sum of a series is 1275, the number of terms 50, and the greater extreme 47¼, what is the less extreme?

607. To find *any given* term, when the *first* term and the common difference are given.

Multiply the common difference by one less than the number of terms required; then if the series be ascending, add the product to the first term; but if it be descending, subtract it.

OBS. The *reason* of this rule may be seen from the manner in which the succeeding terms of a series are formed. (Art. 600.)

10. If the first term of an ascending series is 7, and the common difference 3, what is the 41st term? *Ans.* 127.

11. If the first term of a descending series is 100, and the common difference 1¼, what is the 54th term?

12. If the first term of an ascending series is 7, and the common difference 5, what is the 100th term?

608. To find any given number of arithmetical *means*, when the *extremes* are given.

Subtract the less extreme from the greater, and divide the remainder by 1 *more than the number of means required; the quotient will be the common difference, which being continually added to the less extreme, or subtracted from the greater extreme, will give the mean terms required. One mean term may be found by taking half the sum of the extremes.* (Art. 598.)

OBS. This rule depends upon the same principle as that in Art. 604.

13. Required 3 arithmetical means between 7 and 35.

14. Required 6 arithmetical means between 1 and 99.

GEOMETRICAL PROGRESSION.

609. Numbers which *increase* by a *common multiplier*, or *decrease* by a *common divisor*, are in *Geometrical Progression.*

The numbers 4, 8, 16, 32, 64, &c., are in geometrical progression; and if each preceding term is multiplied by 2, the product will be the succeeding term; thus, $4\times2=8$; $8\times2=16$, &c.

Again, if the order of this series be inverted, the proportion will still be preserved and the common multiplier become a common divisor. Thus, in the series 64, 32, 16, 8, &c., $64\div2=32$; $32\div2=16$, &c.

Note.—If the *first* term and *ratio* are the *same*, the progression is simply series of powers; as 2; 2×2; $2\times2\times2$; $2\times2\times2\times2$, &c.

OBS. 1. *Geometrical Progression* is *geometrical proportion* continued. It is therefore sometimes called continual proportionals, or progression by *quotients.* If the series increases it is called *ascending;* if it decreases, *descending.*

2. The numbers which form the series, are called the *terms* of the progression. The *common multiplier*, or *divisor*, is called the *ratio.* For most purposes, however, it will be more simple to consider the ratio as always a *multiplier*, either *integral* or *fractional.* Thus, in the series 64, 32, 16, &c., the ratio is either 2 considered as a divisor, or $\frac{1}{2}$ considered as a multiplier.

3. In *Geometrical* as well as in *Arithmetical progression*, there are five parts to be considered, viz: *the first term, the last term, the number of terms, the ratio,* and *the sum of all the terms.* These parts have such a relation to each other, that if any *three* of them are given, the other *two* may be easily found.

610. To find the *last* term, when the *first* term, the ***ratio***, and the *number* of terms are given.

*Multiply the first term into that power of the ratio whose **index** is 1 less than the number of terms, and the product will be **the** last term required.*

OBS. 1. The *reason* of this process may be seen by adverting to the manner in which each successive term is formed. (Art. 609.) Thus, in the series 4, 8, 16, 32, &c., the 2d term $8=4\times2$; $16=4\times2\times2$, or 4×2^2; $32=4\times2^3$, &c.

2. It will be seen that the several *amounts* in compound interest, form a *geometrical series* of which the principal is the 1st term; the amount of $1 for 1 year the ratio; and the number of years+1 the number of terms. Hence the required *amount* of compound interest may be found in the same way as the *last term* of a geometrical series.

1. If the first term of a geometrical progression is 2, and the ratio 4, what is the 5th term? *Ans.* 512.

2. The first term is 64, and the ratio $\frac{1}{2}$: what is the 5th term?

3. The first term is 2, and the ratio 3: what is the 8th term?

4. The first term is 7, and the ratio 5: what is the 10th term?

5. A farmer hired a man for a year, agreeing to give him $1 for the 1st month, $2 for the 2d, $4 for the 3d, and so on, doubling his wages each month: how much did he give the last month?

6. What is the amount of $250, at 6 per cent., for 5 years compound int.? Of $500, at 7 per ct., for 6 years? Of $1000, at 5 per ct., for 10 years?

611. To find the *sum* of the series, when the *ratio* and the *extremes* are given.

Multiply the greatest term into the ratio, from the product subtract the least term, and divide the remainder by the ratio less 1.

OBS. 1. When the *first* term, the *ratio*, and the *number* of terms are given, to find the sum of the series we must first find the last term, then proceed as above.

2. The *sum* of an *infinite series* whose terms decrease by a common divisor, may be found *by multiplying the greatest term into the ratio, and dividing the product by the ratio less* 1. The *least* term being infinitely small, is of no comparative value, and is therefore neglected.

7. What is the sum of the series, whose extremes are 5 and 1215, and the ratio 3? *Ans.* 1820.

8. The extremes of a series are 1 and 512, and the ratio 2 what is the sum of the series?

9. The extremes of a series are 1024 and $15274\frac{1}{4}$, and the ratio is $1\frac{1}{2}$: what is the sum of the series?

10. A merchant hired a clerk for a year, and agreed to pay him 1 mill the 1st month; 1 cent the 2d; 10 cents the 3d, and so on, increasing in a tenfold ratio for each successive month: what was the amount of his wages?

11. What is the sum of the infinite series $1+\frac{1}{2}+\frac{1}{4}+\frac{1}{8}$, &c.; that is, the descending series whose first term is 1 and the ratio 2?

Ans. 2.

12. What is the sum of the infinite series $1+\frac{1}{3}+\frac{1}{9}+\frac{1}{27}+\frac{1}{81}$, &c.

612. To find the *ratio*, when the *extremes* and *number* of terms are given.

Divide the greater extreme by the less, and extract that root of the quotient whose index is 1 *less than the number of terms.*

13. The extremes of a series are 3 and 192, and the number of terms 7: what is the ratio? *Ans.* 2.

14. What is the ratio of a series of 5 terms, whose extremes are 7 and 567?

Note.—Other formulas in arithmetical and geometrical progression might be added, but they involve principles with which the student is supposed as yet to be unacquainted. For a fuller discussion of the subject, see Thomson's Day's Algebra.

ANNUITIES.

613. The term *annuity* properly signifies a sum of money, payable annually, for a certain length of time, or forever.

OBS. 1. Payments made semi-annually, quarterly, monthly, &c., are also called annuities. Annuities therefore embrace pensions, salaries, rents, &c.

2. When annuities remain unpaid after they are due, they are said to be *forborne*, or *in arrears*. The *sum* of the annuities in arrears, added to the *interest* due on each, is called the *amount.*

The *present worth* of an annuity is the sum, which being put at interest, will exactly pay the annuity.

3. When an annuity does not commence till a given time has elapsed, it is called an annuity *in reversion;* when it continues *forever*, a *perpetuity.*

4. In finding the *amount* of annuities in *arrears*, it is customary to reckon compound interest on each annuity from the time it is due to the time of payment. The process therefore is the same as finding the sum of an ascending geometrical series. (Art. 611.) Hence,

614. To find the amount of an annuity in arrears.

Make the annuity the first term of a geometrical series, the amount of \$1 *for* 1 *year the ratio, and the given number of years the number of terms; then find the sum of the series, and it will be the amount required.* (Arts. 610, 611.)

OBS. When the payments are not *yearly*, for the amount of \$1 for 1 year, use its amount for the time between the payments; and instead of the *number* of *years*, use the number of payments that have been omitted, and proceed as before.

1. What is the amount of an annuity of \$100 which has not been paid for 3 years, at 6 per cent. compound interest?

Solution.—$100\times(1.06)^2=112.36$; and $(112.36\times1.06)-100\div.06=$ \$318.36.

TABLE, *showing the amount of annuity of* \$1, *or* £1, *at* 5, 6, *and* 7 *per cent. for any number of years from* 1 *to* 20.

Yrs.	5 per ct.	6 per ct.	7 per ct.	Yrs.	5 per ct.	6 per ct.	7 per ct.
1	1.00000	1.00000	1.00000	11	14.20678	14.97164	15.7836
2	2.05000	2.06000	2.07000	12	15.91712	16.86994	17.8884
3	3.15250	3.18360	3.21490	13	17.71298	18.88213	20.1406
4	4.31012	4.37461	4.43994	14	19.59863	21.01506	22.5504
5	5.52563	5.63709	5.75073	15	21.57856	23.27596	25.1290
6	6.80191	6.97532	7.15329	16	23.65749	25.67252	27.8880
7	8.14201	8.39383	8.65402	17	25.84036	28.21287	30.8402
8	9.54911	9.89746	10.2598	18	28.13238	30.90565	33.9990
9	11.02656	11.49131	11.9799	19	30.53900	33.75999	37.3789
10	12.57789	13.18079	13.8164	20	33.06595	36.78559	40.9954

Note.—Multiply the given annuity by the amt. of \$1, for the given number of years found in the Table, and the product will be the amount required.

3. What will an annual rent of \$75 amount to in 9 years, at 5 per cent.?
4. What is the amount of \$200 forborne for 9 years, at 6 per cent.?
5. What is the amount of \$350 forborne for 10 years, at 7 per cent.?
6. What is the amount of \$1000 forborne for 20 years, at 6 per cent.?

615. To find the present worth of an annuity.

Find the amount of \$1 *annuity for the given time as before; then divide this amount by the amount of* \$1 *at compound interest for the same time, multiply the quotient by the given annuity, and the product will be the present worth. If the annuity is a perpetuity, or to continue forever, multiply it by* 100, *divide the product by the given rate, and the quotient will be the present value required.*

OBS. For the amount of \$1 at compound interest, see Table, p. 271.

7. What is the present worth of an annuity of \$40 to continue 5 years, at 5 per cent. compound interest? *Ans.* \$173.178.

8. What is the present worth of an annuity of $80 to continue forever, at 6 per cent.?

616. To find the present worth of an annuity *in reversion.*

Find the present worth of the annuity from the present time till its termination; also find its present worth for the time before it commences; the difference between these two results will be the present worth required.

9. What is the present worth of $79.625 at 5 per cent., to commence in 4 years and continue 6 years? *Ans.* $332.50.

PERMUTATIONS AND COMBINATIONS.

617. By *Permutations* is meant the changes which may be made in the *arrangement* of any given number of things.

The term *combinations,* denotes the taking of a *less* number of things out of a *greater,* without regard to their order or position.

618. To find how many permutations or changes may be made in the arrangement of any given number of things.

Multiply together all the terms of the natural series of numbers from 1 *up to the given number, and the product will be the answer.*

1. How many changes may be rung on 5 bells? *Ans.* 120.
2. How many different ways may a class of 8 pupils be arranged?
3. How many different ways may a family of 9 children be seated?
4. How many ways may the letters in the word *arithmetic,* be arranged?
5. A club of 12 persons agreed to dine with a landlord as long as he could seat them differently at the table: how long did their engagement last?

619. To find how many combinations may be made out of any given number of different things by taking a given number of them at a time.

Take the series of numbers, beginning at the number of things given, and decreasing by 1 *till the number of terms is equal to the number of things taken at a time; the product of all the terms will be the answer required.*

6. How many different words can be formed of 9 letters, taking 3 at a time?
Solution.—$9\times8\times7=504$. *Ans.* 504 words.

7. How many numbers can be expressed by the 9 digits, taking 5 at a time?
8. How many words of 6 letters each can be formed out of the 26 letters of the alphabet, on the supposition that consonants will form a word?

SECTION XIX.

APPLICATION OF ARITHMETIC TO GEOMETRY.

620. In the preceding sections *abstract* numbers have been applied to *concrete substances*, or to objects in general, considered arithmetically. On the same principle, *geometrical magnitudes* may be *compared* or *measured* by means of the numbers representing their dimensions. (Arts. 7, 516. Obs. 3.)

OBS. The measurement of magnitudes is commonly called *mensuration*

MENSURATION OF SURFACES.

621. In the measurement of *surfaces*, it is customary to assume a *square* as the *measuring unit*, whose side is a *linear unit* of the *same name*. (Leg. IV. 4. Sch. Art. 257. Obs. 2.)

Note.—For the demonstration of the following principles, see references.

622. To find the area of a parallelogram, also of a square.

Multiply the length by the breadth. (Art. 285, Leg. IV. 5.)

OBS. When the *area* and *one side* of a rectangle are given, *the other side* is found by dividing the *area* by the *given side.* (Art. 156.)

1. How many acres in a field 240 rods long, and 180 rods wide?
2. How many acres in a square field the length of whose side is 340 rods?
3. If the diagonal of a square is 100 rods, what is its area?
4. A rectangular farm of 320 acres, is ½ a mile wide: what is its length?

623. To find the area of a rhombus. (Leg. I. Def. 18. IV. 5.)

Multiply the length by the altitude or perpendicular height.

5. Find the area of a rhombus whose length is 20 ft., and its altitude 18 ft.

624. To find the area of a trapezium. (Leg. IV. 7.)

Multiply half the sum of the parallel sides by the altitude.

6. Find the area of a trapezium the lengths of whose parallel sides are 17 ft. and 31 ft., and whose altitude is 15 ft.

625. To find the area of a triangle. (Leg. IV. 6.)

Multiply the base by half the altitude or perpendicular height.

7. Find the area of a triangle whose base is 50 ft., and its altitude 44 ft.

626. To find the area of a triangle, the three sides being given.

From half the sum of the three sides subtract each side respectively; then multiply together half the sum and the three remainders, and extract the square root of the product.

9. What is the area of a triangle whose sides are 20, 30, and 40 ft.?
10. How many acres in a triangle whose sides are each 40 rods?

627. To find the circumference of a circle from its diameter.

Multiply the diameter by 3.14159. (Leg. V. 11. Sch.)

Note.—The *circumference* of a circle is a curve line, all the points of which are equally distant from a point within, called the *centre.* The *diameter* of a circle is a straight line which passes through the centre, and is terminated on both sides by the circumference. The *radius* or *semi-diameter* is a straight line drawn from the centre to the circumference.

11. What is the circumferenee of a circle, whose diameter is 20 ft.?
12. What is the circumference of a circle, whose diameter is 45 rods?

628. To find the diameter of a circle from its circumference.

Divide the circumference by 3.14159.

OBS. The *diameter* of a circle may also be found by dividing the *area* by 7854, and extracting the square root of the quotient.

13. What is the diameter of a circle, whose circumference is 314.159 ft.?

629. To find the area of a circle. (Leg. V. 11.)

Multiply half the circumference by half the diameter; or, multiply the square of the diameter by the decimal .7854.

15. What is the area of a circle, whose diameter is 50 rods?
16. Find the area of a circle 200 ft. in diameter, and 628.318 ft. in circum.

630. To find the side of the greatest square that can be inscribed in a circle of a given diameter.

Divide the square of the given diameter by 2, *and extract the square root of the quotient.* (Art. 581. Obs. 1.)

17. The diameter of a round table is 4 ft.; what is the side of the greatest square table which can be made from it?

631. To find the side of the greatest equilateral triangle that can be inscribed in a circle of a given diameter.

Multiply $\frac{1}{2}$ *the given diameter by* 1.73205. (Leg. V. 4. Sch.)

18. Required the side of an equilateral triangle inscribed in a circle of 20 ft. diameter.

MEASUREMENT OF SOLIDS.

632. In the measurement of solids it is customary to assume a *cube* as the *measuring unit,* whose sides are squares of the same name. (Art. 258. Obs. 2.)

633. To find the solidity of bodies whose sides are perpendicular to each other.

Multiply the length, breadth, and thickness together. (Art. 286.)

OBS. When the *contents* of a solid body and *two* of its *sides* are given, the *other side* is found by dividing the *contents* by the *product* of the two given sides. (Art. 159.)

1. What are the contents of a stick of timber 4 ft. square, and 85½ ft. long?
2. What is the capacity of a cubical vessel, 14 ft. 8 in. deep?

634. To find the solidity of a prism.

Multiply the area of the base by the height. (Leg. VII. 12.)

OBS. This rule is applicable to all *prisms*, triangular, quadrangular, pentagonal, &c., also to all *parallelopipedons*, whether rectangular or oblique.

3. Find the solidity of a prism 46½ ft. high, whose base is 7½ ft. square?

635. To find the lateral surface of a right prism.

Multiply the length by the perimeter of its base. (Leg. VII. 5.)

OBS. If we add the areas of both ends to the lateral surface, the sum will be the whole surface of the prism.

4. What is the surface of a triangular prism, whose sides are each 3 ft., and its length 12 ft.?

636. To find the solidity of a pyramid and cone.

Multiply the area of the base by ⅓ of the height. (Leg. VII. 18.)

5. What is the solidity of a pyramid 100 ft. high, whose base is 40 ft. square?
6. What is the solidity of a cone 150 ft. high, whose base is 15 ft. in diameter?

637. To find the lateral or convex surface of a regular pyramid, or cone. (Leg. VII. 16, VIII. 3.)

Multiply the perimeter of the base by ½ the slant-height.

7. What is the lateral surface of a regular pyramid, whose slant-height is 15 ft., and base is 30 ft. square?
8. What is the convex surface of a right cone, whose slant-height is 94 ft. and the perimeter of its base 37 ft.?

638. To find the solidity of a frustum of a pyramid and cone.

To the sum of the areas of the two ends, add the square root of the product of these areas; then multiply this sum by $\frac{1}{3}$ of the perpendicular height. (Leg. VII. 19. Sch., VIII. 6.)

9. If the two ends of the frustum of a pyramid are 3 ft. and 2 ft. square, and the height is 12 ft., what is its solidity?

639. The convex surface of a frustum of a pyramid and cone is found *by multiplying half the sum of the circumferences of the two ends by the slant-height.* (Leg. VII. 17, VIII. 5.)

10. If the circumferences of the two ends of the frustum of a cone are 18 ft. and 14 ft., and its slant-height 11 ft., what is its convex surface?

640. To find the solidity of a cylinder.

Multiply the area of the base by the height. (Leg. VIII. 2.)

11. Find the solidity of a cylinder 10 ft. in diameter, and 35 ft. high.
12. Find the solidity of a cylinder 100 ft. in circumference, and 150 ft. high.

641. To find the convex surface of a cylinder.

Multiply the circumference of the base by the height. (Leg. VIII. 1.)

13. Find the convex surface of a cylinder 5 yds. in diameter, and 5 yds. long.

642. To find the convex surface of a sphere or globe.

Multiply the circumference by the diameter. (Leg. VIII. 9.)

14. What is the surface of a globe 18 inches in diameter?
15. If the diameter of the moon is 2162 miles, what is its surface?

643. To find the solidity of a sphere or globe.

Multiply the surface by $\frac{1}{6}$ of the diameter. (Leg. VIII. 11.)

16. Find the solidity of a globe 15 inches in diameter.
17. The diameter of the moon is 2162 miles: what is its solidity?

MEASUREMENT OF LUMBER.

644. The area of a board is found *by multiplying the length into the mean breadth.* (Arts. 622, 623.)

The solid contents of hewn or square timber are found *by multiplying the length into the mean breadth and depth.*

The solid contents of round timber are found *by multiplying the length by $\frac{1}{4}$ the mean girt or circumference.*

OBS. 1. The mean breadth of a tapering board is found by measuring it in the middle, or by taking $\frac{1}{2}$ the sum of the breadths of the two ends.

2. The *mean* dimensions of square and round timber are found in a similar manner.

3. The method for finding the solidity of round timber makes an allowance of about $\frac{1}{5}$ for waste in hewing. (Arts. 640, 258. Obs. 3.)

18. Find the area of a board 12 ft. long, and the ends 14 in. and 12 in. wide.

19. Find the solidity of a joist 16 ft. long, the ends being 8 in. and 4 in. sq.

20. Find the solidity of a log 50 ft. long, the circumferences of the ends being 6 ft. and 4 ft.

GAUGING OF CASKS.

645. The process of finding the contents or capacities of *casks* and other *vessels* is called GAUGING.

646. *The contents of casks are found by multiplying the square of the mean diameter into the length; then this product multiplied by* .0034 *will give the wine gallons, and multiplied by* .0028 *will give the beer gallons.*

OBS. The *mean diameter* of a cask is found by adding to the head diameter .7 of the difference between the head and bung diameters when the staves are *very much* curved; or by adding .5 when *very little* curved; and by adding .55 when they are of a *medium* curve.

21. How many wine gallons in a cask but little curved, whose length is 45 in., its bung diameter 40 in., and its head diameter 36 in.?

22. How many beer gallons in a cask much curved, whose length is 64 in., its bung diameter 52 in., and head diameter 46 in.?

TONNAGE OF VESSELS.

647. *Government Rule.*—I. If the vessel be *double-decked*, take the length from the fore part of the main stern to the after part of the stern-post, above the upper deck; then the breadth at the broadest part above the main wales, half of which breadth shall be accounted the depth of such vessel; from the length deduct three-fifths of the breadth, multiply the remainder by the breadth and the product by the depth; divide the last product by 95, and the quotient shall be deemed the true tonnage of the vessel.

II. If the vessel be *single-decked*, take the length and breadth as above directed, deduct from the length three-fifths of the breadth, and take the depth from the under side of the deck plank to the ceiling in the hold, then multiply and divide as before, and the quotient shall be deemed the tonnage.

Carpenter's Rule.—The continued product of the length of the keel, the breadth at the main beam, and the depth of the hold in feet, divided by 95 will give the tonnage of a *single-decked* vessel. For a *double-decker*, instead of the depth of the hold, take half the breadth at the beam.

23. What is the government tonnage of a double-decker, whose length is 150 ft., the breadth 35 ft., and the depth 25 ft.?

24. What is the carpenter's tonnage of the same vessel?

MECHANICAL POWERS.

648. The *Mechanical powers* are six, viz: the *lever*, the *wheel and axle*, the *pulley*, the *inclined plane*, the *screw*, and the *wedge*.

649. When the *power* and *weight* act *perpendicularly* to the arms of a *straight lever*, the *power* is to the *weight*, as the *distance* from the fulcrum to the *weight* is to the *distance* from the fulcrum to the *power*.

1. If the power is 100 lbs., the long arm 10 ft., and the short arm 2 ft., what weight can be raised?
2. The arms of a lever are 15 ft. and 4 ft., and the weight raised 500 lbs.: what is the power?

650. When a weight is sustained by a lever resting on *two props*,

The long arm : the *short arm* :: the *weight* supported by the short arm : the *weight* supported by the long arm. Hence,

The *whole length* : *short arm* :: *whole weight* : *weight on l. a.* (Leg. III. 16.)

3. A and B carry 256 lbs. suspended upon a pole 5 ft. from A and 3 ft. from B: how many pounds does each carry?
4. A and B carry 90 lbs. upon a lever 12 ft. long: where must it be placed that B may carry ⅓ of it?

651. The *wheel* and *axle* operate on the same principle as the *lever*; the semi-diameter of the *wheel* answers to the long arm, and the semi-diameter of the *axle* to the short arm.

5. If the diameter of a wheel is 6 ft., and that of the axle 1 ft., what weight will 100 lbs. raise?
6. A wheel is 8 ft. diameter, an axle 1⅓ ft.: what weight will 200 lbs. raise

652. In the application of *movable pulleys*,

The POWER : *the* WEIGHT :: 1 : *twice the* NUMBER *of pulleys*.

7. What weight can a power of 200 lbs. raise with 4 movable pulleys?
8. What power with 8 pulleys will raise a pillar of granite weighing 10 tons?

653. *The perpendicular height of an inclined plane is to its length, as the power to the weight.*

9. What power will draw a train of cars weighing 100000 pounds up an inclined plane which rises 60 ft. to a mile?

654. The screw acts upon the principle of the inclined plane. Hence,

The distance between the threads is to the circumference of a circle described by the power, as the power is to the weight.

10. What weight can be raised by a power of 1000 lbs. applied to a screw whose threads are 1 inch apart, at the end of a lever 12 ft. long?

655. The *power* applied to the head of a wedge is to the *weight*, as half the thickness of the head is to the length of its side. In the use of the wedge, not less than *half* the power is lost by friction against the sides.

MISCELLANEOUS EXAMPLES.

1. The sum of two numbers is 980, and their difference 62: what are the numbers?

2. The product of two numbers is 4410, and one is 63: what is the other?

3. What number multiplied by $28\frac{5}{7}$, will produce 145?

4. What number multiplied by $6\frac{1}{2}$, will be equal to $7\frac{1}{4}$ multiplied by $5\frac{1}{2}$?

5. If an army of 24000 men have 720000 lbs. of bread, how long will it last them, allowing each man $1\frac{1}{2}$ lbs. per day?

6. What is the interest of $5256 for 60 days, at 7 per cent.?

7. What is the amount of $16230 for 4 months, at $6\frac{1}{2}$ per cent.?

8. What is the bank discount on $1200 for 90 days, at 6 per cent.?

9. For what sum must a note be made, payable in 4 months, the proceeds of which shall be $1800, discounted at a bank at 7 per cent.?

10. A capitalist sent a broker $25000 to invest in cotton, after deducting his commission of $2\frac{1}{2}$ per cent.: what amount of cotton ought he to receive?

11. A merchant bought 500 yards of cloth for $1800: how must he retail it by the yard to gain 25 per cent.?

12. A man bought 640 bbls. of beef for $5000, and sold it at a loss of 12 per cent.: how much did he get a barrel?

13. If a man buys 1000 geographies, at $37\frac{1}{2}$ cents apiece, and retails them at 50 cents, what per cent. will he make?

14. A grocer bought 180 boxes of lemons for $360, and sold them at 10 per cent. less than cost: what did he lose?

15. How many dollars, each weighing $412\frac{1}{2}$ grs., can be made from 16 lbs. 5 oz. of silver?

16. How many eagles, weighing 258 grs. apiece, will 21 lbs. 10 oz. make?

17. How long a thread can be spun from 1 ton of flax, allowing 5 oz. will make 100 rods of thread?

18. How many revolutions will the hind wheel of a carriage 5 ft. 6 in. in circumference, make in 2 miles 4 furlongs?

19. How many revolutions will the fore wheel of a carriage 4 ft. 7 in. in circumference, make in the same distance?

20. Bought 1500 doz. buttons for $187.50: what was that per gross?

21. A man paid $132 for 40 bbls. of cider: what is that a quart?

22. A man paid $150 for 10 rods of land, what was that per acre?

23. A man having $2500, laid out $\frac{7}{8}$ of it in flour, at $5 per barrel: how many barrels did he buy?

24. The commander of an exploring expedition found that $\frac{4}{7}$ of his provisions were exhausted in 28 months: how much longer would they last?

25. What cost $15\frac{2}{5}$ lbs. of cheese, at $\$8\frac{5}{8}$ per hundred?

26. How many yards of carpeting $\frac{3}{4}$ yd. wide will it take to cover a floor 18 ft. long and 15 ft. wide?

27. If $\frac{3}{5}$ yard of calico cost $\frac{7}{12}$s., what will $\frac{7}{8}$ of an ell English cost?

28. How long will 468256 lbs. of beef last an army of 8245 soldiers, allowing each man $1\frac{1}{4}$ lb. per day?

29. How long would the same quantity of beef last the army, if reinforced by 2500 men, allowing each man $1\frac{1}{4}$ lb. per day?

30 Bought $\frac{5}{8}$ of a pipe of wine for $126: what was that per gallon?

31. If a man can walk 17 miles in 5 hours, 12 minutes, 31 seconds, how far can he walk in 3 hours, 40 minutes, 36 seconds?

32. If a man traveling 14 hours per day, performs half his journey in 9 days, how long will it take him to go the other half traveling 10 hours a day?

33. If you lend a man $700 for 90 days, how long ought he to lend you $1200 to requite the favor?

34. A milkman's measure was deficient half a gill to a gallon: how much did he cheat his customers in selling 8720 gallons?

35. If 85 yds. of calico cost $10.20, what will 1500 yds. cost?

36. If 1650 lbs. of sugar cost $206.25, what will 87 lbs. cost?

37. If 1424 gals. of oil cost $1062, what will 210 gals. cost?

38. If wind moves $2\frac{3}{4}$ miles per hour, how long is it in moving from the pole to the equator, a distance of 6214 miles?

39. If $\frac{3}{16}$ of a barrel of flour costs $\frac{3}{5}$ of a dollar, what will $\frac{5}{8}$ of a bbl. cost?

40. If $\frac{3}{4}$ of a ton of chalk cost £$\frac{6}{7}$, what will $\frac{7}{8}$ of a ton cost?

41. If $\frac{3}{5}$ of a bushel of wheat cost $$\frac{6}{9}$, how much will $\frac{5}{6}$ of a bushel cost?

42. If $\frac{5}{8}$ of a ship cost $16000, what will $\frac{3}{16}$ of her cost?

43. If $16\frac{1}{8}$ bbls. of mackerel cost $$65\frac{3}{4}$, what will $48\frac{1}{2}$ bbls. cost?

44. If $28\frac{1}{2}$ gals. of oil cost $31.25, how much will $250 buy?

45. At $$3\frac{1}{2}$ for 40 doz. eggs, what will $460\frac{1}{2}$ doz. cost?

46. The President's salary is $25000 per annum: how much can he spend per day, and lay up $10000 of it?

47. If one person lies in bed 9 hours per day, and another 6 hours, how much time will the one gain over the other in 20 years?

48. A cistern has three faucets; the 1st will empty it in 10 min., the 2d in 20 min., the 3d in 30 min.: in what time will they all empty it?

49. A man and his wife drink a barrel of beer in 30 days, and the man alone can drink it in 40 days: how long will it last the wife?

50. A teacher being asked how many scholars he had, replied $\frac{1}{4}$ study Arithmetic, $\frac{1}{5}$ study Latin, $\frac{3}{10}$ study Algebra, $\frac{1}{20}$ study Geometry, and 24 study French: how many scholars had he?

51. A man having spent $\frac{1}{2}$ and $\frac{1}{3}$ of his money, had £$48\frac{2}{3}$ left: how much had he at first?

52. A man bequeathed $\frac{1}{3}$ of his property to his wife, $\frac{1}{4}$ to his son, $\frac{1}{6}$ to his daughter, and the remainder, which was $1500, to the Bible Society: what did his whole property amount to?

53. What is that number $\frac{7}{8}$ of which exceeds $\frac{4}{5}$ of it by 45?

54. Pewter is composed of 112 parts of tin, 15 of lead, and 6 of brass: how much will it take of each ingredient to make 6650 pounds of pewter?

55. Two travelers start at the same time from Boston and Washington to meet each other; one goes 5 miles an hour, the other 7 miles; the whole distance is 436 miles: how far will each travel?

56. A grocer divided a barrel of flour into 2 parts, so that the smaller contained $\frac{3}{5}$ as much as the other: how many pounds were there in each?

57. A, B, and C, build a ship together; A advanced \$1000, B \$12000, and C \$13000; they gain \$5000: what is the gain of each?

58. A, B, and C. entered into partnership; A furnished \$600, B and C together \$1800; they gained \$960, of which B took \$280: how much did A and C gain; and B and C put in respectively?

59. The liabilities of a bankrupt are \$63240, and his assets \$12648: what per cent. can he pay?

60. A bankrupt compromises with his creditors for $37\frac{1}{2}$ per cent.: how much will he pay on a claim of \$3656?

61. How much will he pay on a debt of \$12680.375?

62. A owns $\frac{3}{8}$ and B $\frac{1}{16}$ of a ship; A's share is worth \$10000 more than B's: what is the value of the ship?

63. A man gave his oldest son $\frac{1}{2}$ of his property less \$50; to the second, he gave $\frac{1}{3}$; and to the youngest he gave the remainder, which was $\frac{1}{3}$ less \$10: what was the amount of his property?

64. A man and boy together can frame a house in 9 days; the man can frame it alone in 12 days: how long will it take the boy to frame it?

65. A cistern has a receiving and a discharging pipe; when both are running it takes 18 hours to fill it; if the latter is closed it requires 15 hours to fill it: if the former is closed, how long will it take the latter to empty it?

66. Four men, A, B, C, and D, spent £255, and agreed that A should pay $\frac{3}{8}$; B $\frac{1}{4}$; C $\frac{1}{2}$; and D $\frac{3}{4}$: how much must each pay?

67. A, B, and C, formed a joint stock of £820, and gained £640, in the division of which A received £5 as often as B did £7, and C £8: how much did each put in and receive?

68. A, B, and C, gained a certain sum, of which A and B received \$640, B and C \$880, and A and C \$800: what was the gain of each?

69. What number is that $\frac{3}{15}$ and $\frac{1}{4}$ of which being multiplied together, will produce the number itself?

70. A club spent £2, 12s. 1d.; on settling, each paid as many pence as there were individuals in the party: how many were there in the party?

71. The sum of two numbers is 120, and the difference of their squares is 4800: what are the numbers?

72. The difference of two numbers is 53, and the difference of the squares is 10759: what are the numbers?

73. The diagonal of a square is 80 ft.: what is its side?

74. The diagonal of a square field is 120 rods: what is its area?

75. Find the side of the greatest square beam which can be hewn from a log 5 ft. in diameter?

76. The mainmast of a ship is 95 ft. long, the diameter of the base is $3\frac{1}{2}$ ft., that of the top $2\frac{1}{2}$ ft.: what is its solidity?

77. A man wished to tie his horse by a rope so that he could feed on just an acre of ground: how long must the rope be?

78. What is the area of a circle 1 mile in circumference?

79. If the diameter of the sun is 887000 miles, what is its surface?

80. If the diameter of Jupiter is 86255 miles, what is its solidity?

81. A conical stack of hay is 20 ft. high, and its base 15 ft. in diameter what is its weight, allowing 5 lbs. to a cubic foot?

82. How many bushels will a cubical bin contain whose side is 9 ft.?

83. How many hogsheads will a cylindrical cistern 10 ft. deep and $6\frac{1}{2}$ ft. diameter contain?

84. How far from the end of a stick of timber 30 ft. long, of equal size from end to end, must a lever be placed, so that 3 men, 2 at the lever, and 1 at the end of the stick, may each carry $\frac{1}{3}$ of its weight?

85. How many different ways may a class of 26 scholars be arranged?

86. If 100 eggs are placed in a straight line a rod apart, how many miles must a person travel to bring them one by one to a basket placed a rod from the first egg?

87. What is the sum of the series 1, $1\frac{1}{2}$, 2, $2\frac{1}{2}$, 3, &c., to 50 terms?

88. A blacksmith agreed to shoe a horse for 1 mill for the first nail in his shoe, 2 mills for the second nail, and so on: the shoes contained 32 nails: how much did he receive?

89. Said a mule to an ass, if I take one of your bags, I shall have twice as many as you, and if I give you one of mine, we shall have an equal number: with how many bags was each loaded?

90. What number taken from the square of 48 will leave 16 times 54?

91. Divide $1000 between A, B, and C, and give A $120 more than C, and C $95 more than B.

92. A person being asked the hour of the day, said, that the time past noon was $\frac{4}{5}$ of the time till midnight: what was the hour?

93. A, B, and C, can trench a meadow in 12 days; B, C, and D, in 14 days; C, D, and A, in 15 days; and D, B, and A, in 18 days. In what time would it be done by all of them together, and by each of them singly?

94. Suppose A, B, and C, to start from the same point, and to travel in the same direction, round an island 73 miles in compass, A at the rate of 6, B of 10, and C of 16 miles per day: in what time will they be next together?

95. At what time between 12 and 1 o'clock do the hour and minute hands of a common clock or watch point in directions exactly opposite?

96. In how many years will the error of the Julian Calendar involve the loss of a day?

97. A man's desk was robbed 3 nights in succession; the first night half the number of dollars were taken and half a dollar more; the second, half the remainder was taken and half a dollar more; the third night, half of what was then left and half a dollar more, when he found he had $50 left: how much had he at first?

THE END.

ANSWERS TO EXAMPLES.

NOTE.—At the urgent request of several distinguished Teachers, who have received Thomson's Higher Arithmetic with favor, the publishers have issued an edition of it, containing the answers in the end of the book. It is hoped that pupils, who may use this edition, will have sufficient regard to their own improvement, never to consult the answer till they have made a *strenuous* and *persevering* effort to solve the problem themselves.

N. B.—The work without the answers is published as heretofore.

ADDITION.—ARTS. 59-61.

Ex.	Ans.	Ex.	Ans.	Ex.	Ans.
1.	$5445.	17.	288011295.	35.	9429190.
2.	41757 bushels.	18.	14303433.	36.	11178170.
3.	11596 pounds.	19.	100611775.	37.	10306156.
4.	$31551.	20.	1805851434.	38.	10662291.
5.	$5583.	21.	337351.	39, 40.	Given.
6.	65440 sq. miles.	22.	7221.	41.	214.
7.	102451 sq. m.	23.	4251988.	42.	253.
8.	528524 sq. m.	24.	3795.	43.	276.
9.	666327 sq. m.	25.	73464390.	44.	19443.
10.	1362742 sq. m.	26.	33604444.	45.	20714.
11.	233890.	27.	15821984.	46.	2476372.
12.	828463.	28.	97059404.	47.	$132085946.
13.	990240.	29.	1038220930.	48.	$107109740.
14.	96181521.	32.	$570805.	49.	2069857 tons.
15.	127215713.	33.	6460458 yards.	50.	$57981492.
16.	869754587.	34.	6657039 pounds.	51.	Given.

SUBTRACTION.—ART. 76.

Ex.	Ans.	Ex.	Ans.	Ex.	Ans.
1.	$7095.	10.	$12280043.	19.	5313439.
2.	28984 bushels.	11.	$23563746.	20.	543679.
3.	$30954.	12.	430143 tons.	21.	2007984.
4.	$46025.	13.	149237.	22.	45103074.
5.	58000000 miles.	14.	3393329.	23.	66729549.
6.	$6327597.	15.	54399581.	24.	72820280.
7.	$26176670.	16.	8825431.	25.	55301760.
8.	$1644737.	17.	4001722.	26.	80200180.
9.	$7977899.	18.	2601900.	27.	95658143.

SUBTRACTION CONTINUED.—ART. **76.**

Ex.	Ans.	Ex.	Ans.	Ex.	Ans.
28.	9000001.	39.	85807625.	50.	925.
29.	99899999.	40.	1598.	51.	1511.
30.	83128433.	41.	4004.	52.	41845.
31.	40592424.	42.	1384.	53.	$46900, W. $69450, H.
32.	55352005.	43.	14061.		
33.	19957466.	44.	12494.	54.	$2410 lost.
34.	77919261.	45.	11547.	55.	171825.
35.	70051563.	46.	3295.	56.	$1674737.
36.	53201371.	47.	1606.	57.	$97.
37.	25311703.	48.	3707.	58.	$3893.
38.	86282745.	49.	2664.	59.	Given.

MULTIPLICATION.—ART. **93.**

1.	$24795.	15.	3931476.	29.	239968374861.
2.	$36099.	16.	415143630.	30.	449148410434.
3.	$56700.	17.	31884470.	31.	289975559744.
4.	90520 miles.	18.	8468670.	32.	294144537440.
5.	74175 pounds.	19.	43506216.	33.	335834314400.
6.	372500 days.	20.	11847672.	34.	18834782688.
7.	960000 rods.	21.	57380625.	35.	109588282650.
8.	20835.	22.	11050155200.	36.	654638320927.
9.	21576.	23.	12810000.	37.	396890151372.
10.	68198.	24.	48288058.	38.	554270292192.
11.	176400.	25.	3473567604.	39.	2985984.
12.	1554768.	26.	88789980848.	40.	57111104051.
13.	5497800.	27.	9313702853.	41.	60435595442394
14.	1674918.	28.	67226401140.	42.	87112343040000

CONTRACTIONS IN MULTIPLICATION.—ARTS. **97—108.**

8.	$1776.	20.	312046700000.	29.	96000 pounds.
9.	$5760.	21.	52690078000000	30.	359400000.
10.	$8100.	22.	6890634570000- 000.	31.	143759940000.
11.	5782 s.			32.	28708635000000
12.	23808 miles.	23.	494603050600- 000000.	34.	123240000.
13.	$11736.			35.	2309760000.
14.	19845 s.	24.	87831206507- 000000000.	36.	26366200000.
15.	$32256.			37.	144447000000.
17.	46500 bushels.	25.	678560051090- 000000000.	39.	31276000000.
18.	365000 days.			40.	374760000000.
19.	1534860000.	28.	18750 pounds.	41.	18054680000000

CONTRACTIONS IN MULTIPLICATION CONTINUED.

Ex.	Ans.	Ex.	Ans.	Ex.	Ans.
42.	664726500000-	74.	4140.	99.	180600000.
	000.	75.	27936.	100.	2722946304.
43.	1075635900000-	76.	154250.	101.	2172069918.
	000.	77.	11348400.	102.	7225.
45.	45514.	78.	34639552.	103.	65536.
46.	68476.	79.	2685942.	104.	104650.
47.	400624.	80.	2801960.	105.	12744790.
48.	907002.	81.	72156000.	106.	31049291000.
50.	132525.	82.	1680000000.	107.	273211606224[illegible]
51.	307664.	83.	2000000000.	108.	222310980000.
52.	2333616.	84.	43644865.	109.	20066857745-
53.	5691627.	85.	81708550.		896.
55.	474309.	86.	401939564.	110.	1256700743298
56.	6027966.	87.	476413195.	111.	37968867755.
57.	7293699.	88.	62220780.	112.	39073118478.
58.	4629537.	89.	637049231.	113.	1021288493520
63.	54530.	90.	406101366.	114.	1421400000000
64.	72819.	91.	42261696.	115.	60302400000-
65.	346896.	92.	504159579.		000.
66.	6624403632.	93.	6724232757.	116.	91300203000-
67.	17651712450.	94.	7306359.		000.
68.	21983532672.	95.	21760506.	117.	680040000000-
71.	625.	96.	39429936.		000.
72.	2916.	97.	2283344802.	118.	4000000000-
73.	5184.	98.	650633256.		000000.

DIVISION.—Art. 127.

1. 45 bu.	13. $5697\frac{67}{74}$.	25. 3679.	35. 826451-
2. 85 bbls.	14. $3823\frac{45}{85}$.	26. 4500.	$\frac{70404}{123456}$.
3. $68\frac{130}{365}$.	15. $4166\frac{64}{96}$.	27. $50830\frac{13}{102}$.	36. 1387805-
4. $3.	16. $21276\frac{27}{47}$.	28. 630.	$\frac{649635}{654321}$.
5. $68\frac{24}{52}$.	17. $12152\frac{9}{29}$.	29. 235.	37. 900900900-
6. $73972\frac{220}{365}$.	18. $191\frac{40}{251}$.	30. 648.	$9009\frac{1}{111}$.
7. $20\frac{120}{144}$ days.	19. 873.	31. $26710\frac{170}{365}$.	38. 90009000-
8. $173\frac{88}{144}$ ds.	20. 48.	32. 563.	$9000\frac{1000}{11111}$.
9. $2773\frac{27}{32}$.	21. $48\frac{3764}{6274}$.	33. 8826211-	39. 90000900-
10. $1139\frac{1}{42}$.	22. $87\frac{74}{145}$.	$\frac{83033}{85593}$.	$009\frac{1}{111111}$.
11. $1443\frac{7}{52}$.	23. 108.	34. 23434402-	
12. $1489\frac{33}{63}$.	24. 45.	$\frac{645}{27789}$.	

CONTRACTIONS IN DIVISION.—Arts. 129—139.

Ex.	Ans.	Ex.	Ans.	Ex.	Ans.	Ex.	Ans.
1,	2. Given.	18.	36.	37.	15.	55.	2283781$\frac{1}{1}$
3.	132$\frac{30}{35}$ a.	19.	68.	38.	16$\frac{76}{175}$.	56.	941501$\frac{36}{45}$.
4.	672.	20.	36$\frac{32}{85}$.	39.	17.	57.	478676$\frac{12}{35}$.
5.	460.	21.	75.	40.	30.	58.	59207.
6.	205.	23.	1207.	41.	250 days.	59.	1826896.
7.	1265.	24.	1690.	42.	950 years.	60.	138791$\frac{38}{45}$.
8.	20; 34; 56d.	25.	6512$\frac{1}{5}$.	43.	10\frac{1}{17}$.	61.	65964$\frac{23}{125}$.
9.	$650; 7650;	26.	8654.	44.	285\frac{20}{28}$.	62.	6162$\frac{6}{75}$.
	$43200.	27.	83$\frac{11}{15}$.	45.	39\frac{29}{119}$.	63.	15831$\frac{214}{225}$.
10.	267, and	28.	76$\frac{13}{35}$.	46.	2\frac{478}{511}$.	64.	21$\frac{13726}{32000}$.
	50000 R.	29.	77$\frac{42}{45}$.	47.	11\frac{48}{341}$.	65.	4134$\frac{114}{175}$.
11.	144, and	30.	142$\frac{43}{55}$.	48.	54\frac{18}{333}$.	66.	3966$\frac{215}{225}$.
	360791 R.	31.	94.	49.	219\frac{153}{253}$.	67.	1658$\frac{262}{275}$.
12	5823, and	32.	194$\frac{10}{25}$.	50.	18$\frac{17}{25}$.	68.	7405$\frac{48}{125}$.
	67180309R	33.	1693$\frac{15}{25}$.	51.	13529$\frac{22}{42}$.	69.	4362$\frac{71}{175}$.
14.	105 b.	34.	3795$\frac{5}{25}$.	52.	12466$\frac{14}{63}$.	70.	3186$\frac{90}{275}$.
15.	184 bbls.	35.	67.	53.	12454$\frac{48}{72}$.	71.	97$\frac{65600}{800000}$.
16.	197$\frac{11}{17}$.	36.	203$\frac{51}{125}$.	54.	13446913$\frac{3}{5}$	72.	920$\frac{4578}{100000}$.

CANCELATION.—Arts. 150, 151.

2.	45.	4.	65.	6,	7. Given.	9.	3$\frac{1}{2}$.
3.	63.	5.	73.	8.	6.	10.	3.

APPLICATIONS OF THE FUNDAMENTAL RULES. Arts. 152—159.

1.	Given.	11.	79. years;	19.	Given.	27.	632.
2.	255 acres.		94 yrs.	20.	48 beggars.	28.	974.
3.	925 bu.	12.	510\frac{1}{2}$ car.	21.	20 flocks.	29.	7124; 5510
4.	Given.		345\frac{1}{2}$ hor.	22.	Given.	30.	13000;
5.	$190.	14.	65 years.	23.	20 years.		12264.
6.	1125 sheep.	15.	175 rods.	24.	10 months.	31.	21151;
8.	$2240.	17.	187825.	25.	1842.		20975.
9.	$3436.	18.	1033062.	26.	1062.	32.	786.

PROPERTIES OF NUMBERS.—Arts. 162, 163.

1—9.	Given.	13.	2024122.	18.	707961.	22.	1614386.
10.	20212331.	14.	1522365.	19.	1036993.	23.	118620366
11.	2350147.	15.	Given.	20.	9753020.	24.	3879090-
12.	1331124.	16,	17. Given.	21.	360913096.		582.

ANALYSIS OF COMPOSITE NUMBERS.—ART. 165.

Ex.	Ans.	Ex.	Ans.	Ex.	Ans.
4.	9=3×3.	28.	2, 3, and 7.	52.	2, and 37.
5.	2, and 5.	29.	2, 2, and 11.	53.	3, 5, and 5.
6.	2, 2, and 3.	30.	3, 3, and 5.	54.	2, 2, and 19.
7.	2, and 7.	31.	2, and 23.	55.	7, and 11.
8.	3, and 5.	32.	2, 2, 2, 2, and 3.	56.	2, 3, and 13.
9.	2, 2, 2, and 2.	33.	7, and 7.	57.	2, 2, 2, 2, and 5.
10.	2, 3, and 3.	34.	2, 5, and 5.	58.	3, 3, 3, and 3.
11.	2, 2, and 5.	35.	3, and 17.	59.	2, and 41.
12.	3, and 7.	36.	2, 2, and 13.	60.	2, 2, 3, and 7.
13.	2, and 11.	37.	2, 3, 3, and 3.	61.	5, and 17.
14.	2, 2, 2, and 3.	38.	5, and 11.	62.	2, and 43.
15.	5, and 5.	39.	2, 2, 2, and 7.	63.	3, and 29.
16.	2, and 13.	40.	3, and 19.	64.	2, 2, 2, and 11.
17.	3, 3, and 3.	41.	2, and 29.	65.	2, 3, 3, and 5
18.	2, 2, and 7.	42.	2, 2, 3, and 5.	66.	7, and 13.
19.	2, 3, and 5.	43.	2, and 31.	67.	2, 2, and 23.
20.	2, 2, 2, 2, and 2.	44.	3, 3, and 7.	68.	3, and 31.
21.	3, and 11.	45.	2, 2, 2, 2, 2, and 2	69.	2, and 47.
22.	2, and 17.	46.	5, and 13.	70.	5, and 19.
23.	5, and 7.	47.	2, 3, and 11.	71.	2, 2, 2, 2, 2, and 3
24.	2, 2, 3, and 3.	48.	2, 2, and 17.	72.	2, 7, and 7.
25.	2, and 19.	49.	3, and 23.	73.	3, 3, and 11.
26.	3, and 13.	50.	2, 5, and 7.	74.	2, 2, 5 and 5.
27.	2, 2, 2, and 5.	51.	2, 2, 2, 3, and 3.	75.	2, 2, 3, 3, and 3.

76. 120=2×2×2×3×5.
144=2×2×2×2×3×3

77. 180=2×2×3×3×5.
420=2×2×3×5×7.

78. 714=2×3×7×17.
836=2×2×11×19.

79. 574=2×7×41.
2898=2×3×3×7×23.

80. 11492=2×2×13×13×17
980=2×2×5×7×7.

81. 650=2×5×5×13.
1728=2×2×2×2×2×2×3×3×3.

82. 1492=2×2×373.
8032=2×2×2×2×2×251.

83. 4604=2×2×1151.
16806=2×3×2801.

84. 71640=2×2×2×3×3×5×199.
20780=2×2×5×1039.

85. 84570=2×3×5×2819.
65480=2×2×2×5×1637.

86. 92352=2×2×2×2×2×2×3×13×37.
81660=2×2×3×5×1361.

GREATEST COMMON DIVISOR.—Arts. 168—171.

Ex. Ans.	Ex. Ans.	Ex. Ans.	Ex. Ans.
1. Given.	6, 7. Given.	12. 1.	18. 12.
2. 3.	8. 15.	13. Given.	19. 18.
3. 7.	9. 14.	14. 3.	20. 35.
4. 5.	10. 111.	15. 16.	21. 6.
5. 2.	11. 39.	17. 15.	22. 28.

LEAST COMMON MULTIPLE.—Arts. 176, 177.

1—3. Given.	8. 720.	14. 144.	19. 360.
4. 90.	9. 12600.	15. 600.	21. 600.
5. 144.	10. 504.	16. 2520.	22. 1440.
6. 180.	11. 1134.	17. 252.	23. 13824.
7. 360.	12. 15015.	18. 1134.	24. 51000.

REDUCTION OF FRACTIONS.—Arts. 195-201

1, 2. Given.	15. $\frac{2}{3}$.	30. $107\frac{197}{425}$.	47. $\frac{5}{192}$.
3. $\frac{2}{5}$.	16. $\frac{3}{4}$.	33. $\frac{71}{4}$.	48. $\frac{1}{270}$.
4. $\frac{3}{7}$.	17. $\frac{47}{191}$.	34. $\frac{77}{3}$.	49. $\frac{14}{325}$.
5. $\frac{4}{5}$.	18. $\frac{431}{469}$.	35. $\frac{342}{7}$.	52. $\frac{5}{15}$.
6. $\frac{3}{4}$.	21. 9.	36. $\frac{707}{10}$.	53. $\frac{1}{21}$.
7. $\frac{7}{9}$.	22. 5.	37. $\frac{1385}{12}$.	54. $\frac{3}{40}$.
8. $\frac{25}{29}$.	23. $3\frac{3}{5}$.	38. $\frac{51405}{60}$.	55. $\frac{2}{9}$.
9. $\frac{33}{50}$.	24. $9\frac{1}{7}$.	39. $\frac{3913}{3}$.	56. $\frac{1}{35}$.
10. $\frac{7}{11}$.	25. 1.	40. $\frac{23626}{5}$.	57. $\frac{13}{102}$.
11. $\frac{13}{31}$.	26. 60.	41. $\frac{4450}{10}$.	58. $\frac{9}{136}$.
12. $\frac{1}{2}$.	27. 21.	42. $\frac{5376}{8}$.	59. $\frac{4}{5}$.
13. $\frac{10}{13}$.	28. 52.	43. $\frac{440450}{115}$.	60. $\frac{4}{15}$.
14. $\frac{7}{9}$.	29. $60\frac{3}{4}$.	44. $\frac{3589375}{625}$.	61, 62. Givea

63. $\frac{280}{420}$; $\frac{252}{420}$; $\frac{105}{420}$; $\frac{300}{420}$.	75. $\frac{10}{14}$; $\frac{5}{14}$.
64. $\frac{324}{540}$; $\frac{60}{540}$; $\frac{360}{540}$; $\frac{405}{540}$.	76. $\frac{10}{36}$; $\frac{30}{36}$; $\frac{28}{36}$; $\frac{21}{36}$.
65. $\frac{120}{270}$; $\frac{135}{270}$; $\frac{216}{270}$; $\frac{180}{270}$.	77. $\frac{18}{24}$; $\frac{20}{24}$; $\frac{21}{24}$.
66. $\frac{1575}{2205}$; $\frac{1260}{2205}$; $\frac{1470}{2205}$; $\frac{882}{2205}$.	78. $\frac{18}{72}$; $\frac{40}{72}$; $\frac{63}{72}$; $\frac{66}{72}$.
67. $\frac{6720}{7560}$; $\frac{5400}{7560}$; $\frac{4536}{7560}$; $\frac{4410}{7560}$.	79. $\frac{54}{72}$; $\frac{45}{72}$; $\frac{64}{72}$; $\frac{6}{72}$.
68. $\frac{4200}{5775}$; $\frac{4950}{5775}$; $\frac{4620}{5775}$; $\frac{2310}{5775}$.	80. $\frac{24}{36}$; $\frac{9}{36}$; $\frac{20}{36}$; $\frac{30}{36}$.
69. $\frac{42000}{92400}$; $\frac{83160}{92400}$; $\frac{62370}{92400}$.	81. $\frac{35}{40}$; $\frac{32}{40}$; $\frac{26}{40}$; $\frac{35}{40}$.
70. $\frac{30000}{65000}$; $\frac{45500}{65000}$; $\frac{135200}{65000}$.	82. $\frac{35}{60}$; $\frac{57}{60}$; $\frac{54}{60}$; $\frac{56}{60}$.
71. $\frac{25000}{56250}$; $\frac{39375}{56250}$; $\frac{22500}{56250}$.	83. $\frac{30}{42}$; $\frac{28}{42}$; $\frac{36}{42}$; $\frac{33}{42}$.
72. $\frac{375000}{607500}$; $\frac{1316250}{607500}$; $\frac{296460}{607500}$.	84. $\frac{48}{60}$; $\frac{50}{60}$; $\frac{57}{60}$; $\frac{18}{60}$.
73, 74. Given.	85. $\frac{588}{1008}$; $\frac{861}{1008}$; $\frac{728}{1008}$; $\frac{168}{1008}$

REDUCTION OF FRACTIONS CONTINUED.—ART. 201.

Ex.	Ans.	Ex.	Ans.
86.	$\frac{18275}{76755}$; $\frac{30345}{76755}$; $\frac{54180}{76755}$; $\frac{54825}{76755}$.	89.	$\frac{868}{1260}$; $\frac{357}{1260}$; $\frac{504}{1260}$; $\frac{432}{1260}$.
87.	$\frac{2880}{6300}$; $\frac{2205}{6300}$; $\frac{980}{6300}$; $\frac{3150}{6300}$.	90.	$\frac{432}{720}$; $\frac{195}{720}$; $\frac{610}{720}$; $\frac{125}{720}$.
88.	$\frac{84}{144}$; $\frac{60}{144}$; $\frac{76}{144}$; $\frac{128}{144}$.	91.	$\frac{486}{540}$; $\frac{195}{540}$; $\frac{441}{540}$; $\frac{190}{540}$.

ADDITION OF FRACTIONS.—ARTS. 202-204.

1--3. Given.
4. $3\frac{19}{60}$.
5. $2\frac{7}{24}$.
6. $2\frac{2}{45}$.
7. $2\frac{309}{728}$.
8. $1\frac{332}{495}$.
9. $2\frac{2}{105}$.
10. $3\frac{653}{728}$.
11. $5\frac{3}{4}$.
12. $2\frac{179}{420}$.
13. $1\frac{1}{15}$.
14. $\frac{61}{240}$.
15. $2\frac{73}{840}$.
16. $21\frac{19}{20}$.
17. $11\frac{13}{21}$.
18. $6\frac{1}{4}$.
19. $61\frac{5}{8}$.
20. $15\frac{41}{120}$.
22. $\frac{55}{700}$, or $\frac{11}{140}$; $\frac{105}{2600}$, or $\frac{21}{520}$
23. $\frac{19}{616}$; $\frac{152}{5607}$.
24. $\frac{15}{442}$; $\frac{278}{86025}$
25. $\frac{125}{136}$; $\frac{91}{153}$.
26. $1\frac{47}{52}$; $\frac{9}{20}$.
27. $1\frac{23}{201}$; $\frac{19}{28}$.
29. $\frac{407}{9}$.
30. $\frac{4807}{15}$.
31. $\frac{4523}{10}$.
32. $\frac{43025}{16}$.
33. $\frac{19764865}{3000}$.
34. $\frac{35033799}{2520}$.
35. $\frac{35006999}{4536}$.
36. $\frac{719890959}{106590}$
37. $\frac{533711}{30}$.
38. $\frac{2066609}{1386}$.
39. $\frac{209038}{180}$.
40. $\frac{9614765}{40}$.
41. $\frac{17377967}{84}$.
42. $\frac{39952281}{616}$.
43. $\frac{21601}{70}$.

SUBTRACTION OF FRACTIONS.—ARTS. 206-208.

4. $\frac{4}{15}$.
5. $\frac{8}{35}$.
6. $\frac{9}{27}=\frac{1}{3}$.
7. $\frac{347}{1880}$.
8. $\frac{785}{3152}$.
9. $\frac{906}{3685}$.
10. $\frac{137}{360}$.
11. $\frac{41}{99}$.
12. 23.
13. $\frac{151}{1260}$.
15. $12\frac{1}{21}$.
16. $69\frac{43}{50}$.
17. $121\frac{37}{90}$.
18. $278\frac{11}{15}$.
21. $125\frac{1}{8}$.
22. $238\frac{1}{12}$.
23. $137\frac{3}{8}$.
24. $466\frac{2}{5}$.
25. $291\frac{7}{9}$.
26. $603\frac{7}{40}$.
27. $974\frac{1}{10}$.
28. $\frac{2}{7}$.
29. $8263\frac{3}{4}$.
30. $71\frac{11}{84}$.

MULTIPLICATION OF FRACTIONS.—CASE I.—ARTS. 211–17

1--3. Given.
4. $\frac{17}{5}=3\frac{2}{5}$.
5. $\frac{25}{2}=12\frac{1}{2}$.
6. $18\frac{9}{94}$.
7. $37\frac{3}{4}$.
8. $48\frac{36}{43}$.
9. $89\frac{7}{37}$.
10. $66\frac{522}{683}$.
11. $167\frac{2}{119}$.
12. $776\frac{8}{67}$.
13. $662\frac{181}{955}$.
14, 15. Given.
16. $735\frac{3}{4}$.
17. $633\frac{1}{3}$.
18. $3715\frac{5}{8}$.
19. $4448\frac{8}{9}$.
20. $2264\frac{6}{11}$.
21. $12519\frac{4}{9}$.
24. 108.
25. 127.
26. 240.
27. 435.
28. 560.
29. $621\frac{1}{4}$.
30. $701\frac{1}{4}$.
31. $76\frac{2}{13}$.
32. $514\frac{1}{2}$.
33. $305\frac{160}{463}$.
35. $10396\frac{4}{5}$.
36. $17460\frac{20}{63}$.
37. $366\frac{23}{128}$.
38. $2067\frac{3}{367}$.
39. $35650\frac{62}{65}$.
40. $235554\frac{13}{25}$.
43. 375.
44. 738.
45. 1178.
46. 3450.
47. 6795.
48. 6897.
49. 15282.
50. 29318.
51. 1280.
52. 279.
53. 4496.
54. 8113.
55. $10413\frac{8}{13}$.
56. $5672\frac{67}{75}$.
57. $5086\frac{8}{13}$.
58. 43452.
59. $74290\frac{2}{13}$.
60. $92280\frac{83}{93}$.

MULTIPLICATION OF FRACTIONS CONTINUED.—CASE II.

ARTS. 219, 220.

Ex.	Ans.	Ex.	Ans.	Ex.	Ans.	Ex.	Ans.
1, 2.	Given.	11.	17.	20.	$3232\frac{50}{333}$.	29.	$99\frac{1}{6}$.
3.	$\frac{35}{80}=\frac{7}{16}$.	12.	$53\frac{1}{7}$.	21.	$5143\frac{487}{1344}$.	30.	$78\frac{3}{7}$.
4.	$\frac{165}{336}=\frac{55}{112}$.	13.	$242\frac{26}{27}$.	22.	$6998\frac{1663}{2881}$.	31.	$47\frac{3}{10}$.
5.	$\frac{44}{115}$.	14.	329.	23.	$167\frac{122}{375}$.	32.	86.
6.	$\frac{1}{2}$.	15.	$1362\frac{5}{6}$.	24.	$53\frac{47}{108}$.	33.	$401\frac{5}{8}$.
7.	$\frac{2775}{4928}$.	16.	$3198\frac{18}{25}$.	25.	$24091\frac{2}{3}$.	34.	$1474\frac{314}{1269}$.
8.	$\frac{136}{475}$.	17.	$451\frac{173}{221}$.	26.	$466\frac{14}{15}$.	35.	\$$972\frac{5}{8}$.
9.	$\frac{35}{66}$.	18.	$834\frac{317}{408}$.	27.	300000.	36.	\$$6323\frac{9}{28}$.
10.	Given.	19.	$2001\frac{91}{111}$.	28.	700.	37.	$5501\frac{3}{10}$ m.

CONTRACTIONS IN MULTIPLICATION OF FRACTIONS.

ARTS. 221—225.

3.	$\frac{3}{8}$.	11.	$\frac{4}{9}$.	21.	57600.	32.	$4762\frac{1}{2}$.
4.	$\frac{1}{42}$.	12.	$\frac{1}{40}$.	22.	99000.	33.	$15937\frac{1}{2}$.
5.	$\frac{1}{9}$.	13.	$\frac{1}{4}$.	23.	$1871\frac{2}{3}$.	34.	$40187\frac{1}{2}$.
6.	$\frac{21}{8}=2\frac{5}{8}$.	14.	$\frac{1}{28}$.	24.	14220.	35.	65450.
7.	$\frac{2}{5}$.	15.	$\frac{5}{81}$.	26.	$2133\frac{1}{3}$.	37.	$1278\frac{1}{3}$.
8.	$\frac{9}{13}$.	16.	$\frac{16}{243}$.	27.	$18466\frac{2}{3}$.	38.	$4083\frac{1}{3}$.
9.	$\frac{1}{4}$.	19.	$493\frac{1}{3}$.	28.	5580.	39.	8150.
10.	26.	20.	$8533\frac{1}{3}$.	29.	430000.	40.	$9383\frac{1}{3}$

DIVISION OF FRACTIONS.—ARTS. 226—241.

1--3.	Given.	18, 19.	Given.	33.	$9\frac{57}{178}$.	51.	$\frac{7}{10}$.
4.	$\frac{2}{23}$.	20.	$\frac{22}{45}$.	34.	$13\frac{137}{251}$.	52.	$\frac{5}{6}$.
5.	$\frac{34}{217}$.	21.	$\frac{2}{27}$.	37.	$13\frac{59}{100}$.	53.	$31\frac{1}{3}$.
6.	$\frac{4}{70}=\frac{2}{35}$.	22.	$23\frac{1}{3}$.	38.	$8\frac{349}{1000}$.	54.	$\frac{1}{27}$.
7.	$\frac{7}{75}$.	23.	$40\frac{1}{2}$.	40.	$220\frac{1}{5}$.	55.	$1\frac{3}{7}$.
8.	$\frac{5}{30}=\frac{1}{6}$.	24.	$\frac{1}{2}$.	41.	$30\frac{21}{50}$.	56.	1.
9.	$\frac{64}{725}$.	25.	$\frac{1}{8}$.	42.	$50\frac{47}{50}$.	58.	$1\frac{13}{15}$.
10--13.	Given.	26.	$3\frac{57}{65}$.	43.	$6\frac{93}{125}$.	59.	$\frac{3}{25}$.
14.	$1\frac{29}{216}$.	29.	135.	46.	$383\frac{67}{100}$.	60.	$\frac{4}{33}$.
15.	$2\frac{13}{16}$.	30.	168.	47.	$54\frac{333}{1000}$.	61.	$5\frac{3}{5}$.
16.	$3\frac{21}{67}$.	31.	$11\frac{7}{43}$.	49.	$2\frac{6}{13}$.	63.	$\frac{1}{2}$.
17.	$\frac{280}{413}$.	32.	$11\frac{67}{73}$.	50.	$\frac{31}{42}$.	64.	$\frac{3}{8}$.

APPLICATION OF FRACTIONS.—ART. 242.

1.	$88\frac{13}{16}$ yds.	4.	\$$14\frac{13}{16}$.	7.	\$$1548\frac{1}{4}$.	10.	$5606\frac{3}{16}$
2.	$162\frac{1}{12}$ lbs.	5.	\$$62\frac{11}{16}$.	8.	\$$5516\frac{1}{2}$.	11.	\$$483\frac{3}{4}$.
3.	\$$54\frac{5}{8}$.	6.	\$$635\frac{7}{16}$.	9.	\$1515.	12.	\$100.

APPLICATION OF FRACTIONS CONTINUED.—ART. 242.

Ex	Ans	Ex.	Ans.	Ex	Ans.	Ex.	Ans.
13.	$\$16111\frac{1}{9}$.	21.	$165\frac{1}{3}$ yds.	29.	$38\frac{2}{11}$ rods.	37.	$\$5\frac{731}{775}$.
14.	$404552\frac{23}{47}$lb	22.	$218\frac{1}{3}$ lbs.	30.	$\$3\frac{269}{456}$.	38.	$42\frac{8}{447}$ tons.
15.	$\$1806\frac{7}{8}$.	23.	$626\frac{2}{3}$ gals.	31.	$\$6\frac{2}{9}$.	39.	$\$1\frac{237}{1286}$.
16.	$3612\frac{1}{2}$ bu.	24.	$20\frac{7}{16}$ lbs.	32.	$584\frac{4}{9}$ bu.	40.	$\$1\frac{1108}{1502}$.
17.	$\$30968\frac{6}{7}$.	25.	$53\frac{1}{7}$ yds.	33.	$4\frac{2}{3}$ doz.	41.	$\$3\frac{508}{950}$.
18.	$6939\frac{9}{10}$ m.	26.	27 boxes.	34.	$6\frac{3}{62}$ cts.	42.	$266\frac{74}{4511}$ d.
19.	5229 m.	27.	$153\frac{3}{5}$ bbls.	35.	$9\frac{181}{198}$ s.	43.	$\$33798\frac{117}{152}$.
20.	$\$9175\frac{1}{2}$.	28.	$29\frac{5}{9}$ suits.	36.	$\$13\frac{27}{152}$.		

REDUCTION.—ART. 282.

2.	68810 far.	28.	8553600 in.	52.	28992 pts.
3.	86768 far.	29.	5280000 yds.	53.	1427 bu. 1 pk.
4.	284079 far.	30.	54 m. 7 fur. 38 r. 2 yds. 2 ft.	54.	130100 qts.
5.	96615 far.			55.	36360 min.
6.	£25, 13s. 6d. 3 far.	31.	9 l. 2 m. 4 fur. 31 r. $1\frac{1}{2}$ yds. 2 ft. 7 in.	56.	31557600 sec.
7.	£433, 1s. 2d. 3 far.			57.	84 wks. 6 hrs. 45 min
8.	266 guin. 18s. 8d.	32.	5031 rods.	58.	65 d. 2 h. 4 m. 40 sec
9.	1448 sixpences.	33.	17 m. 20 r.	59.	31556928 sec.
10.	6050 threepences.	34.	132105600 ft.	60.	946728000 sec.
11.	170472 grs.	35.	2560 na.	61.	10 yrs.
12.	9000 pwts.	36.	5000 qrs.	62.	397200″.
13.	1010047 grs.	37.	6396 yds. 2 qrs. 1 na	63.	1350000.″
14.	2 lbs. 1 oz. 10 pwts. 16 grs.	38.	9302 F. e. 4 qrs. 3 na	64.	2126°, 11′, 54″.
		39.	10156 na.	65.	555555s. 16°, 40
15.	177 lbs. 9 oz. 12 pwts	40.	7116 qts.	66.	470660 sq. ft.
16.	1596 lbs.	41.	693 gals.	67.	$4366073\frac{1}{4}$ sq. ft.
17.	564000 oz.	42.	26528 gi.	68.	32640858360 sq. in.
18.	104300 lbs.	43.	48 bar. 20 gals.	69.	582 A. 1 R. 3 r $269\frac{1}{4}$ sq. ft.
19.	71680000 drs.	44.	117 pi. 1 hhd. 46 g. 3 qts. 1 pt. 2 gi.		
20.	10 cwt. 16 lbs.			70.	259200 cu. in.
21.	133 T. 12 cwt. 35 lbs	45.	102128 gi.	71.	4551552 cu. in.
22.	1 T. 202 lbs. 1 oz.	46.	12960 pts.	72.	10877760 cu. in.
23.	9120 drs.	47.	87 bar. 26 gals.	73.	49 cu. ft. 1 cu. in
24.	37440 sc.	48.	630 hhds. 44 gals.	74.	306 C. 48 cu. ft.
25.	64 lbs. 11 oz. 5 drs.	49.	19520 pts.	75.	4492800 cu. in.
26.	88 lbs. 4 oz. 7 drs. 2 sc	50.	488 qts.	76.	52 T. 40 cu. ft. 180 cu. in.
27.	142560 ft.	51.	24440 qts.		

APPLICATIONS OF REDUCTION.—ARTS. 283—284.

1.	Given.	4.	$177\frac{1}{7}$ lbs. Troy, or $145\frac{187}{245}$ lbs. avoir.	7.	58 lbs. 4 oz. Troy,
2.	576 lbs. avoir.			8.	$21\frac{763}{1152}$ lbs. Troy
3.	691 lbs. 10 oz. $5\frac{43}{175}$ drams.	5.	$265\frac{5}{7}$ lbs. Troy, or $218\frac{158}{245}$ lbs. avoir.	9.	271 lbs. 3 oz.
				10.	Given.

APPLICATIONS OF REDUCTION CONTINUED.—ARTS. 285—293.

Ex.	Ans.	Ex.	Ans.	Ex.	Ans.
11.	360 sq. ft.	30.	3456 wine gals.	49.	14 hhds. $48\frac{10}{11}$ g.
12.	14 A. 10 sq. rds.	31.	$8640\frac{1}{7}$ w. gals.	51.	$51\frac{57}{94}$ beer gals.
13.	108 sq. y. 8 sq. ft.	32.	5184 beer gals.	52.	$45\frac{27}{47}$ wine gals.
14.	446 A. 1 R.	33.	6912 b. gals. 2 qts	53.	$24\frac{32}{77}$ w. gals.
15.	40 A.	34.	$80\frac{5}{14}$ bu.	54.	1598 w. gals.
16.	36 sq. yds.	35.	100 bu.	55.	$2207\frac{61}{77}$ w. gals.
17.	66 sq. yds.	36.	800 bu.	56.	$3125\frac{61}{77}$ qts.
18.	$111\frac{1}{9}$ sq. yds.	37.	$897\frac{51}{77}$ w. gals.	57.	$2734\frac{22}{47}$ gals.
20.	$56\frac{7}{8}$ cu. ft.	38.	$712\frac{232}{539}$ bar.	59.	8 min. 36 sec.
21.	126 cu. ft.	39.	$902867\frac{447}{539}$ hhds.	60.	39 min.
22.	86 C. 2 cu. ft.	41.	$622\frac{2}{9}$ cu. ft.	61.	1 hr. 8 m. 40 sec.
23.	748 cu. ft.	42.	$1244\frac{4}{9}$ cu. ft.	62.	33 min. 48 sec.
24.	756 cu. ft.	43.	$8\frac{27}{64}$ cu. ft.	63.	12 h. 28 m. 12 s
25.	72 cu. yds.	44.	$210\frac{35}{64}$ cu. ft.	64.	Given.
26.	160 cu. ft.	45.	$842\frac{3}{16}$ cu. ft.	65.	4° 45′.
27.	1800 cu. ft.	47.	$42\frac{83}{256}$ bu.	66.	12° 46′.
29.	17280 bu.	48.	$46\frac{6}{11}$ gals.	67.	13° 23′.

COMPOUND NUMBERS REDUCED TO FRACTIONS.—ART. 296

1–4.	Given.	12.	$\frac{7}{9}$ yd.	20.	$\frac{1}{806400}$.	29.	$\frac{313}{860}$.
5.	£$\frac{11}{48}$.	13.	$\frac{247}{960}$ m.	22.	$\frac{3}{16}$.	30.	$\frac{61}{16000}$.
6.	£$\frac{77}{1920}$.	14.	$\frac{91}{320}$ A.	23.	$\frac{60}{143}$.	31.	$\frac{453}{32400}$.
7.	£$\frac{7}{1920}$.	15.	$\frac{28}{121}$ sq. r.	24.	$\frac{3}{8}$.	32.	$\frac{16291}{1296000}$.
8.	$\frac{7}{12}$ lb. Troy.	16.	$\frac{1}{5}$ gal.	25.	$\frac{1}{32}$.	33.	$\frac{5}{12}$.
9.	$\frac{43}{640}$ lb. Troy	17.	$\frac{1}{9}$ hhd.	26.	$\frac{5}{33}$.	34.	$\frac{51}{120}$.
10.	$\frac{35}{64}$ lb. avoir.	18.	$\frac{7}{192}$ d.	27.	$\frac{55}{504}$.	35.	$\frac{1}{4}$.
11.	$\frac{283}{400}$ T.	19.	$\frac{1}{108}$ hr.	28.	$\frac{45}{136}$.	36.	$\frac{3}{10}$.

FRACTIONAL COMPOUND NUMBERS

REDUCED TO WHOLE NUMBERS OF LOWER DENOMINATIONS.—ARTS. 297, 298

3.	17s. 6d.	13.	2 qts. 1 pt. $1\frac{1}{3}$ gi.	25.	$6\frac{78}{87}$ hrs.
4.	7d. $\frac{4}{5}$ far.	14.	55 gals. 1 pt.	26.	2688 min.
5.	5 oz. 2 p. $20\frac{4}{7}$ g.	16.	3 pks. 1 qt. $1\frac{1}{5}$ pts.	27.	$8\frac{8}{89}$ na.
6.	12 pwts. 12 grs.	17.	46 min. 40 sec.	28.	$17\frac{39}{57}$ qts.
7.	10 oz. $10\frac{2}{3}$ drs.	18.	21 hrs. 36 min.	29.	$174\frac{102}{137}$ qts.
8.	57 lbs. 2oz. $4\frac{4}{7}$ drs.	19.	$22\frac{1}{2}$ sec.	30.	$4\frac{32}{37}$ oz.
9.	1250 lbs.	20.	17′ $8\frac{4}{7}$″.	31.	66 pwts.
10.	2 ft. $4\frac{4}{5}$ in.	22.	$\frac{240}{417}$ d.	32.	$7\frac{37}{109}$ r.
11.	6 ft. $2\frac{1}{4}$ in.	23.	$\frac{64}{217}$ oz.	33.	$\frac{3}{10}$ sq. ft.
12.	177 r. 12 ft. 10 in.	24.	$\frac{20}{29}$ r.	34.	70″.

COMPOUND ADDITION.—Art. 300.

Ex.	Ans.	Ex.	Ans.	Ex.	Ans.
3.	£106, 3s. 1d.	10.	109 l. 2 m. 6 fur. 1 ft.	16.	240 gals.
4.	£188, 13s. ½d.	11.	114 yds. 3 qrs.	17.	181 hhds. 59 gals 1 pt. 1 gi.
5.	9 T. 8 cwt. 17 lbs.	12.	387 yds. 1 qr.	18.	115 w. 15 h. 25 m.
6.	45 T. 4 cwt. 57 lbs. 2 oz.	13.	138 A. 114 sq. r. 80 sq. ft.	19.	322 bu. 1 pk. 5 qts.
7.	107 lbs. 7 o. 8 p. 1 g.	14.	468 A. 1 R. 33 sq. r.	20.	135 qrs. 3 bu. 3 pks. 2 qts.
8.	330 lbs. 2 o. 3 p. 5 g.	15.	43 sq. yds. 5 sq. ft. 125 sq. in.		
9.	4 fur. 13 r. 13 ft. 3 in.				

COMPOUND SUBTRACTION.—Arts. 302, 303.

1.	Given.	9.	35 bu. 2 pks. 6 qts.	19.	25° 3′ 15″.
2.	£9, 2s. 8d. 3 qrs.	10.	19 qrs. 6 bu. 2 pks.	20.	35° 3′ 30″.
3.	£60, 4s. 7d. 3 qrs.	11.	55 yds. 2 qrs. 3 na.	21.	10° 26′.
4.	£499, 13s. 4d. 2 qrs.	12.	44 yds. 1 qr. 3 na.	22.	54 yrs. 2 mos. 2 wks. 6 d. 2 hrs. 45 min. 6 s.
5.	8 cwt. 1 qr. 6 lbs. 10 oz.	13.	6 gals. 2 qts. 1 pt.	23.	Given.
6.	24 T. 1 cwt. 71 lbs.	14.	Given.	24.	67 yrs. 9 mos. 22 d.
7.	19 m. 289 r. 2 ft.	15.	85 A. 119 r.	25.	———
8.	1 l. 1 m. 7 fur. 10 r. 12½ ft.	16.	235 A. 48 r.	26.	1 yr. 5 mos. 11 d.
		17.	56 C. 90 cu. ft.	27.	3 yrs. 9 mos. 22 d.
		18.	339 cu. ft. 26 in.		

COMPOUND MULTITPLICATION.—Art. 305.

1, 2.	Given.	14.	2044 l. 1 m. 4 fur. 30 r.	23.	15891° 13′ 30″
3.	£247, 6s. 1d.	15.	8962 bu. 16 qts.	24.	204° 10′.
4.	£24, 9d.	16.	2968 qrs. 5 bu. 2 pks. 6 qts.	25.	4581 bu. 8 qts.
5.	17 T. 55 lbs.	17.	7821 A. 20 r.	26.	2453 bu. 4 qts.
6.	403 T. 17 cwt. 55 lbs	18.	25172 A. 1 R. 3 r.	27.	£5, 16s. 10½d.
7.	689 lbs. 8 oz. 16 pwts	19.	24645 cu. ft. 930 cu. in.	28.	£679, 3s. 4d
8.	6 lbs. 10 oz. 10 pwts	20.	96350 C. 50 cu. ft.	29.	£297.
9.	3039 hhds. 39 gals. 1 qt. 1 pt.	21.	12783 d. 11 h. 28 m.	30.	£507, 16s. 3d.
10.	5668 pi. 32 gals.	22.	1199 yrs. 9 mos. 3 wks. 1 d.	31.	36 C. 74 cu. ft. 944 in.
11.	2358 yds.			32.	865 lbs. 12 oz.
12	5375 yds.			33.	25418 lbs. 12 oz.
13	14778 m. 1 fur. 32 r.			34.	8662 gals. 2 qts.

COMPOUND DIVISION.—Art. 307.

1-3.	Given.	8.	£4, 17s. 3d. $\frac{1}{3}$ qr.
4	51 lbs. 3 oz. 10 pwts. 15¾ grs.	9.	10 yds. 3 qrs. 1$\frac{2}{5}$ na.
5.	31 bu. 14$\frac{4}{5}$ qts.	10.	9 yds. 1 qr. $\frac{11}{27}$ na.
6.	25 bu. 1½ pts.	11.	83 m. 2 fur. 26 r. 11 ft.
7.	£20, 1s. 6d.	12.	214 m. 2 fur. 27 4 ft. $\frac{6}{7}$ in.

COMPOUND DIVISION CONTINUED.—ART. 307.

Ex.	Ans.	Ex.	Ans.
13.	1 gal. 2 qts. 1 pt. $1\frac{13}{18}$ gi.	18.	1s. 17° 52′ $21\frac{39}{41}$″.
14.	44 hhds. 29 gals. 1 pt. $1\frac{7}{9}$ gi.	19.	9 C. 84 ft. $1016\frac{8}{17}$ in.
15.	24 d. 8 hrs. 42 min. 40 sec.	20.	6 C. 92 ft. $850\frac{11}{61}$ in.
16.	10 yrs. 35 d. 1 hr. 13 min. $11\frac{9}{11}$ sec.	21.	6s. $10\frac{1}{2}$d.
		22.	7s. 11d. $3\frac{3}{7}$ qrs.
17.	1° 48′ $41\frac{1}{5}$″.	23.	10s. 11d. $2\frac{2}{73}$ qrs.

ADDITION OF DECIMALS.—ART. 320.

1, 2. Given.
3. 428.1739.
4. 103.8523.
5. 14.747274.
6. 60.149.
7. 332.1249.
8. 501.15998.
9. 857.005.
10. 1097.84143.
11. 1408.25559.
12. 127.05034.
13. 33.3182746.
14. 15674.1613.
15. 1.807.
16. 2.471092.
17. 0.0711824.
18. 0.3532637.
19. 0.807711.
20. 0.1627165.
21. 0.996052.
22. 0.329773.

SUBTRACTION OF DECIMALS.—ART. 322.

1, 2. Given.
3. 1427.633782.
4. 20.987651.
5. 72.5193401.
6. 81.16877.
7. 0.066721522.
8. 0.01.
9. 9.999999.
10. 64.0317753.
11. 24680.12377.
12. 24.75.
13. 2.291.
14. 9.9999999.
15. 8.000001.
16. 4635.5346.
17. 541.787.
18. 46.43606.
19. 0.0000999.
20. 0.0000396.
21. 31.99968.
22. 44.99955.
23. 98.99999901.
24. 0.000999.
25. 699.93.
26. 28999.908.
27. 255999999.744.
28. 0.414.
29. 0.0041.
30. 0.000000000999.
31. 0.002873789.
32. 0.062156.
33. 0.71699.
34. 0.0000843174.

MULTIPLICATION OF DECIMALS.—ART. 324.

1. 681.45 ft.
2. 25020 miles.
3. 2055.375 gals.
4. 136.125 nails.
5. 788.0125 sq. yds.
6. 43560 sq. ft.
7. 2465.375 sq. rods.
8. 0.250325.
9. 18.93978.
10. 14.78091.
11. 0.613836.
12. 0.0320016.
13. 36.740232.
14. 919.82036.
15. 0.000000072.
16. 0.00105175.
17. 390.657556.
18. 275.230594.
19. 148.64244532.
20. 73.25771882.
21. 52.17977576.
22. 0.0306002448.
23. 4701.169144360.
24. 536.660075952.
25. 0.00164389993.
26. 160.86701632806.
27. 0.06288405909156.
28. 2.5067823.
29. 64.327106105314.
30. 0.0000118260069.
31. 11027.40199543716.
32. 94167471.869654039.
33. .00000006676542672.

CONTRACTIONS IN MULTIPLICATION OF DECIMALS.
ARTS. 325–327.

Ex.	Ans.	Ex.	Ans.	Ex.	Ans.
1.	Given.	9.	75000.	17–20.	Given.
2.	429302.13401.	10.	6.5.	21.	0.09484.
3.	106723.50123.	11.	48.	22.	1.262643.
4.	608340.17.	12.	2480.	23.	0.0769.
5.	304672.14067.	13.	381.	24.	0.0389254.
6.	44632140.32.	14.	65.04.	25.	0.00876.
7.	2134567.82106.	15.	834000.	26.	0.002516.
8.	500.	16.	10.	27.	0.001789.

DIVISION OF DECIMALS.—ART. 330.

1–3. Given.	12. 79098.8235+.	21. 83671000.
4. 13 boxes.	13. 0.6344+.	22. 255.1210+.
5. 8 suits.	14. 1210.2344+.	23. 0.000005.
6. 4.98347+days.	15. 0.03.	24. 60.2589.
7. 82.9997+loads.	16. 134 8805+.	25. 211.076.
8. 27.7173+days.	17. 59.4060+.	26. 400000.
9. 150.25 bales.	18. 24.8266+.	27. 60000000.
10. 5.9291+.	19. 4320.67.	28. 4000000.
11. 6.632.	20. 0.02.	29. 311.487360+.

CONTRACTIONS IN DIVISION OF DECIMALS.—ARTS. 331-33

1, 2. Given.	7. 0.000012300456.	12. 1.611.
3. 67234.567.	8. 0.0000020076346.	13. 0.04026.
4. 103.42306.	9. Given.	14. 0.0954776.
5. 0.42643621.	10. 0.1274.	15. 2.0208.
6. 6.72300045.	11. 0.09471.	16. 0.980439.

DECIMALS REDUCED TO COMMON FRACTIONS.—ART. 335.

1, 2. Given.	6. $\frac{23}{40}$.	10. $\frac{3}{250}$.	14. $\frac{139}{2500}$.
3. $\frac{1}{8}$.	7. $\frac{82}{125}$.	11. $\frac{1}{400}$.	15. $\frac{76}{625}$.
4. $\frac{19}{20}$.	8. $\frac{51}{250}$.	12. $\frac{1001}{10000}$.	16. $\frac{401}{2000}$.
5. $\frac{87}{200}$.	9. $\frac{3}{40}$.	13. $\frac{461}{2500}$.	17. $\frac{3}{2000}$.

COMMON FRACTIONS REDUCED TO DECIMALS.
ARTS. 337—344.

1–3. Given.	10. 0.6.	17. 0.625.	25. Terminate.
4. 0.5.	11. 0.8.	18. 0.75.	26. Terminate.
5. 0.25.	12. 0.5.	19. 0.875.	27. Interminate.
6. 0.5.	13. 0.125.	20, 21. Given.	28. Terminate.
7. 0.75.	14. 0.25.	22. Terminate.	31. $0.\dot{3}$.
8. 0.2.	15. 0.375.	23. Terminate.	32. $0.\dot{6}$.
9. 0.4.	16. 0.5.	24. Interminate	33. $0.1\dot{6}$

COMMON FRACTIONS REDUCED TO DECIMALS CONTINUED.

Ex. Ans.	Ex. Ans.	Ex. Ans.	Ex. Ans.
34. $0.\dot{3}$.	41. $0.\dot{7}1428\dot{5}$.	48. $0.\dot{6}$.	56. 0.0048828-125.
35. $0.\dot{6}$.	42. $0.\dot{8}5714\dot{2}$.	49. $0.\dot{7}$.	
36. $0.8\dot{3}$.	43. $0.\dot{1}$.	50. $0.\dot{8}$.	57. $0.58\dot{3}$.
37. $0.\dot{1}4285\dot{7}$.	44. $0.\dot{2}$.	51. 0.1875.	58. $0.\dot{0}7692\dot{3}$.
38. $0.\dot{2}8571\dot{4}$.	45. $0.\dot{3}$.	52. $0.\dot{0}7692\dot{3}$	59. $0.0\dot{1}0489\dot{5}$.
39. $0.\dot{4}2857\dot{1}$.	46. $0.\dot{4}$.	53. 0.024.	60. $0.\dot{4}683544303797\dot{}$ 4683544303797.
40. $0.\dot{5}7142\dot{8}$.	47. $0.\dot{5}$	54. 0.0112.	
		55. 0.275.	

COMPOUND NUMBERS REDUCED TO DECIMALS.—ART. 346.

1, 2. Given.	6. $0.41\dot{6}$ s.	10. $0.258\dot{3}$ hr.	14. 0.875 bu.
3. £0.5375.	7. $0.541\dot{6}$ s.	11. $0.12708\dot{3}$ d.	15. 0.5625 pk
4. £0.825.	8. 0.115625 m.	12. 0.0525 cwt.	16. 1.125 gals.
5. $£0.8791\dot{6}$.	9. 0.25625 m.	13. 0.46875 lb.	

DECIMAL COMPOUND NUMBERS REDUCED TO WHOLE ONES.—ART. 348.

2. 14s. 6d.	7. 6 oz. 15.36 drs.	11. 1 qt. 1 pt. 3.4432 gi.
3. 2s. 7d. 3.2 qrs.	8. 88 rods.	12. 10 h. 13 m. 9.12 sec.
4. 1d. 2 qrs.	9. 7 ft 0.51 in.	13. 50 min. 42 sec.
5. 9d. 3.6 qrs.	10. 11 gals. 1 qt. 1 pt. 3.7184 gills.	
6. 12 lbs. 8 oz.		

REDUCTION OF CIRCULATING DECIMALS —ARTS. 355—61.

1, 2. Given.	8. $\frac{45}{999}$, or $\frac{5}{111}$.	17. $\frac{1}{44}$.	$0.33\dot{3}$.
3. $\frac{18}{99}$, or $\frac{2}{11}$.	9. $\frac{1}{7}$.	18. $\frac{251}{550}$.	$0.04\dot{5}$.
4. $\frac{123}{999}$, or $\frac{41}{333}$.	10. $\frac{1}{13}$.	19. $\frac{2333}{4950}$.	24. $4.\dot{3}21\dot{3}$.
5. $\frac{297}{999}$, or $\frac{11}{37}$.	14. $\frac{8}{15}$.	20. $\frac{849715}{9999990000}$, or $\frac{83}{9768}$.	$6.4\dot{2}6\dot{3}$.
6. $\frac{72}{99}$, or $\frac{8}{11}$.	15. $\frac{5925}{9990}$, or $\frac{16}{27}$.		$0.6\dot{0}0\dot{0}$.
7. $\frac{9}{99}$, or $\frac{1}{11}$.	16. $\frac{7}{12}$.	23. $0.27\dot{7}$.	

ADDITION OF CIRCULATING DECIMALS.—ART. 362.

2. $179.\dot{2}74556\dot{3}$.	5. $594.6\dot{9}\dot{1}$.	8. $1380.0\dot{6}4819\dot{3}$.
3. $476.65\dot{1}2\dot{9}$.	6. $112.72\dot{2}\dot{4}$.	9. $5974.1037\dot{1}$.
4. $47.8\dot{6}68\dot{3}$.	7. $223.5\dot{1}0774\dot{4}$.	10. $339.626\dot{1}7744\dot{3}$

SUBTRACTION OF CIRCULATING DECIMALS.—ART. 363.

1, 2. Given.	5. $4.\dot{7}8\dot{9}$.	8. $218.6\dot{0}$.
3. $391.552\dot{4}$.	6. $400.91\dot{5}$.	9. $0.613\dot{6}4073\dot{1}$.
4. $3.818\dot{2}\dot{4}$.	7. $3.90\dot{4}\dot{6}$.	10. $2451.\dot{3}8\dot{6}$.

MULTIPLICATION OF CIRCULATING DECIMALS.—Art. 364.

Ex. Ans.	Ex. Ans.	Ex. Ans.
1, 2. Given.	5. $389.\dot{1}8\dot{5}$.	8. $31.\dot{7}9\dot{1}$.
3. $0.\dot{0}8\dot{2}$.	6. $77\dot{8}.1\dot{4}$.	9. $34998.41\dot{9}900\dot{3}$
4. 1.8.	7. $750730.\dot{5}1\dot{8}$.	10. $2.\dot{2}9\dot{7}$.

DIVISION OF CIRCULATING DECIMALS.—Art. 365.

1, 2. Given.	5. $7.\dot{7}\dot{2}$.	9. $62.3\dot{2}3834196$-
3. $55.\dot{6}\dot{9}$.	6. $8574.\dot{3}$.	$89\dot{1}$.
4. $5.\dot{4}146\dot{3}$.	7. $3.\dot{5}0649\dot{3}$.	10. $\dot{1}.4229249011$-
	8. $3.1\dot{4}\dot{5}$.	$8577075098\dot{8}$.

ADDITION OF FEDERAL MONEY.—Art. 374.

1. Given.	7. $3531.432.	13. $8765.12.	18. $1945.258.
2. $265.04.	8. $12200.524	14. $16989.	19. $82110.17.
3. $581.128.	9. $185.285.	15. $378.383.	20. $71774.75.
4. $560.56.	10. $74.33.	16. $300.166.	21. $27860.74.
5. $1795.34.	11. $350.32.	17. $256.213.	22. $81800.63.
6. $1431.50.	12. $6491.05.		

SUBTRACTION OF FEDERAL MONEY.—Art. 375.

1. Given.	6. $183.22.	11. $9947.788.	16. $2.937.
2. $12.13.	7. $323.47.	12. $61119.364	17. $32.056.
3. $84.82.	8. $373.82.	13. $18.981.	18. $10890.07.
4. $247.15.	9. $10870.75.	14. $88.11.	19. $89989.90.
5. $918.48.	10. $1699.49.	15. $189.92.	

MULTIPLICATION OF FEDERAL MONEY.—Arts. 377, 378.

3. $83.60.	9. $84.875.	14. $2.84375.	19. $28.125.
4. $517.625.	10. $193.75.	15. $909.375.	20. $220.50.
6. $39.59375.	11. $205.625.	16. $2.70.	21. $142.50.
7. $1440.75.	12. $326.25.	17. $14.0625.	22. $2331.875.
8. $40.59375.	13. $2.0925.	18. $15.78375.	23. $14084.125

DIVISION OF FEDERAL MONEY.—Arts. 379-381.

1. Given.	5. Given.	10. 543.518+ yds.
2. $4.50.	6. 8.207 coats.	11. 991.421+ doz.
3. $0.06.	7. 7.871+ times.	12. 360 skeins.
4. $3.13.	9. 308.035+ gals.	13. $3.524+.

DIVISION OF FEDERAL MONEY CONTINUED.—ART. 381.

Ex.	Ans.	Ex.	Ans.	Ex.	Ans.
14.	$1.50.	19.	$0.04049+.	24.	113.56377+ tons
15.	$6.25.	20.	$0.02709+.	25.	$0.595238+.
16.	$1.973+.	21.	$1.78008+.	26.	245.517+ acres.
17.	$3.615+.	22.	$1.5435+.	27.	500 cows.
18.	$0.084+.	23.	1714.285+ bu.	28.	150 carriages.

APPLICATIONS OF FEDERAL MONEY.—ARTS. 382-85.

1. Given.	13. $5885.	28. $0.0072.
2. $800.	14. $10538.625.	29. $0.0064.
3. $511.50.	15, 16. Given.	30. $13.4719+
4. $780.	17. $6.33375.	per cwt.;
5. $780.	18. $104.55.	$0.134719+
6. $1350.	19. $114.198.	per lb.
7. $1020.	20. $59.5856.	31. $12.88506 cwt.
8. $864.50.	21. $505.3775.	$0.1288506 lb.
9. $2418.	22. $1901.75.	32. $129.625.
10. $4440.	23. $5.40625.	33. $208.838.
11. $1424.75.	24. $52.126.	34. $1734.875.
12. $2691.875.	25. $437.645.	35. $13703.78.

PERCENTAGE.—ART. 388.

5, 6. Given.	14. $43.13 rec'd.	22. 375 sheep.	30. $90.4824.
7. $7.6875.	$819.43 paid.	23. $1568.	31. $844.08.
8. $8.7526.	15. $402.05.	24. 187.5 lost;	32. $4724.775.
9. $3.4608	16. $134.	1312.5 saved.	33. $1250.
10. $8.7078	17. $32.625.	25. $8.125.	34. $12000.
11. $114.1070.	18. $34.03575.	26. $6.316.	35. $21900, 1st
12. $10.50.	19. $62.50.	27. $84.52016.	$14600, 2d.
13. $219.	20. $146.666+.	28. $250.	36. $200.
	21. $8.771875.	29. $750.	37. $0.95.

APPLICATIONS OF PERCENTAGE.—ARTS. 395-97.

1. Given.	10. $619.887.	19. $761904.761.	30. $8840.70.
2. $12.507	11. $44.32.	20. $4126.55.	31. $7072.
3. $58.878.	12. $673.75.	21. $1413.975	32. $3552.50.
4. $73.159.	13. $57.29.	22. $46.50.	33. $1350 rec'd.
5. $115.203.	14. $415.831.	23. $22.113.	$180 lost.
6. $615.	15. $106.831 A.	24. $9.375.	34. $7490.50.
7. $583.842.	$2029.798 O.	25. $318.975.	35. $960.
8. $52.834.	17. $21078.431.	28. $3692.50.	36. $4427.50.
9. $155.875.	18. $3439.613.	29. $2250.	37. $9028.50.

INTEREST.—Art. 404.

Ex. Ans.	Ex. Ans.	Ex. Ans.	Ex. Ans.
1. $29.61.	10. $8.103.	19. $889.44.	28. $3312.209.
2. $43.255.	11. $6.853.	20. $1.135.	29. $5278.162.
3. $40.367.	12. $19.14.	21. $1.409.	30. $16.158.
4. $51.20.	13. $60.27.	22. $1.898.	31. $206.718, at
5. $60.263.	14. $89.40.	23. $102.125.	360 days;
6. $44.414.	15. $958.41	24. $154.216.	$203.886, at
7. $194.58.	16. $657.45.	25. $704.083.	365 days.
8. $17.803.	17. $1006.833.	26. $2.975.	32. $66778.64.
9. $28.206.	18. $1585.018.	27. $76.131.	

SECOND METHOD.—Arts. 409—413.

4. $8.50.	15. $82.078.	26. $307.65.	38. $137.288.
5. $1.065.	16. $39.179.	27. $227.994.	39. $481.016.
6. $70.151.	17. $320.833.	28. $8.	40. $391.062.
7. $97.28.	18. $9.8437.	29. $0.07.	41. $1531.25.
8. $30.78.	19. $85.207.	31. $15.60.	42. $3425.655.
9. $398.287.	20. $400.	32. $21.09.	43. $16320.528.
10. $1177.50.	21. $1638.442.	33. $1.272.	45. $2.145.
11. $1113.024.	22. $144.	34. $4.778.	46. $74.392.
12. $10.05.	23. $90.	35. $46.35.	47. $10.835.
13. $11.0025.	24. $12666.075.	36. $129.15.	48. $398.055.
14. $988.761.	25. $16360.996.	37. $168.552.	49. $14.532

APPLICATIONS OF INTEREST.—Arts. 415—419.

2. $5.25.	7. $36.08.	12. $6547.20.	20. £8, 18s. 9d.
3. $3.15.	8. $91.085.	14. $499.034.	21. £12, 10s.
4. $17.	9. $107.854+.	15. $498.595.	22. £1898, 10s.
5. $60.	10. $533.867.	16. $4149.689.	4¾d.
6. $45.014.	11. $25729.166+	19. £19, 5s. 10½d	23. £2900.

PROBLEMS IN INTEREST.—Arts. 422-424.

1, 2. Given.	10. 5 per cent.	19. $30000.	28. 14 y. 3 mo.
3. 6 per cent.	11. 5 per cent.	20. $20833⅓.	13 d. nearly.
4. 6 per cent.	13. $1800.	22. 4 years.	29. 14 y. 3 mo.
5. 8 per cent.	14. $5400.	23. 6 months.	13 d. nearly.
6. 7½ per cent.	15. $10000.	24. 1 y. 3 mos.	30. 10 years.
7. 5½ per cent.	16. $8000.	1 d. nearly.	31. 8 y. 4 mo.
8. 7 per cent.	17. $14285.7143.	25. 1 y. 6 mo.	32. 9 y. 6 mo. 8 d.
9. 6 per cent.	18. $20000.	27. 16 y. 8 mo.	33. 28 years.

COMPOUND INTEREST.—Arts. 426, 427.

1, 2. Given.	5. $4590.09.	8. $1551.328.	11. $16035.675.
3. $507.213.	6. Given.	9. $877.506.	12. $149744.
4. $2177.426.	7. $1888.464.	10. $3491.395.	

DISCOUNT.—Art. 430.

Ex. Ans.	Ex. Ans.	Ex. Ans.	Ex. Ans.
1, 2. Given.	5. $88.461+.	8. $6208.955+.	10. $9950.248+.
3. $934.579+.	6. $83.52+.	9. $3404.347+.	11. $36.636.
4. $1488.687+.	7. $4729.064+.		

BANK DISCOUNT.—Arts. 433, 434.

12, 13. Given.	20. $24.822.	27. $456.785.	34. $3821.883.
14. $14 1825.	21. $48.3237.	28. $1126.523.	35. $4355.102.
15. $16.605.	22. $37.595.	29. Given.	36. $63717.884.
16. $26.98.	23. $43.694.	30. $414.507.	37. $10416.666.
17. $5.495.	24. $6381.59.	31. $966.101.	38. $51194.539.
18. $2034.1213.	25. $1495.625.	32. $1252.70.	39. $46638.655.
19. $2774.655.	26. $80.	33. $2514.247.	40. $8301.342.

INSURANCE.—Arts. 437—442.

1. Given.	8. $1875.	16. 1 per cent.	25. $8365.482.
2. $20.70	9. $487.50.	17. 1$\frac{1}{4}$ per cent.	26. $13876.288.
3. $94.20	10. $243.125	19. $52000.	27. $27027.027.
4. $63.75	11. $19278.	20. $65600.	29. $48.60.
5. $104.	12. $3375.	21. $65000.	30. $373.75.
6. $70.50.	14. 2$\frac{1}{2}$ per cent.	22. 57333\frac{1}{3}$.	31. $10000, ins.
7. $900.	15. 2$\frac{1}{2}$ per cent.	23. 3416\frac{2}{3}$.	$12250,prem

PROFIT AND LOSS.—Arts. 444—447.

1-3. Given.	10, 11. Given.	19. 15$\frac{5}{9}$ per cent	27. $2622.222.
4. $218.	12. $156.804.	20. 100 per cent.	28. $2736.
5. $680.	13. $4238.50.	21. 20$\frac{94}{173}$ per ct.	29. $13043.478.
6. $935.25.	14. $5926.85.	22. 2$\frac{29}{30}$ per cent.	30. $6317.391.
7. $1366.75.	15. $29504.875.	23, 24. Given.	31. $17806.122
8. $68730.28.	17. 23$\frac{1}{13}$ per ct.	25. $460.869.	32. $42654.028
9. $12500 lost.	18. 4$\frac{1}{6}$ per cent.	26. $205.882.	33. $42160.

DUTIES.—Arts. 451-453.

1. Given.	7. $3784.	13. $717.40.	19. $12642.40.
2. $370.80.	8. $345.744.	14. $492.	20. $2807.10.
3. $163.20.	9. $679.14.	15. $1051.71.	21. $11172.30.
4. $1323	10. $1882.406.	16. $715.75.	22. $17328.75.
5. $546.	11. Given.	17. $1230.	23. $15770.70.
6. $1235.22.	12. $248.	18. $15884.75.	

ASSESSMENT OF TAXES.—Arts. 456, 457.

1, 2. Given.	5. $\frac{4}{5}$ of 1 per cent., or	7. $121.92, B's tax.
3. $54.15, B's tax.	8 mills on $1.	8. $283.68, C's tax
4. $80.50, C's tax.	6. $80, A's tax.	10. $8854.166

ASSESSMENT OF TAXES CONTINUED.—ARTS. **459, 460.**

Ex.	Ans.	Ex.	Ans.	Ex.	Ans.
11.	$16125.654.	20.	$314.50, J. F's.	27.	$370.50, F. M's.
12.	$17342.105.	21.	$621.90, T. G's.	28.	$458.20, C. P's.
13.	$34051.815.	22.	$526.40, W. H's.	29.	$480.50, J. S's.
16.	$73, G. A's.	23.	$263.30, L. J's.	30.	$541, R. W's.
17.	$116, H. B's.	24.	$631.00, W. L's.	32.	$13.36.
18.	$451.50, W. C's.	25.	$196.90, J. K's.	33.	$3.45.
19.	$481.22, E. D's.	26.	$404.90, G. L's.	34.	$13.40.

ANALYSIS.—ARTS. **462—470.**

1, 2.	Given.	12.	0.039\frac{1}{16}$.	22.	$7.98.	32.	Given.
3.	$300.	13.	$64.	23.	$6.03.	33.	360 lbs.
4.	$320.	14.	$1080.	24.	$160.	34.	1500 lbs.
5.	12.33\frac{1}{3}$.	15.	$480.	25.	3.70\frac{2}{7}$.	35.	95.2 cords.
6.	$10.50.	16.	$8.	26.	$3430.	36.	100 pair.
7.	1.68\frac{3}{4}$.	17.	60 days.	27.	$119.918.	38.	$450, A's.
8.	$2640.	18.	29$\frac{13}{28}$ mos.	28.	$636.479.		$750, B's.
9.	$24.80.	19.	1088 days.	29.	Given.	39.	$450, A's.
10.	$0.055.	20.	$0.56.	30.	2$\frac{2}{5}$ hours.		$600, B's.
11.	0.29\frac{1}{6}$.	21.	$3.	31.	2$\frac{10}{13}$ days.		$750, C's.

40.	763.63\frac{7}{11}$, A's.	49.	40 tons. A's.	65.	188$\frac{4}{7}$ lbs. at 8d.
	654.54\frac{6}{11}$, B's.		80 tons, B's.		17$\frac{1}{7}$ lbs. " 12d.
	981.81\frac{9}{11}$, C's.		120 tons, C's.		17$\frac{1}{7}$ lbs. " 18d.
41.	150.95\frac{265}{313}$, A's.	50.	25 per cent.		17$\frac{1}{7}$ lbs. " 22d.
	164.53\frac{241}{313}$, B's.	51.	33$\frac{1}{3}$ per cent.	67.	9 horses.
	185.70\frac{90}{313}$. C's.		$30000, loss.	68.	38$\frac{2}{5}$ days.
	123.80\frac{60}{313}$, D's.	53.	5s. per gal.	69.	278160 men.
42.	66$\frac{2}{3}$ cts. on $1.	54.	5$\frac{5}{7}$s. per lb.	70.	15$\frac{1}{17}$ months.
	266.66\frac{2}{3}$, 1st.	55.	9 cts. per lb.	71.	$54.60.
	333.33\frac{1}{3}$, 2d.	56.	19$\frac{1}{3}$ cts. per lb.	72.	$459.90.
	$400.000, 3d.	57.	91$\frac{2}{3}$ cts. per gal.	74.	$600.
43.	70 cts. on $1.	58–60.	Given.	75.	$3600.
44	25 per cent.	61.	1 part 16 car.	76.	$630.
45.	$2990.00, A's.		1 " 18 car.	77.	72.
	$4197.50, B's.		2$\frac{1}{2}$ " 23 car.	78.	360.
	$4312.50, C's.		1 " 24 car.	79.	120.
46.	66$\frac{2}{3}$ per cent.	63.	100 gals. at 80cts.	80.	240.
47.	37$\frac{1}{2}$ per cent.		40 " 30cts.	81.	68$\frac{4}{7}$ feet.
48.	10 per cent.		40 " 40cts.	85.	$239.

ANALYSIS CONTINUED.—ART. 471.

Ex	Ans.	Ex.	Ans.	Ex.	Ans.	Ex.	Ans.
86.	$1170.	100.	£266.	114.	$288.	127.	$140.
87.	$900.	101.	£131$\frac{1}{4}$.	115.	$\$43\frac{3}{4}$.	128.	$1560.
88.	$1125.	102.	£353.	116.	$814.	129.	$180.
89.	$367.50.	105.	$\$250\frac{2}{3}$.	117.	$640.	130.	$630.
90.	$442.	106.	$231.	118.	$3000.	131.	$180.
91.	$201.	107.	$\$119\frac{7}{32}$.	119.	$200.	132.	$1281.
92.	$350.	108.	$186.	121.	$625.	133.	$800.
93.	$240.	109.	$280.	122.	$480.	134.	$12.66
94.	$754.	110.	$1170.	123.	$808.	135.	$45.
95.	$1080.	111.	Given.	124.	$420.	136.	$45.
96.	$630.	112.	$378.	125.	$690.	137.	$90.
99.	£206$\frac{2}{3}$.	113.	$810.	126.	$\$877\frac{1}{2}$.	138.	$150.

RATIO.—ARTS. 480—488.

1, 2.	Given.	14.	$8\frac{10}{99}$.	26.	120.	40.	45 to 72.
3.	2.	15.	$\frac{1}{5}$.	27.	60.	41.	Equal.
4.	4.	16.	$\frac{1}{4}$.	28.	$\frac{2}{21}$.	42.	936 to 560.
5.	9.	17.	$\frac{1}{4}$.	29.	$\frac{3}{80}$.	43.	G. inequality
6.	6.	18.	$\frac{1}{5}$.	30.	$1\frac{1}{3}$.	44.	L. inequality
7.	6.	19.	$\frac{1}{4}$.	31.	$\frac{1}{4}$.	45.	Equality.
8.	8.	20.	$\frac{1}{7}$.	32.	240.	46.	60 : 12=5
9.	9.	21.	3.	35.	8; $\frac{1}{7}$.	47.	1.
10.	9.	22.	7.	36.	4; 8.	48.	$\frac{3}{4}$.
11.	9.	23.	112 avoir.	37.	4; $\frac{1}{4}$.	49.	$\frac{27}{40}$.
12.	9.	24.	4.	38.	$\frac{1}{7}$; 9.	50.	$\frac{147}{190}$.
13.	4.	25.	6.	39.	72 to 8.	51.	$\frac{5}{8}$.

SIMPLE PROPORTION.—ARTS. 502—506.

1.	12.	17.	$51\frac{1}{5}$ lbs.	35.	$1925\frac{1}{4}$lbs.cop.	47.	480.
2.	3.	18.	$1640.64.		$641\frac{3}{4}$ lbs. tin.	48.	375 sheep.
3.	16.	19.	$7066.40.	36.	1520 lbs. n.	49.	20 days.
4.	3.	22.	3 far.		280 lbs. c.	50.	400 rods.
5.	Given.	23.	$792.		200 lbs. sul.	51.	$8\frac{4}{7}$ weeks.
6.	20.	24.	$2768.	37.	980.5155 lbs.	52.	£1, 3s. 6d.
7.	$55\frac{4}{5}$.	25.	435 miles.	38.	$1350.		$1\frac{7}{17}$ far. 1st.
8.	120.	26.	252 days.	39.	£45.		£1, 1s. 2d.
9, 10.	Given.	28.	$2.70.	40.	$3375.		$\frac{8}{17}$ far. 2nd.
11.	$903.	29.	3s. 3d. $2\frac{34}{49}$ q.	41.	$2562.50.		£0, 18s. 9d.
12.	$1309.50.	30.	$3.15.	42.	$16480.		$3\frac{9}{17}$ far. 3rd.
13.	$225.	31.	$8.555.	43.	70400 times.		£0, 16s. 5d.
14.	775 miles.	32.	$26.40.	44.	57600 times.		$2\frac{10}{17}$ far. 4th
15.	20 tons.	34.	30 bu. oats;	45.	170.	53.	$888\frac{8}{9}$ oz. ox.
16.	2156 lbs.		70 bu. corn.	46.	200.		$111\frac{1}{9}$ oz. hg.

COMPOUND PROPORTION.—ARTS. 508—511.

Ex.	Ans.	Ex.	Ans.	Ex.	Ans.	Ex.	Ans.
3.	10 horses.	9.	24 days.	13.	$140.	17, 18.	Given.
4.	$19\frac{1}{5}$ days.	10.	144 days.	14.	$768.	19.	56 yds. Can.
5.	1314 gals.	11.	1125 miles.	15.	$600.	20.	127 b. N. O.
6.	27 laborers.	12.	$225.	16.	32 days.	21.	16 rupees.

DUODECIMALS.—ART. 516.

1, 2. Given.
3. 28 sq. ft. 6′ 10″.
4. 59 cu. ft. 3′ 8″.
5. 268 cu. ft. 6′ 11″.
6. 235 sq. ft.
7. 734 sq. ft. 0′ 9″.
8. 105 ft. 5′ 4″ 5‴ 5⁗ 4′′′′′.
9. 154 ft. 3′ 1″ 5‴ 4⁗ 6′′′′′ 8′′′′′′.
10. 85 ft. 1′ 11″ 0‴ 5⁗ 2′′′′′ 6′′′′′′.
11. 195 ft. 4′ 1″ 3‴ 8⁗ 0′′′′′ 6′′′′′′.
12. 23 C. 111 ft. 3′.
13. 3840 ft. 0′ 5″.
14. $15.819+.
15. 33750 bricks.

EQUATION OF PAYMENTS.—ART. 521.

3 6 months. | 4. 6 months. | 5. 3 years. | 6. 62 days.

PARTNERSHIP.—ART. 523.

1. Given.
2. $240, A's gain.
   $320, B's gain.
   $400, C's gain.
3. 274.21\frac{723}{737}$, A's.
   373.40\frac{420}{737}$, B's.
   212.37\frac{331}{737}$, C's.
4. $1178.947, A's.
   $2259.649, B's.
   $3340.351, C's.
   $4421.053, D's.
5. $850, A's.
   $800, B's.
   $700, C's.
   $650, D's.
7. $1655.172, X's.
   $1448.276, Y's.
   $1396.552, Z's.
8. $22.486, A's.
   $21.024, B's.
   $16.490, C's.
9. $3492.06, A's.
   $4761.91, B's.
   $6746.03, C's.

EXCHANGE OF CURRENCIES.—ARTS. 533—537.

3. $4116.42.
4. $850.63.
5. $414.667.
6. $969.815.
7. $2041.59+.
8. $4841.089+.
9. $7746.082+.
10. $60652.55+.
11. $208683.819+.
12. $330661.605+.
13. $242840.369+.
14. $257791.397+.
15. $369716.864+.
16. $284412.622+.
17. $4840000.
19. £82.
20. £90.
21. £181, 10¼d.
22. £261, 8s. 7½d.
23. £446, 7s. 8¼d.
24. £201, 11s. 7¾d.
25. £883, 5s. 8¼d.
26. £1095, 3s. 11¾d.
27. £5220, 9¾d.
28. £8568, 3s. 7½d.
29. £10384, 18s. 4d.
30. £20661, 3s. 1½d.
32. £135.
33. £227.
34. £315, 9d.
35. £375.
37. $534.166.
38. $614.1875.
39. $986.083.
40. $7714.285.
41. $20000.

EXCHANGE.—ART. 548.

2. $4791.60.
3. $25391.084+.
4. $284.58.
5. $10152.527+.
6. $707.
7. $1881.60.
8. $15418.509.
9. $20665.20.
10. $36480.755.

ARBITRATION OF EXCHANGE.—ART. **549.**

Ex. Ans.	Ex. Ans.	Ex. Ans.
1. $2\frac{1}{2}$ florins.	2. $45 gain.	3. 180 milrees, circu.

ALLIGATION.—ARTS. **552—556.**

2. 0.87\frac{1}{2}$.	**7.** 10 oz. 16 car. fine.	10. 40 gals. at 15s.
3. 5s. 4d. $1\frac{45}{53}$ qr.	5 oz. 18 "	40 gals. at 17s.
5. 3 grs. at 18 car. fine.	5 oz. 22 "	40 gals. at 18s.
1 gr. " 20 "	**8.** 133 lbs. at 20 cts.	200 gals. at 22 s.
1 gr. " 22 "	95 lbs. at 30 cts.	11. 28 gals. water;
3 grs. " 24 "	190 lbs. at 54 cts.	98 gals. wine.

INVOLUTION.—ART. **562.**

12. Given.	17. 3125.	22. 6.25.	27. $\frac{625}{1156}$.
13. 15129.	18. 279936.	23. .000001728.	28. $\frac{27000}{1000000}$.
14. 2460375.	19. 117649.	24. .0000015625.	29. $20\frac{1}{4}$.
15. 8294400.	20. 65536.	25. $\frac{4}{5}$.	30. $54\frac{25}{64}$.
16. 10000.	21. 387420489.	26. $\frac{64}{125}$.	31. $1480\frac{369}{1936}$.

SQUARE ROOT.—ARTS. **574, 575.**

3. 51.	12. 9.848+.	21. 792.	30. 186.9951+.
4. 73.	13. 2.6457+.	22. 1.7810+.	31. 12345.
5. 28.	14. 13.78404+.	23. 3216.	32. 345761.
6. 9.327+	15. 209.	24. $\frac{5}{7}$.	33. 31.05671.
7. 69.	16. 217.	25. $\frac{11}{16}$.	34. 19.104973174
8. 84.	17. 23.8.	26. .79056+.	35. 1.41421356-
9. 99.	18. 2.71.	27. 4.1683+.	237+.
10. 167.	19 .9044+.	28. 28.181.	36. 1.732050807
11. 31.	20. 34.2.	29. 14.4116+.	5688772.

APPLICATIONS OF THE SQUARE ROOT.—ARTS. **581—585.**

1. Given.	7. 18.	14. $\frac{8}{27}$.	20. 320 rods.
2. 32 feet.	8. 36.	15. $\frac{40}{60}$.	21. 480, length;
3. 166.709+m.	9. 40.	16. $\frac{84}{144}$.	160, breadth.
4. 240rds. side.	10. 66.	17. $\frac{78}{168}$.	22. 148 in rank;
339.4112 r. d.	11. 168.	18. 63 rods.	74 in file.
5. Given.	12. 11.2.	19. 160 rods.	24. 25 and 40.
6. 10.	13. 67.5.		25. 18 and 47.

EXTRACTION OF THE CUBE ROOT.—ARTS. **590-92.**

4. 45.	11. 0.623.	18. 3.5463+.	26. $379\frac{11}{16}$ lbs.
5. 52.	12. 3.332222+.	19. $3\frac{2}{3}$.	27. 24 and 72
6. 83.	13. 1.817121+.	20. 1.25992104.	28. 128 and 256
7. 136.	14. 7.217652+.	21. .64365958974.	29. 60 and 300.
8. 217.	15. 8.315517+.	22. 68 ft.	30. 160 and 640.
9. 22.6.	16. $\frac{5}{7}$.	24. 3.1748+yds.	31. 426 and 2556
10. 2.74.	17. $\frac{14}{45}$.	25. 4 lbs.	32. 747 and 6723

ROOTS OF HIGHER ORDERS.—Arts. 593-95.

Ex.	Ans.	Ex.	Ans.	Ex.	Ans.	Ex.	Ans.	Ex.	Ans.
2.	2.	5.	6.	8.	7.	12.	2.4872+.	16.	1.080059
3.	16.	6.	26.	9.	3.	13.	414.5+.	17.	1.004074.
4.	376.	7.	5.	10.	2.	15.	1.104089.	18.	1.047128.

ARITHMETICAL PROGRESSION.—Arts. 603—608.

. Given.	4. Given.	9. 3¾.	13. 14, 21, & 28
2. 5050.	5. 33.	11. 33¾.	14. 15, 29, 43, 57 71, & 85.
3. 78 strokes.	7. 44.	12. 502.	

GEOMETRICAL PROGRESSION.—Arts. 610-12.

2. 4.	$750.3651759245, amt. of $500.	8. 1023.
3. 4374.	$1628.894614622-37890625, amt. of $1000.	9. 43774¾.
4. 13671875.		10. $111111111.111.
5. $2048.		12. 1⅓.
6. $334.5563944, amt. of $250.		14. 3.

ANNUITIES.—Arts. 614, 615.

1, 2. Given.	4. $2298.262.	6. 36785.59.	8. $1333.333.
3. $826.992.	5. $4835.74.	7. Given.	9. Given.

PERMUTATIONS AND COMBINATIONS.—Arts. 618, 619.

2. 40320 ways.	4. 3628800 ways.	7. 15120 numbers.
3. 362880 ways.	5. 479001600 days.	8. 165765600 words.

MENSURATION OF SURFACES.—Arts. 622—631.

1. 270 acres.	7. 1100 sq. ft.	13. 100 ft.
2. 722⅓ acres.	9. 290.4737 sq. ft.	15. 12 A. 43.49375 r.
3. 31¼ acres.	10. 4 A. 52.82 rods.	16. 31415.9 sq. ft.
4. 320 rods, or 1 m.	11. 62.8318 ft.	17. 2 ft. 9.94 in.
5. 360 sq. ft.	12. 141.37155 rods.	18. 17.3205 ft.
6. 435 sq. ft.		

MENSURATION OF SOLIDS.—Arts. 633—647.

1. 1364 cu. ft.	11. 2748.89125 cu. ft.	18. 13 sq. ft.
2. 3154 ft. 11′ 6″ 8‴.	12. 119366.25 cu. ft.	19. 4 cu. ft.
3. 2615 cu.ft. 1080 in.	13. 78 yds. 4 ft. 123-.1128 in.	20. 62⅓ cu. ft.
4. 115 ft. 114.368 in.		21. 220 gals. 3 qts. 1 pt. 1.824 gi.
5. 53333⅓ cu. ft.	14. 7 sq. ft. 9.87516 in.	
6. 8835.75 cu. ft.	15. 14684558.20796 sq. miles.	22. 451 gals. 2 qts. 0.729344 pt.
7. 900 sq. ft.		
8. 1739 sq. ft.	16. 1767.14437 cu. in.	23. 831.71526+tons.
9. 76 cu. ft.	17. 5291335807.60158 cu. m.	24. 967.10521+tons.
10. 176 sq. ft.		

MECHANICAL POWERS.—ARTS. 648—655.

Ex.	Ans.	Ex.	Ans.	Ex.	Ans.	Ex.	Ans.
1.	500 lbs.		160 lbs. B.	5.	600 lbs.	8.	1250 lbs.
2.	$133\frac{1}{3}$ lbs.	4.	4 ft. from A.;	6.	$1066\frac{2}{3}$ lbs.	9.	1136.3636 lb.
3.	96 lbs. A.;		8 ft. from B.	7.	1600 lbs.	10.	904777.92 lb.

MISCELLANEOUS EXAMPLES.

1.	459 less.	34.	136 g. 1 q.		\$440, C's g.	79.	247170562
	521 greater	35.	\$180.		\$700, B's s.		2710 s. m.
2.	70.	36.	\$10.875.		\$1100, C's s.	80.	33600914-
3.	$5\frac{10}{201}$.	37.	\$156.615.	59.	20 per cent.		2264006.2-
4.	$6\frac{7}{52}$.	38.	94 d. 3 h.	60.	\$1371.		3104 c. m.
5.	20 days.		38m. $10\frac{10}{11}$s.	61.	\$4755.141.	81.	5890.5 lbs.
6.	\$61.32.	39.	\$2.	62.	\$32000.	82.	585.80357 b
7.	\$16581.65.	40.	£1.	63.	\$360.	83.	39.401 hhd
8.	\$18.60.	41.	\$$\frac{25}{27}$.	64.	36 days.	84.	$7\frac{1}{2}$ ft.
9.	\$1843.003.	42.	\$4800.	65.	90 hours.	85.	403291461
10.	\$24390.243	43.	\$197.759.	66.	£51, A's.		126605635
11.	\$4.50.	44.	228 gals.		£34. B's		584000000.
12.	\$6.875.	45.	\$40.29$\frac{3}{8}$.		£68, C's	86.	31 m. 180 r.
13.	$33\frac{1}{3}$ per ct.	46.	\$41.095.		£102, D's.	87.	$662\frac{1}{2}$.
14.	\$36.	47.	2 y. $182\frac{1}{2}$ d.	67.	£160, A's.	88.	\$4294967-
15.	\$$229\frac{13}{55}$.	48.	$5\frac{5}{11}$ min.		£224, B's.		.295.
16.	$487\frac{19}{43}$. E.	49.	120 days.		£256, C's.	89.	5 bags, A
17.	2000 miles.	50.	120 schol.		£205, A's.		7 bags, M
18.	2400 times.	51.	£292.		£287, B's.	90.	1440.
19.	2880 times.	52.	\$6000.		£328, C's.	91.	\$230, B's.
20.	\$1.50 per g.	53.	600.	68.	\$520, D's.		\$325, C's.
21.	$2\frac{13}{21}$ cts.	54.	5600 lbs. t.		\$280, A's.		\$445, A's.
22.	\$2400.		750 lbs. l.		\$360, B's.	92.	5 o'clock,
23.	$437\frac{1}{2}$ bbls.		300 lbs. b.	69.	20.		20 min.
24.	21 months.	55.	$254\frac{1}{3}$ miles.	70.	25 persons.	93.	$10\frac{220}{349}$ d. all
25.	\$1.328.	56.	$78\frac{2}{5}$ lbs.	71.	40 and 80.		$47\frac{67}{79}$ d. A.
26.	40 yds.		$117\frac{3}{5}$ lbs.	72.	75 and 128.		$38\frac{84}{97}$ d. B.
27.	1s. $3\frac{1}{24}$ qrs.	57.	\$$192.307\frac{9}{13}$	73.	56.5685 ft.		$27\frac{27}{139}$ d. C.
28.	$37\frac{21317}{24735}$ d.		A's gain.	74.	7200 rods.		$111\frac{3}{17}$ d. D.
29.	$34\frac{46374}{53725}$ d.		\$2307.692	75.	3.535519 ft.	94.	$36\frac{1}{2}$ days.
30.	\$1.60.		$\frac{4}{13}$, B's g.	76.	677.73475f.	95.	12 o'clock
31.	12 miles.		\$2500.000,	77.	7.13645 r.		$32\frac{8}{11}$ min.
32.	$12\frac{3}{5}$ days.		C's g.	78.	50 A. 3 R.	96.	$128\frac{4}{7}$ yrs.
33.	$52\frac{1}{2}$ days.	58.	\$240, A's g.		28.7399+r.	97.	\$407.

A CATALOGUE

OF

Newman and Ivison's Publications,

To which we invite the attention of Teachers and the friends of Education generally. The retail prices are attached to each Book, from which we make a discount by the quantity. We are in the habit of making very favorable terms for the first introduction of our School Books, and those Teachers who desire to introduce and establish an uniform series of the best Text Books, will do well to apply to us, post paid.

Copies of any of our School or Music Books, for examination, will be sent to any one by Mail, post paid, who will send us the price of the Book in P. O. Stamps or money.

NEWMAN & IVISON, Publishers,
178 Fulton Street, New York.

SCHOOL AND COLLEGE TEXT-BOOKS.

Agriculture for Schools ;

Or, Lessons in Modern Farming. Containing Scientific Exercises for Recitation, and elegant Extracts from Rural Literature for Academic and Family reading. By Rev. JOHN L. BLAKE. $1 00.

Alabama Readers,

In four parts. Prepared expressly for the Schools in the Southern States, and are in general use in Alabama, Georgia, and Mississippi. They are as follows:

FIRST PART: a Primary Primer. By C. W. SANDERS. 12½ cents.

SECOND PART: designed for children who are too young to read in Porter's Rhetorical Reader. By T. D. P. STONE. 25 cents

THIRD PART: consisting of Exercises in Reading and Speaking, for the use of middle classes in Schools. 37½ cents.

FOURTH PART: consisting of Instructions for regulating the Voice, with a Rhetorical Notation, and a course of Rhetorical Exercises. Designed for the use of High Schools and Academies. By Dr. PORTER, late of Andover Theological Seminary. 62½ cents.

Barrington's Physical Geography;

Being a treatise on the subject, comprising Hydrology, Geognosy, Geology, Meteorology, Botany, Zoology, Anthropology. By A. BARRINGTON. Edited by CHARLES BURDETT. $1 00.

Butler's Analogy of Religion;

Or, the Analogy of Religion, Natural and Revealed, to the constitution and course of Nature. By JOSEPH BUTLER, LL.D., late Lord Bishop of Durham. With an Introductory Essay by Rev. ALBERT BARNES. 1 vol. 12mo.—20th edition. 62½ cents.

The same Work,

With an Introductory Essay by Bishop HALIFAX. To which are added Copious Analytical Questions for the examination of Students.. By Rev. JOSEPH MCKEE, A.M. Academy and School edition. 62½ cents.

De Sacy's General Grammar;

Being the principles of General Grammar, adapted to the capacity of youth, and proper to serve as an introduction to the study of Languages. By A. J. SYLVESTER DE SACY, member of the Royal Council for Public Institutions. Translated and fitted for American use by DAVID FOSDICK, Jr. 37½ cents.

Elements of Political Economy.

By SAMUEL P. NEWMAN, late Professor in Bowdoin College. 75 cents.

Fasquelle's French Course;

Or, a New Method for Learning to Read, Write, and Speak the French Language, on the plan of Woodbury's "New Method with German." $1 25.

Key to, the Exercises in Fasquelle's French

Course, for the use of Teachers. 75 cents.

Fasquelle's Colloquial French Reader.

75 cents.

Fasquelle's Telemaque ;

Les Aventures de Telemaque, Fils d'Ulysse. Par M. FENELON. A new edition, with Notes. By LOUIS FASQUELLE, L.L.D., Prof. of Modern Languages in the University of Michigan. The Text carefully prepared from the most approved French Editions. 75 cents.

Gray's Chemistry ;

Or, Elements of Chemistry. Containing the Principles of the Science, both Experimental and Theoretical. Intended as a Text-Book for Academies, High Schools, and Colleges. Illustrated with numerous Engravings. By ALONZO GRAY, A.M., formerly Professor of Chemistry and Natural Philosophy in Phillips Academy, Andover, Mass. 60th edition, newly revised and greatly enlarged. $1 00.

Gale's Natural Philosophy ;

Or, Elements of Natural Philosophy. Embracing the General Principles of Mechanics, Hydrostatics, Hydraulics, Pneumatics, Acoustics, Optics, Electricity, Magnetism, Galvanism, and Astronomy. Illustrated by Several Hundred Engravings. Designed for the use of Schools. By LEONARD GALE. 20th edition. 62½ cents.

Hitchcock's Geology ;

Elements of Geology. By EDWARD HITCHCOCK, D.D., L.L.D., President of Amherst College, and Professor of Natural Theology and Geology. A new edition, revised, enlarged, and adapted to the present advanced state of the Science. With an Intro ductory Notice by JOHN PYE SMITH, D.D., F.R.S., and F.G.S. $1 25.

Hallock's Grammar :

A Grammar of the English Language, for the use of Schools, Academies, and Seminaries. By E. J. HALLOCK, A.M., Principal of the Castleton Seminary, Vermont. 2d edition. 62½ cents.

Kuhner's Elementary Grammar of the Greek

Language. Containing Exercises for the Writing of Greek, and the requisite Vocabularies. By RAPHAEL KUHNER. Translated by S. H. TAYLOR, of Andover. 12mo. $1 25.

Kendrick's Primary Greek Book;

Or, the Child's Book in Greek; being a Series of Elementary Exercises in the Greek Language. By ASAHEL C. KENDRICH, Prof. of the Greek Language and Literature in the Madison University. 16mo. 31 cents.

Kendrick's Greek Introduction;

Or, an Introduction to the Greek Language; containing an outline of the Grammar, with appropriate Exercises for the use of Schools and private learners. By ASAHEL C. KENDRICK. Enlarged edition. 62½ cents.

Kiddle's Astronomy;

Being a Manual of Astronomy and the use of the Globes. For Schools and Academies. By HENRY KIDDLE. Just published. 31 cents.

MacGregor's Book-keeping:

A Practical Treatise on Book-keeping, by Single and Double Entry, on a new plan; containing General Book-keeping for the use of Farmers, Mechanics, Professional Men, and other non-mercantile persons—Retailers' Book-keeping, and Merchants' Book-keeping. With an Appendix of Definitions, Directions, and Practical Forms. For the use of Seminaries and Self-Instructors. By P. MACGREGOR. 75 cents.

McElligott's Analytical Manual:

A Manual, Analytical and Synthetical, of Orthography and Definition. By JAMES N. MCELLIGOTT, L.L.D., Associate Principal of the Collegiate School, N. Y., and late Principal of the Mechanics' Society School. 75 cents.

McElligott's Young Analyzer;

Being an Easy Outline of the course of instruction in the English Language presented in the Analytical Manual. Designed to serve the double purpose of Spelling Book and Dictionary in the younger classes in Schools. 31 cents.

Newman's Rhetoric,—150th edition;

Being a Practical System of Rhetoric, or the principles and rules of style, inferred from examples of writing. To which is added an Historical Dissertation on English Style. By SAMUEL P. NEWMAN, late Professor of Rhetoric in Bowdoin College. 75 cents.

Parley's Universal History;

On the basis of Geography. Illustrated by Maps and Engravings. Especially designed for the use of Schools. $1 00.

Porter's Rhetorical Reader.

Consisting of Instructions for Regulating the Voice. With a Rhetorical Notation, illustrating Inflection, Emphasis, and Modulation; and a course of Rhetorical Exercises. Designed for the use of Academies and High Schools. By EBENEZER PORTER, D.D., late President of the Theological Seminary, Andover. New and enlarged edition. 62½ cents:

Woodbury's New Method of Learning the

German Language; embracing the Analytic and Synthetic modes of instruction. By the author of

WOODBURY'S SHORTER COURSE WITH GERMAN.
" ELEMENTARY GERMAN READER.
" ECLECTIC GERMAN READER, and
" NEUE METHODE, or NEW METHOD FOR GERMANS TO LEARN ENGLISH.

SANDERS' SERIES OF READERS.

Sanders' Series comprises eight books. It is the most thorough, correct, complete, and regularly progressive series of Reading-Books now before the public. It has received the unqualified approval of nearly every Teacher and School Committee who have examined it, and is in extensive use in nearly every State in the Union, in California, Oregon, and the West India Islands, and some of the South American States. The Series consists of

Sanders' Spelling-Book;

Designed to teach the Orthography and Orthoëpy of Dr. Webster. 12½ cents.

Sanders' Primary School Primer.

Paper covers. 6¼ cents.

Sanders' Primary School Primer.

Stiff covers. 8 cents.

Sanders' Pictorial Primer.

Bound. Green covers. 12½ cents.

Sanders' First Reader. 12½ cents.

Sanders' Second Reader. 18¾ cents.

Sanders' Third Reader. 37½ cents.

Sanders' Fourth Reader. Full Sheep binding. [illegible] cents

Sanders' Fifth Reader. Full Sheep and Embossed. 75 cts.

Sanders' Elocutionary Chart;

Designed as an accompaniment to Sanders' Series of Reading Books, for the use of Primary Schools, Academies, Institutes, Seminaries, Colleges, &c. By C. W. SANDERS, A.M., and Prof. E. W. MERRILL, A.M. $2 50.

This Chart is intended to give a knowledge of the Elementary Sounds of the English Language; to apply those Elements in Reading; to remedy defects in Articulation, Modulation, Inflection, &c.; also to give force, smoothness, and compass to the voice. It is colored and varnished, and mounted on rollers, being in size six feet by four and a half, and intended to hang up in the Recitation Room, to which it is a very handsome appendage.

Sawyer's Moral Philosophy;

Being the Elements of Moral Philosophy, on the basis of the Ten Commandments; containing a complete system of Moral Duties. By LEICESTER H. SAWYER, A.M., President of Central College, Ohio. 75 cents.

Schuster's Practical Drawing-Book,

For Schools and Self-Instruction; containing Heads and Figures, Landscapes, Flowers, Animals, and Ornamental Drawings, as well as some very useful instructions for their imitation, with a Historical Sketch of the Arts of Painting, Drawing, and Sculpture, and an Exposition of the New and Celebrated Method of M. DUPUIS, highly interesting for teachers, By SIGISMOND SCHUSTER, Professor of Drawing and Painting in New York City. $1 50.

Smith's Natural Philosophy:

A new and valuable Text-Book for the use of Schools and Academies. Illustrated by numerous Examples and appropriate Diagrams. New and enlarged edition. By HAMILTON L. SMITH, A.M. 75 cents.

Spencer's English Grammar;

Combining the Analytical and Synthetical Principles. Illustrated by Exercises for Grammatical Analysis; with numerous Examples of False Syntax; adapted to all classes of learners. By GEORGE SPENCER, A.M., late Principal of the Utica Academy 37½ cents.

Stone's Child's Reader;

For children who are too young to read in the Rhetorical Reader. Prepared at the request of Dr. PORTER, late President of the Andover Theological Seminary. By T. D. P. STONE, A.M 10th edition. 25 cents.

Scripture School Reader;

Consisting of selections of Sacred Scripture, for the use of Schools. Compiled and arranged by W. W. EVARTS, A.M., author of the "Bible Manual," and "Pastor's Hand Book," and W. H. WYKOFF, A.M., late Principal of the Collegiate School, N. Y. 75 cents.

THOMSON'S SERIES OF ARITHMETICS.

INTRODUCTORY TO DAY AND THOMSON'S SERIES OF MATHEMATICS.

All that can be said in praise of the best works which treat of the science of Numbers may be truly said of Thomson's Arithmetics The unusual favor with which they have been received since their publication, is a pretty fair recommendation of their merits. The Series is as follows:

Thomson's Arithmetical Tables,

and Exercises for Primary Schools. 18mo. Half Bound. 12½ cents

Thomson's Mental Arithmetic;

Or, First Lessons in Numbers, for Children. 18mo. New edition, enlarged. 12½ cents.

Thomson's Slate and Black-Board Exercises,

Or, First Lessons in Written Arithmetic. For Beginners. 20 cents

Thomson's Practical Arithmetic;

Uniting the Inductive with the Synthetic modes of Instruction, also illustrating the Principles of Cancellation. 110th edition. Revised, and greatly enlarged. 37½ cents.

Key to the same.

37½ cents.

Thomson's Higher Arithmetic;

Or, the Science and Application of Numbers. Combining the Analytic and Synthetic modes of Instruction. 75 cents.

Key to the same.

50 cents.

Thomson's Trigonometry;

Being a Treatise on Plain Trigonometry, and the Mensuration of Heights and Distances. To which is prefixed a summary view of the Nature and Use of Logarithms. Adapted to the method of Instruction in Schools and Academies. By JEREMIAH DAY, D.D., L.L.D, of Yale College. $1 00.

Thomson and Day's Surveying.

In Press.

WILLSON'S HISTORIES.

NEW AND REVISED EDITIONS.

Willson's Juvenile American History:

For Primary Schools. 31 cents.

Willson's History of the United States.

For the use of Schools and Academies. The History is brought down to the Election of Franklin Pierce to the Presidency; and embraces a Summary of the History of Mexico, Texas, and the Canadas. The author has also added to this edition the Constitution of the United States, with valuable notes and questions; the whole making the most complete and reliable History for the School-room that has yet appeared in this country. 75 cents.

Willson's American History,

(School edition,) comprising Historical Sketches of the Indians; a Description of American Antiquities, with an inquiry into their Origin, and the Origin of the Indian Tribes; History of the United States, with Appendices, showing its connection with European History; History of the present British Provinces; History of Mexico, and History of Texas, brought down to the time of its Admission into the American Union. Large 12mo. $1 25.

Willson's American History

Comprising Historical Sketches of the Indian Tribes ; a Description of American Antiquities, with an inquiry into their Origin, and the Origin of the Indian Tribes; History of the United States, with Appendix, showing its connection with European History; History of the present British Provinces ; History of Mexico and Texas, brought down to the time of its admission into the American Union. By MARCUS WILLSON. Library edition. Cloth $2 00. Sheep $2 50.

Willson's Comprehensive Chart of American

History. On Rollers, and Varnished. $6 00.

This is an elegant Chart of our Own Country's History, neatly engraved, colored, varnished, and mounted on rollers, and measuring about four feet by five and a half. It is arranged on a plan essentially different from any other historical chart; and yet is so simple that an intelligent child can readily understand it. It embraces the History of all the Countries, Colonies, States, and Provinces of North America, from the first discovery and settlement down to the present time. The Chart, with an accompanying Epitome and Questions, is designed for practical use in the business of instruction, and not for reference merely. Competent teachers who have used it, express the opinion, that it will be found as serviceable in teaching History, as maps are in teaching Geography. No description can give an adequate idea of its value.

Willson's Outlines of General History ;

Or, Universal History, is in course of preparation, and will, when completed, be the most perfect book of the kind ever published. The Work will be on the same General Plan and Arrangement with Mr. Willson's other Histories.

MUSIC FOR CHOIRS AND CHURCHES.

Psalmista, or Choir Melodies :

An extensive collection of new and available Church Music, together with some of the choicest selections from the former publications of the Authors, for Choirs and Congregational use. By THOMAS HASTINGS and WM. B. BRADBURY. 75 cents.

The Mendelssohn Collection.

By THOMAS HASTINGS and WM. B. BRADBURY. Containing original

and selected Music from the best European composers, consisting of Metrical Tunes, Anthems, Set Pieces, Motets, Sentences, and Chants, and an Appendix containing a selection of the most approved old Standard Church Tunes, for Congregation and Family use. 75 cents.

The New York Choralist:

A new and valuable Collection of Tunes in all the Metres, with an entire new collection of Anthems and Set Pieces for the use of Choirs, Congregations, Singing Schools, and Music Societies. By THOMAS HASTINGS and WM. B. BRADBURY. 75 cents.

The Psalmodist:

A Choice Collection of Psalm and Hymn Tunes, chiefly new, together with Chants, Anthems, Motets, and Set Pieces. By THOMAS HASTINGS and WM. B. BRADBURY. 75 cents.

The Christian Lyre:

A Collection of Hymn Tunes, adapted for Social Worship, Prayer Meetings, and Revivals of Religion. By JOSHUA LEAVITT. 26th edition, revised. 75 cents.

SECULAR MUSIC.

The Alpine Glee Singer.

Vol. 1. The most complete and choice Glee-Book ever published in this country. The music is new, popular, and beautiful, and the arrangements are easy of execution. $1 00.

The Metropolitan Glee-Book, or Alpine Glee

Singer. Vol. 2. An entirely new collection of Glees, Choruses, Part Songs, &c. $1 00.

The Social Singing-Book:

A collection of Glees, Part Songs, Rounds, Madrigals, &c., chiefly from European masters. With an Introductory Course of Elementary Exercises and Solfeggios, designed for singing-classes and schools of Ladies and Gentlemen. By W. B. BRADBURY. 50 cents.

TEMPERANCE MUSIC.

The Temperance Lyre:

A Collection of Temperance Songs, adapted to popular Melodies, and designed for the use of Temperance Meetings, Conventions, and Celebrations. By Mrs. M. S. B. Dana. 12½ cents.

The Crystal Fount:

A Temperance Song-Book, beautifully arranged with Hymns, Songs, and Music, entirely new. By Thomas Hastings. This book is admirably adapted for Temperance Choirs and Celebrations, being arranged in parts for Bass and Treble voices. 25 cents.

JUVENILE SINGING-BOOK.

The Singing-Bird, or Progressive Music Reader:

a new Collection of Juvenile Music arranged on an entirely new plan, designed to facilitate the introduction of Vocal Music as a study in Schools. By Wm. B. Bradbury. 37½ cents.

A New Juvenile Oratorio,

Entitled "The Seasons." Just published, in four parts.

1. Spring (in press). 25 cents.
2. Summer. 25 cents.
3. Autumn (in press). 25 cents.
4. Winter. 25 cents. (In press.)

Musical Gems for School and Home:

A new and complete collection of Music for the young. By Wm. B. Bradbury. 37½ cents.

Bradbury's Sabbath School Melodies:

A complete Singing-Book for all Sabbath School occasions. 16 cents.

The Young Melodist:

A new and rare collection of Social, Moral, and Patriotic Songs, designed for Schools and Academies. Composed and arranged for one, two, and three voices. By Wm. B. Bradbury. 25 cents

Flora's Festival:

A Musical Recreation, for Juvenile Singing-Classes, &c., together with Songs, Duets, Trios, Solfeggios, and plain Tunes for singing by note, in Thirteen Keys. By W. B. Bradbury. 25 cents

The Young Choir;

Or, School Singing-Book. Original and Selected. By W. B. BRADBURY and C. W. SANDERS. 25 cents.

The School-Singer, or Young Choir's Com

panion: a Choice Selection of Music, original and selected, for Juvenile Singing-Schools, Sabbath-Schools, Select Classes, &c. containing some of the most popular German Melodies. Also, a complete course of instruction in the Elements of Vocal Music By W. B. BRADBURY and C. W. SANDERS. 37½ cents.

THEOLOGICAL AND MISCELLANEOUS.

American Biblical Repository:

The First Twelve Volumes of this Valuable Work, while under the editorial charge of Dr. Robinson. Bound in half calf. $30 00

Appleton's Works.

The Works of Rev. JESSE APPLETON, D.D., late President of Bowdoin College, embracing his Course of Theological Lectures, his Academic Addresses, and a Choice Selection from his Sermons with a Memoir of his Life and Character. 2 vols. 8vo. $4 00.

Assembly's Shorter Catechism:

A New and neat Edition. 18mo. Paper covers. Per Hundred $1 50.

Beman on the Atonement.

Christ the only Sacrifice; or, the Atonement in its Relations to God and Man.. By N. S. S. BEMAN, D.D. With an Introductory Chapter, by Dr. Cox, of Brooklyn. 2d edition, enlarged. 50 cents.

Bush's Notes on the Old Testament.

Critical and Practical; designed as a general help to Biblical Reading and Instruction. By GEORGE BUSH, Prof. of Hebrew and Oriental Literature, in the New York City University. In 7 vols., as follows:

NOTES ON GENESIS. 2 vols. $1 75.
NOTES ON EXODUS. 2 vols. $1 50.
NOTES ON LEVITICUS. 1 vol. 75 cents.
NOTES ON JOSHUA. 1 vol. 75 cents.
NOTES ON JUDGES. 1 vol. 75 cents.

www.ingramcontent.com/pod-product-compliance
Lightning Source LLC
LaVergne TN
LVHW021143110826
845150LV00005B/1113

9781425546939